The Theory of Auger Transitions

The Theory of Auger Transitions

D. CHATTARJI
Department of Physics
Visva-Bharati University
Santiniketan
India

1976

Academic Press
London New York San Francisco

A Subsidiary of Harcourt Brace Jovanovich, Publishers

ACADEMIC PRESS INC. (LONDON) LTD.
24/28 Oval Road,
London NW1

United States Edition published by
ACADEMIC PRESS INC.
111 Fifth Avenue
New York, New York 10003

Library of Congress Catalog Card Number: 75 19624
ISBN: 0-12-169850-5

Printed in Great Britain by
William Clowes & Sons Limited
London, Colchester and Beccles

Preface

Now in the circulating torrent of the stars
Certain events are drawn correct and clear—Charles Madge

In the years 1925–26, P. Auger irradiated inert gases contained in a Wilson cloud-chamber with a beam of X-rays and observed paired electron tracks originating from some of the ionized atoms. In each of these pairs, one track represented the photo-electron and had a variable length depending on the energy of the incident ionizing radiation. The other track had a constant length and represented the radiationless reorganization of the atom now named after Auger. In 1927, G. Wentzel gave the non-relativistic theory of the Auger effect, explaining it as an autoionization process resulting from the electrostatic interaction between two electrons in an atom which is already singly ionized in an inner shell. Evidently, this was one of the earliest applications of the then newly-discovered tools of quantum mechanics.

Since then, a great deal has happened in the area of physics broadly labelled as inner-shell ionization phenomena. In particular, there has been a remarkable revival of interest in the Auger effect and allied processes. To an extent, this is part of the renewed growth of atomic physics. At the same time, it is also related to technical developments in other areas like nuclear and solid state physics. For example, the availability of solid-state detectors and high resolution electron spectrometers with high transmission has undoubtedly helped the measurement of fluorescence and Auger yields. The usefulness of Auger spectra in the analysis of surface impurities has, in recent years, necessitated the study of environmental effects on these spectra. Auger transitions are also important to the understanding of mesic atoms, nuclear internal conversion and orbital capture of electrons, ion-atom collisions, etc.

Because of its bearing on these diverse branches of pure and applied physics, the theory of the Auger effect deserves some attention at this point of time. Over and above, involving as it does both discrete and continuous states of the atomic system, it can be used as a good testing ground for fundamental atomic theory, i.e. the various atomic models and their concomitant wave functions, the symmetries inherent in these models and the simplifications resulting from them. In other words, it can be a good play-room for atomic theorists of all degrees of maturity. More than anything else, this was the idea behind the writing of this book. Of course the interplay between theory and experiment can hardly be overemphasized and an up-to-date survey of experiments and their interpretation covering the different applications would be both useful and interesting. Fortunately, excellent reviews designed to meet these needs are now available. The object of this book is to supplement them with a review of Auger theory as such, a review which would provide some perspective by relating it to the broad spectrum of atomic and other physical theory. At the same time, it is hoped that this book will be of some help to the graduate student about to specialize in atomic physics and the more experienced immigrant from a neighbouring discipline. Therefore, this book spells out in some detail techniques of calculation using different atomic models. Inevitably, some portions of this book will be of interest primarily to the theoretician. For this I must apologize to the committed experimentalist (we all know his kind indulgence of our vagaries) and hope that he will still find something of interest.

The last chapter makes a conscious departure from the emphasis on theory. The application of Auger spectroscopy to surface chemical analysis is perhaps the most important technical use of this field made so far and can hardly be the province of a theoretician who is no more than a curious onlooker, a reminder here of the well-known behavioural contrast between fools and angels. Our excuse, if there can be one, is that a dialogue between surface scientists and atomic physicists would be worth their while and somewhere, at some point, the ice needs to be broken. Our effort will be rewarded if at least a few more competent theorists get interested in this interface between physics and technology.

One of the pleasures of writing a preface is to be able to acknowledge the many facets of one's indebtedness to professional colleagues. I must start out by stating that the idea of writing this book in its present form was conceived during my visit to Professor E. H. S. Burhop at the University College, London, in the summer of 1973. With his characteristic kindness and hospitality, Professor Burhop encouraged me to pursue the idea by carefully going through the first few chapters as I wrote them during that summer, by giving me his critical comments on the completed manuscript, by drawing my attention to important items in the literature and by getting the publishers interested in my project. In short, his active interest was crucial at every stage.

Work on the manuscript was completed during a year's stay at the University of Freiburg in 1974–75 and I am grateful to Professor Werner Mehlhorn for his ungrudging help with his own time as well as institutional facilities during this period. He was kind enough to read critically most of the chapters as they were being put into their final form and I shall always remember the long hours of animated discussion which followed. Professor Mehlhorn's laboratory was an ideal place for this kind of work and it gave me the opportunity to learn some good physics.

This work could not have been completed without the cooperation of many colleagues who helped me with comments and information about their work and permitted me to use copyrighted material. While the latter is acknowledged in the text, I should like to thank everyone for the prompt attention I received. Dr. E. J. McGuire helped me understand his work on Auger transitions involving partially filled shells. Drs J. A. D. Matthew, T. E. Gallon, M. Prutton and D. W. O. Heddle of the University of York very kindly introduced me to the mysteries of Auger spectroscopy for solids during a visit to their laboratory. Drs C. C. Chang and N. J. Taylor sent me material on their surface Auger work. Professor B. Crasemann, Dr E. J. Callan and Dr S. Aksela kept me abreast of their research during the writing of the book.

My visits to the University College, London, and the University of Freiburg were made possible by the British Council and the Alexander von Humboldt Foundation respectively. I am grateful to this Foundation and the Department of Atomic Energy of the Government of India for partial financial support towards the preparation of the manuscript. Finally, I am pleased to acknowledge the expert assistance of Miss R. Kempter and Mr. P. Banerjee in typing two different versions of the manuscript and of Mrs. I. Batt in preparing the figures.

In spite of all this help, I am sure there will be many lapses and shortcomings. These must, alas, be traced to the author.

Dipankar Chattarji

Santiniketan, India
November, 1975

To Professor E. H. S. Burhop, F.R.S.,
with kind regards

Contents

Preface v

1. Symmetry and Invariance—Some Preliminary Notions 1

1. Introduction 1
2. The group concept 2
3. Lie groups and Lie algebras 3
4. Symmetry and invariance 7
5. The direct product group 9

2. Theory of the Auger Process 13

1. Introduction 13
2. Mechanism of the Auger process. Real and virtual photons 14
3. The Møller formula 20
4. Generalized derivation of the Møller formula 22
5. Some related concepts and processes 27

3. The Coulomb Field and Coulomb Wave Functions 30

1. Introduction 30
2. Symmetry properties of the Coulomb field 32
3. The non-relativistic spin-$\frac{1}{2}$ (NRS) particle in a Coulomb field 36
4. Some Coulomb field calculations of Auger transitions 39
5. Screening parameters 56
6. The hydrogenic atom—some dividends 58

4. Symmetry-breaking and the Classification of States 64

1. Introduction 64
2. The *LS* coupling scheme (Russell–Saunders coupling) 68

3. The *jj* coupling scheme 72
4. The intermediate coupling scheme 76
5. Configuration interaction 79
6. The mixed coupling scheme 87

5. Some Central-field Calculations 91

1. Introduction 91
2. Wigner algebra and selection rules 92
3. Auger transition probabilities in *jj* coupling 93
4. Transition rates with *LS* coupling 104
5. Transition probabilities with intermediate coupling 109
6. Electron energies of the *K*-Auger spectrum 119

6. The Many-Electron Atom 126

1. Introduction 126
2. Symmetries of the many-electron system 128
3. Techniques for the many-electron system 130
4. Nonrelativistic self-consistent-field calculations—Theory and Experiment 138
5. The Relativistic Hartree–Fock method 142
6. The Thomas–Fermi–Dirac method 148
7. Relativistic approaches to the many-electron atom—Theory and Experiment 151

7. Some Recent Advances in Auger Theory 158

1. Satellite spectra 158
2. Auger transitions in partially filled shells 163
3. Effects of electron correlation 171

8. Auger Electron Spectroscopy and its Application to Surface Science 180

1. Introduction 180
2. Auger electron energies—general features 188
3. Auger electron spectroscopy as a surface technique—historical background 193

4. The secondary electron spectrum 195
5. Auger electron spectra of atoms in solids 199
6. Experimental technique 218
7. Surface chemical analysis with AES 238

References 251

Index 259

Chapter One

Symmetry and Invariance—Some Preliminary Notions

1. Introduction

The symmetry properties of atomic systems have been known so long that they are almost likely to be forgotten in the course of day-to-day work. The general feeling is that the basic law of force, namely the electromagnetic interaction, is known and all that has to be done is to plod through with the various coupling schemes and arrive at detailed results. There still remains some reverence among experimentalists for things like recoupling coefficients, 9-j symbols, etc. although even these are regarded as no more than technical tools. Familiarity often leads to a kind of indifference to the intrinsic beauty of this oldest branch of quantum physics where group-theoretic ideas were first tested and elucidated.

In a field like particle physics, symmetries and invariance properties assume a more crucial role because they provide a dependable way of classifying states and of looking for those properties of the interaction which are not otherwise amenable to analysis. Even the partial breaking of a symmetry provides valuable clues to the interaction which may have caused it. In atomic physics, there is an analogous breaking of symmetry from the textbook case of the hydrogen atom to the many-electron atom. Because electrostatic interaction between electrons and spin-orbit coupling are well-defined perturbations, symmetry-breaking is not usually talked about. Yet, that the resulting physical situation is far from being closed and completely understood can be seen from the fact that techniques like configuration mixing began to be applied to an active field like Auger transitions only in recent years. Studies on partially filled shells are only now being made. Even the Coulomb field has lately been the subject of considerable interest, particularly in relation to the study of non-invariance groups. As in most branches of science, completeness of knowledge seems to be an illusion even in atomic physics and the study of symmetries may still play a role.

Apart from the fundamental elegance of the symmetry approach, its pedagogic importance cannot be overemphasized. For example by regarding the coupling schemes as mere technique, the dynamical nature of the interactions which motivate them tends to be forgotten. One result of this is that the more general perturbation scheme, namely intermediate coupling, is somewhat underemphasized in the classroom. Then there is the question of "mixed" coupling, where two different coupling schemes are used to describe the initial and final states of an atomic system. Such a combination does seem to provide good agreement with experiment in some cases. Whether this is permitted in principle can be decided only on the basis of a clear understanding of the dynamical conditions which these coupling schemes are supposed to embody.

In the ultimate analysis, whenever a residual interaction to explain the discrepancy between theory and experiment is invoked, in effect, some kind of symmetry-breaking is being looked for. The fact that generally there is no need to search for a new type of fundamental interaction should not affect the appreciation of the power of group-theoretic techniques which help classify the states of the perturbed system. Unfortunately, formal group theory is yet to become popular among physicists, particularly among experimentalists, and a formal approach will not be adopted in the following chapters. However, this book will try to develop the theory of Auger transitions, keeping in mind the progressive violation of the simple symmetry of the Coulomb field as it approaches a more realistic description of heavier atoms. While the principal endeavour will be to understand these transitions and some of their applications, simple considerations of symmetry will be used as far as possible without making the discussion too esoteric.

This chapter introduces without any claim to originality some of the elementary tools of group theory that may be needed in the following chapters. There is incompleteness and lack of rigour in the treatment, but those who are interested in details may look up one or more of the excellent sources given in the references.

2. The group concept

When a physical system undergoes some transformation, such as a rotation about some axis, it may or may not look different at the end of this transformation. A right circular cylinder has one axis—its symmetry axis—about which it can be rotated without looking different. A rubber ball has any number of such axes—it is completely "degenerate" with respect to rotations. However, if the ball is slightly deformed by pressing at the "poles", part of this degeneracy is lifted. Even then, it may retain the corresponding polar axis as an axis of symmetry—rotations about this axis can be distinguished from other types of rotation. In fact, it may be used as a reference axis to describe the rotation of the deformed system. This simple form of symmetry-breaking is not entirely alien to the kind of situation that shall be encountered in the following chapters.

In classical physics, the transformations undergone by a system are described in terms of its generalized coordinates. In quantum mechanics, the state of a system is described by its state vector or wave function Ψ, and the coordinates appear as arguments of Ψ. Therefore, when we talk of transformations, we are really interested in the changes which occur to Ψ and in those dynamical features of the system which may remain unchanged in course of a transformation. The latter, the so-called constants of the motion, serve to characterize the states of the system. Of special interest is the Hamiltonian operator H, because it gives the time-development of the system.

A set of transformations t_a, t_b, t_c, etc. are said to form a group G under the following conditions.

1. An associative "multiplication" is defined for the elements of the group. This provides a way of combining any two elements t_a and t_b to yield a product $t_a t_b$. The meaning of such a product is made explicit by the associative law: $t_a(t_b t_c) = (t_a t_b) t_c$ for all t_a, t_b and t_c in the group.
2. Given that t_a and t_b belong to the group G, the product $t_a t_b$ also belongs to G. This is the property of closure.
3. There exists an identity transformation I in the group such that $t_a I = I t_a = t_a$ for all t_a in the group.
4. Every transformation t_a in the group has an inverse t_a^{-1} also belonging to the group such that $t_a t_a^{-1} = t_a^{-1} t_a = I$.

It is not necessary that $t_a t_b = t_b t_a$, i.e. multiplication need not commute. However, if it does commute, the group G is said to be Abelian.

If any subset of the group elements have the group properties, then they are said to form a subgroup.

The most familiar groups of transformations in classical physics are geometrical in nature, i.e. translations, rotations, reflections, etc. For a system like an equilateral triangle or a cube, there are a small number of rotations, i.e. rotations by 120° or 90° and multiples thereof, which return them to their original position. These are the well-known examples of finite groups. We shall, however, be concerned with transformations which are parametrized by one or more continuous variables.

3. Lie groups and Lie algebras

Consider, for example, the rotation of a rigid body about an arbitrary z-axis fixed in space. Let (x, y) be a point in the body before rotation and let (x', y') denote the coordinates of the same point after the body has been rotated through an angle θ. In classical physics, we describe the transformation by writing

$$\begin{pmatrix} x' \\ y' \end{pmatrix} = \begin{pmatrix} \cos\theta & \sin\theta \\ -\sin\theta & \cos\theta \end{pmatrix} \begin{pmatrix} x \\ y \end{pmatrix}, \tag{1.1}$$

where the matrix $\begin{pmatrix} \cos\theta & \sin\theta \\ -\sin\theta & \cos\theta \end{pmatrix}$ embodies the effect of the transformation on a vector in the xy plane. In quantum mechanics, we write the transformed wave function

$$T\Psi(x,y) = \Psi(x',y') = \Psi(x\cos\theta + y\sin\theta, -x\sin\theta + y\cos\theta). \tag{1.2}$$

Let us assume for the moment that θ is small, so that $\cos\theta \simeq 1$ and $\sin\theta \simeq \theta$. Then

$$\begin{aligned}\Psi(x',y') &= \Psi(x + y\theta, y - x\theta) = \sum_{n=0}^{\infty} \frac{\theta^n}{n!}\left[y^n\left(\frac{\partial}{\partial x}\right)^n - x^n\left(\frac{\partial}{\partial y}\right)^n\right]\Psi(x,y) \\ &= \sum_{n=0}^{\infty} \frac{(i\theta)^n}{n!}\frac{1}{i^n}\left[y^n\left(\frac{\partial}{\partial x}\right)^n - x^n\left(\frac{\partial}{\partial y}\right)^n\right]\Psi(x,y) \\ &= e^{iJ_z\theta/\hbar}\Psi(x,y). \end{aligned} \tag{1.3}$$

Here we have made a Taylor series expansion of $\Psi(x + y\theta, y - x\theta)$ about (x,y) and used the usual representation

$$J_z = \frac{\hbar}{i}\left[y\frac{\partial}{\partial x} - x\frac{\partial}{\partial y}\right].$$

Obviously, $e^{iJ_z\theta/\hbar}$ represents an infinitesimal rotation about the z-axis. By multiplying a number of such infinitesimal rotations, we can generate any finite rotation about the z-axis and the group property is evident. If we write $\mathcal{O}(\theta) = e^{iJ_z\theta/\hbar}$, then $\mathcal{O}(-\theta) = e^{-iJ_z\theta/\hbar}$ is its inverse element. The identity element is $\mathcal{O}(0)$. Because all such rotations are only different functions of the single operator J_z, they commute, i.e. $\mathcal{O}(\theta_1)\mathcal{O}(\theta_2) = \mathcal{O}(\theta_2)\mathcal{O}(\theta_1)$, and this is an Abelian group.

Now, since all rotations about the z-axis are functions of a single operator J_z, we could conveniently talk about J_z rather than the infinite number of elements of the group. We can do this because there is a unique prescription for obtaining J_z from the elements of the group:

$$J_z = -i\hbar\left[\frac{\partial\mathcal{O}(\theta)}{\partial\theta}\right]_{\theta=0}. \tag{1.4}$$

Equation (1.4) is based on the fact that $\mathcal{O}(\theta)$ is an analytic function of θ near $\theta = 0$, i.e. in some neighbourhood of the identity element. In other words, not only is $\mathcal{O}(\theta)$ a continuous function of θ, but it is also analytic in this neighbourhood. This is a rather special property. Continuous groups whose elements are analytic functions of the group parameters are called "Lie groups". The one-parameter group of rotations $\mathcal{O}(\theta)$ is a very simple example. We can of course generalize these statements to the three-dimensional rotation group O_3 whose elements are $\mathcal{O}(\theta_1, \theta_2, \theta_3) = e^{i\mathbf{J}\cdot\boldsymbol{\theta}/\hbar}$, where $\mathbf{J} = (J_x, J_y, J_z)$ and $\boldsymbol{\theta} = (\theta_1, \theta_2, \theta_3)$. The operators J_x, J_y, J_z are called the generators of the rotation group. Here the rotations do not commute in general and the generators satisfy the well-known commutation relations

$$[J_i, J_j] = \hbar i J_k \quad (i, j, k = x, y, z \text{ taken in a cyclic order}). \tag{1.5}$$

The rank of a Lie group is given by the number of mutually commuting generators it has. From our knowledge of angular momentum, we can see that the rotation group has rank one, because no two generators commute. The question may of course be asked: since the generators are defined in terms of the properties of the Lie group only in some neighbourhood of the identity element, can they uniquely specify the group everywhere? The answer to this question involves topological considerations and we need not go into it in detail. However, one of the important properties of a Lie group is *homogeneity*, i.e. if we think of the group elements as points in a group space G, then for any two points x and y, there exists a mapping f such that $f(x) = y$. Thus, even if we know only the local properties of the group in the neighbourhood of a single point, e.g. the identity element, we can in principle derive the same properties at any other point by utilizing this mapping (Kleima *et al.*, 1968).

The local properties of a Lie group, i.e. the properties of infinitesimal transformations, are completely defined by its generators. For the general Lie group, these generators satisfy commutation relations of the type

$$[A_i, A_j] = \sum_{k=1}^{n} C_{ijk} A_k, \tag{1.6}$$

where n is the number of parameters characterizing the group and the C_{ijk} are called structure constants. Equation (1.5) is a special case of eqn (1.6). By demanding that the infinitesimal transformations of the Lie group may be integrated to yield a finite transformation, it may be shown that the generators must satisfy the Jacobi identity (Kleima *et al.*, 1968)

$$[[A_i, A_j], A_k] + [[A_j, A_k], A_i] + [[A_k, A_i], A_j] = 0. \tag{1.7}$$

Thus the generators form a closed set under commutation. The set of all possible linear combinations of the generators is called the Lie algebra of the group. There exists, for every Lie group, a Lie algebra. The advantage in using the Lie algebra rather than the group is, as we have already pointed out, one of economy.

Consider, for example, the 2×2 Pauli matrices

$$\sigma_1 = \begin{pmatrix} 0 & 1 \\ 1 & 0 \end{pmatrix}, \quad \sigma_2 = \begin{pmatrix} 0 & -i \\ i & 0 \end{pmatrix}, \quad \sigma_3 = \begin{pmatrix} 1 & 0 \\ 0 & -1 \end{pmatrix}. \tag{1.8}$$

These are the generators of the group SU_2, i.e. the two-dimensional group of unitary transformations having unit determinant. To see this, we write such a unitary transformation in the form

$$U = e^{iH}. \tag{1.9}$$

Unitarity implies that

$$UU^+ = e^{iH}e^{-iH^+} = e^{i(H-H^+)} = 1, \tag{1.10}$$

so that we must have $H - H^+ = 0$, i.e. H must be Hermitian. Now consider a transformation of the type $e^{i\boldsymbol{\sigma}\cdot\boldsymbol{\lambda}}$ generated by the Pauli matrices, $\boldsymbol{\lambda}$ being a real three-dimensional vector. Here

$$H = \boldsymbol{\sigma}\cdot\boldsymbol{\lambda} = \begin{pmatrix} \lambda_3 & \lambda_1 - i\lambda_2 \\ \lambda_1 + i\lambda_2 & -\lambda_3 \end{pmatrix} \tag{1.11}$$

is obviously Hermitian, so that $e^{i\boldsymbol{\sigma}\cdot\boldsymbol{\lambda}}$ is unitary. Additionally, note that $\boldsymbol{\sigma}\cdot\boldsymbol{\lambda}$ is traceless. Now there must exist a unitary transformation S which diagonalizes the Hermitian matrix $\boldsymbol{\sigma}\cdot\boldsymbol{\lambda}$. Let α_1 and α_2 be the eigenvalues of the transformed matrix H'. Then

$$H' = SHS^{-1} = \begin{pmatrix} \alpha_1 & 0 \\ 0 & \alpha_2 \end{pmatrix}, \quad U' = SUS^{-1} = \begin{pmatrix} e^{i\alpha_1} & 0 \\ 0 & e^{i\alpha_2} \end{pmatrix}$$

and

$$\det U' = e^{i\alpha_1}e^{i\alpha_2} = e^{i(\alpha_1+\alpha_2)} = e^{iTrH'}.$$

But the trace of H is invariant under a unitary transformation, so that

$$TrH' = 0 \text{ and } \det U' = \det U = 1.$$

The commutation relations for the Pauli matrices are

$$[\sigma_i, \sigma_j] = 2i\sigma_k \qquad (i, j, k = 1, 2, 3, \text{ in cyclic order}), \tag{1.12}$$

or,

$$[\tfrac{1}{2}\sigma_i, \tfrac{1}{2}\sigma_j] = i\tfrac{1}{2}\sigma_k. \tag{1.12a}$$

By comparing eqns (1.5) and (1.12a), we can see that the groups SU_2 and O_3, both of rank 1, have the same Lie algebra. Here we have an example of two Lie groups sharing a Lie algebra. Because the Lie algebra is defined only by the local properties of the group, Lie groups having different topological properties can have the same Lie algebra. To the extent that these topological differences may be ignored, we can say that a Lie group is uniquely characterized by its Lie algebra.

4. Symmetry and invariance

Consider a spinless particle of mass μ in a spherically symmetric potential $V(r)$. For such a system, the orbital angular momentum $\mathbf{L}$ and its z-component are constants of the motion and we write

$$\mathbf{L}^2\, Y_l^m(\theta,\varphi) = l(l+1)\hbar^2\, Y_l^m(\theta,\varphi),$$
$$L_z\, Y_l^m(\theta,\varphi) = m\hbar\, Y_l^m(\theta,\varphi). \tag{1.13}$$

The spherical harmonic $Y_l^m(\theta,\varphi)$ constitutes the angular part of the wave function regardless of the explicit form of $V(r)$. The operators L_x, L_y and L_z satisfy the commutation relations (1.5) and we say that the system has O_3 as its symmetry group. We shall use these elementary ideas to introduce some of the basic concepts relating to symmetries and invariance.

1. When we say that the symmetry group of this system is O_3, we mean that the Hamiltonian H commutes with all rotations of the type

$$\mathcal{O}(\boldsymbol{\theta}) \equiv \mathcal{O}(\theta_1,\theta_2,\theta_3) = e^{i\mathbf{L}\cdot\boldsymbol{\theta}/\hbar},$$
$$\text{i.e. } [H,\mathcal{O}] = 0. \tag{1.14}$$

2. Now consider that part of the Hilbert space of the system which is spanned by the orthogonal vectors $Y_1^1(\theta,\varphi)$, $Y_1^0(\theta,\varphi)$ and $Y_1^{-1}(\theta,\varphi)$. These vectors are said to form an invariant subspace of the Hilbert space of the system in the sense that each of the operators L_x, L_y, L_z transforms any one of the vectors $Y_1^m(\theta,\varphi)(m = \pm 1, 0)$ only into another member of the set and into none outside it. Since all rotations in configuration space are generated by these three operators, any such rotation will also leave this subspace invariant. Furthermore, it contains no smaller invariant subspace, i.e. no subspace with the same property but having a smaller number of orthogonal vectors. Such an invariant subspace is said to be *irreducible*. In the language of group theory, it is sometimes called a *multiplet*.‡ If we now add to this set the vector $Y_0^0(\theta,\varphi)$, we can see that the resulting subspace is no more irreducible. In fact the enlarged invariant subspace may be reduced to a triplet and a singlet, namely $(Y_1^1(\theta,\varphi)$, $Y_1^0(\theta,\varphi)$, $Y_1^{-1}(\theta,\varphi))$ and $Y_0^0(\theta,\varphi)$, each of which is irreducible. Actually, this is a general property of invariant subspaces of any semisimple Lie group.§ There is a theorem which states that every finite-dimensional invariant space of a semisimple Lie group can be completely decomposed into a number of non-overlapping multiplets.

‡ In atomic physics, the term multiplet has a specific connotation, namely, a set of states labelled by the total orbital angular momentum L and total spin S.

§ Let G be a group with an Abelian subgroup G'. Let a be any element of G' and b some element of the full group G. If $bab^{-1} = a'$, where a' is some other element of G', then G' is said to be an invariant subgroup of G. If G possesses no such invariant Abelian subgroup, it is said to be *semisimple*.

3. By operating on any one of the members of a multiplet with the general rotation operator $\mathcal{O}(\boldsymbol{\theta})$, we can generate all the other members of the multiplet, but no vector outside it. Now, because the Hamiltonian H commutes with $\mathcal{O}(\boldsymbol{\theta})$, we have, for example,

$$\begin{aligned} H\mathcal{O}(\theta)\, Y_1^0(\theta,\varphi) &= \mathcal{O}(\boldsymbol{\theta})\, H Y_1^0(\theta,\varphi) \\ &= E_1\, \mathcal{O}(\boldsymbol{\theta})\, Y_1^0(\theta,\varphi), \end{aligned} \tag{1.15}$$

where E_1 is the energy eigenvalue corresponding to $Y_1^0(\theta,\varphi)$. Thus the vector $\mathcal{O}(\boldsymbol{\theta})\, Y_1^0(\theta,\varphi)$ is also an eigenvector belonging to the same eigenvalue E_1. In other words, members of a multiplet form a set of degenerate energy eigenstates of the system.

4. Invariant operators: The spherical harmonic $Y_l^m(\theta,\varphi)$ is an eigenvector of the operators $\mathbf{L}^2$ and L_z. Unlike L_z, $\mathbf{L}^2$ is not a generator of the group O_3. It is a function of the generators L_x, L_y, L_z and commutes with all of them. Hence it commutes with all elements of O_3. It is an invariant of the group in the sense that if we rotate the coordinate system, then in the new frame $\mathbf{L}'^2 = \mathcal{O}(\boldsymbol{\theta})\mathbf{L}^2\mathcal{O}^{-1}(\boldsymbol{\theta}) = \mathbf{L}^2$. Such an operator is called a Casimir invariant of the group. Obviously, it also commutes with the Hamiltonian and shares with it the property that its degenerate eigenvectors form multiplets of the symmetry group O_3. Thus its eigenvalues uniquely label the multiplets. There is a theorem (Racah, 1951) which states that for every semisimple Lie group of rank l there exists a set of l invariant operators. These are functions $C_i(L_1,\ldots,L_n)$ $(i = 2,3,\ldots,l+1)$ of the generators which commute with every operator of the group, and whose eigenvalues uniquely characterize the multiplets of the group.

For a given symmetry group not only are these invariants unique, but they are also complete, i.e. any other operator commuting with the group elements must be functions of the C_i. Thus, once the Casimir invariants of the symmetry group S of a system are known, all the eigenstates of the Hamiltonian H can be classified in terms of the eigenvalues of these invariants. Each multiplet is then characterized by the same set of eigenvalues of these operators, i.e. the members of the multiplet form a degenerate set. This is the state of affairs so long as S remains the symmetry group of the system. If we now turn on some perturbation H', so that S is no more the symmetry group, i.e. $[S, H+H'] \neq 0$, then we have to search for a fresh symmetry group S'. If there exists such a group, all the above statements will apply to its multiplets. If it so happens that S' is a subgroup of S, then the multiplets of S may split into a number of multiplets of S'. A familiar example in atomic physics is that of the triplet $(Y_1^1(\theta,\varphi),\ Y_1^0(\theta,\varphi),\ Y_1^{-1}(\theta,\varphi))$ splitting into three individual energy levels corresponding to $m = 1$, 0 and -1 under the influence of a magnetic field. This is what we mean when we say that the degeneracy has been removed by a perturbation. In a situation like this, one needs new quantum numbers, i.e.

new Casimir invariants, to characterize states which belonged to the same multiplet of the larger group S, but now belong to different multiplets of the more restricted group S'. The magnetic quantum number m, corresponding to the invariant operator J_z, serves this purpose in the above example. It tells us that the system is still symmetric with respect to rotations about the z-axis, i.e. the direction of the magnetic field. The group of rotations about a fixed axis in space is O_1 and its operators are of the form $\mathcal{O}(\theta) = e^{iJ_z\theta/\hbar}$ already discussed. This is a subgroup of O_3.

5. The direct product group

In the last paragraph, we have considered a situation where the symmetry of a system is reduced by the introduction of a perturbation. There is another kind of situation and it occurs quite often in nature, where a system possesses several symmetries at the same time. It is not difficult to see that here the overall symmetry is larger than the individual symmetry groups in question (McVoy, 1965).

To see exactly what this larger symmetry means, we might proceed as follows. Consider a system with a symmetry group S_1 of which the Casimir invariants are known, say, $C_i(i = 1,2,\ldots)$. If we now discover that it has an additional symmetry S_2, we ask the following question: Do the generators of S_2 all commute with the C_i? If they do not, they will connect different multiplets of S_1, i.e. the multiplets of S_1 will no longer form invariant subspaces of the larger group. Thus different multiplets of S_1 may have to be combined together to form multiplets of the symmetry group of the new system. This is what happens if we add to the generator J_z of the O_1 group in the example above two more generators J_x and J_y of two more O_1 groups, i.e. the transformations $e^{iJ_x\theta_1/\hbar}$ and $e^{iJ_y\theta_2/\hbar}$. In this case all three states Y_1^1, Y_1^0 and Y_1^{-1} taken together form a triplet of O_3.

The other possibility is, of course, that the added generators commute with the Casimir invariants C_i of S_1. In this case, the multiplets of S_1 continue to be multiplets of the enlarged symmetry group, because the C_i continue to be Casimir operators of the latter. We shall only consider the simple case where the generators of S_2 commute with all the elements of S_1 and not just its invariants. In this case one says that S_1 and S_2 are compatible. The larger group S which describes the total symmetry of the system is called the direct product group and its elements are obtained by taking all possible products of the elements of S_1 and S_2. This relationship is symbolically written

$$S = S_1 \otimes S_2. \tag{1.16}$$

Just as in the case of a single Lie group, the generators of S are obtained by taking functional values of the partial derivatives of the group elements in the neighbourhood of the identity element. For example let $S_1 = \{X(\boldsymbol{\alpha})\}$ and $S_2 = \{Y(\boldsymbol{\beta})\}$, $\boldsymbol{\alpha}$ and $\boldsymbol{\beta}$ being vectors in the parameter spaces of the two groups, i.e. $\boldsymbol{\alpha} = (\alpha_1, \alpha_2, \ldots, \alpha_m)$ and $\boldsymbol{\beta} = (\beta_1, \beta_2, \ldots, \beta_n)$. Then the generators of S are given by

$$\frac{\partial}{\partial \alpha_i} X(\boldsymbol{\alpha})\, Y(\boldsymbol{\beta})|_{\alpha=0,\,\beta=0} = \frac{\partial}{\partial \alpha_i} X(\boldsymbol{\alpha}) Y(0)|_{\alpha=0}\, (i = 1, \ldots, m) \tag{1.17}$$

and

$$\frac{\partial}{\partial \beta_j} X(\boldsymbol{\alpha})\, Y(\boldsymbol{\beta})|_{\alpha=0,\,\beta=0} = \frac{\partial}{\partial \beta_j} Y(\boldsymbol{\beta})\, X(0)|_{\beta=0}\, (j = 1, \ldots, n).$$

Here, $X(0)$ and $Y(0)$ are identity elements of S_1 and S_2. Because the generators of S_1 commute with the generators of S_2, the total number of commuting generators of S is

$$l = l_1 + l_2, \tag{1.18}$$

where l_1 and l_2 are the number of commuting generators of S_1 and S_2 respectively. In other words, the rank of the direct product group S is just the sum of the ranks of S_1 and S_2. Moreover, because S_1 has l_1 Casimir invariants and S_2 has l_2 of them, S will have l invariants given by eqn (1.18). Each of them commutes with all the elements of S and, taken together, these l invariants serve to label the multiplets of S.

Finally, it remains to obtain the multiplets of S. Again we start with our knowledge of quantum mechanics, namely, that the basis vectors spanning the space of S are written as linear combinations of products of the basis vectors of S_1 and S_2:

$$\Psi(\boldsymbol{\alpha}, \boldsymbol{\beta}) = \sum \sum C_{ij}\, \varphi_{1i}(\boldsymbol{\alpha})\, \varphi_{2j}(\boldsymbol{\beta}). \tag{1.19}$$

Here φ_{1i} stands for the ith basis vector of S_1, etc. Thus the elements of S operate in the space of the product wave functions Ψ. This leads us to believe that the multiplets of S will be formed by taking all possible products of the type $\varphi_{1i}\varphi_{2j}$, where φ_{1i} and φ_{2j} are members of corresponding multiplets of S_1 and S_2 respectively. In other words, if we take a multiplet M_1 of S_1 with m_1 members and a multiplet M_2 of S_2 with m_2 members, then we get a multiplet M of S with $m_1 m_2$ members by taking all possible products of these members.

Equation (1.19) assumes a familiar meaning if we consider S_1 and S_2 to be the same group relating to two different constituents of a system. For example, let us consider two electrons such that S_1 and S_2 represent the rotational symmetry of their wave functions, both of them being in a common central potential. The elements of S_1 and S_2 will have the general form $e^{i\mathbf{J}\cdot\boldsymbol{\theta}/\hbar}$. The vector parameter $\boldsymbol{\theta}$ represents an ordered triad of rotations θ_1, θ_2, θ_3 about the x, y and z axes respectively. For simplicity, let us call this a rotation $\boldsymbol{\theta}$. The operator $e^{i\mathbf{J}\cdot\boldsymbol{\theta}/\hbar}$ causes such a rotation of the unit vector $\hat{r}$ whenever it operates on any function of $\hat{r}$. We can therefore write

$$\begin{aligned} e^{i\mathbf{J}\cdot\boldsymbol{\theta}/\hbar}[Y_{l_1}^{m_1}(\hat{r})\, Y_{l_2}^{m_2}(\hat{r})] &= Y_{l_1}^{m_1}(\hat{r}')\, Y_{l_2}^{m_2}(\hat{r}') \\ &= [e^{i\mathbf{J}\cdot\boldsymbol{\theta}/\hbar} Y_{l_1}^{m_1}(\hat{r})]\,[e^{i\mathbf{J}\cdot\boldsymbol{\theta}/\hbar} Y_{l_2}^{m_2}(\hat{r})]. \end{aligned} \tag{1.20}$$

Here we have of course implied that $\hat{r}'$ is the unit vector obtained by rotating $\hat{r}$ through $\boldsymbol{\theta}$. Now notice that $e^{i\mathbf{J}\cdot\boldsymbol{\theta}/\hbar}$ operating on $Y_{l_1}^{m_1}(\hat{r})$ yields only another member of the same multiplet characterized by the orbital angular momentum l_1. Similarly, $e^{i\mathbf{J}\cdot\boldsymbol{\theta}/\hbar}\, Y_{l_2}^{m_2}(\hat{r})$ also belongs to the multiplet labelled by l_2. Thus we have the result that the rotation operator operating on a product of the type $Y_{l_1}^{m_1} Y_{l_2}^{m_2}$ yields another product of the same set. Hence, a product of multiplets of S_1 and S_2 defines an invariant subspace of the direct product group S. Because S is a semi-simple Lie group, we can at once see that the product $M_{l_1} \otimes M_{l_2}$ will decompose into a sum of non-overlapping multiplets of S. From our knowledge of angular momentum algebra, we can write, in this simple case,

$$M_{l_1} \otimes M_{l_2} = M_{|l_1-l_2|} \oplus M_{|l_1-l_2|+1} \oplus \cdots \oplus M_{l_1+l_2}. \tag{1.21}$$

The dimensionality of each multiplet on the right-hand side is $(2L+1)$, L being the subscript labelling it. So long as we talk about the rotation group, this is just a statement of the triangle rule. It can be put in the more explicit form

$$Y_{l_1}^{m_1}(\hat{r})\, Y_{l_2}^{m_2}(\hat{r}) = \sum_{L=|l_1-l_2|}^{l_1+l_2} \sum_{M=-L}^{L} C_{m_1 m_2 M}^{l_1 l_2 L}\, Y_L^M(\hat{r}), \tag{1.22}$$

where $C_{m_1 m_2 M}^{l_1 l_2 L}$ is a Clebsch–Gordan coefficient.

Equation (1.22) may be inverted to yield

$$Y_L^M(\hat{r}) = \sum_{\substack{m_1+m_2=M \\ \Delta(l_1 l_2 L)}} C_{m_1 m_2 M}^{l_1 l_2 L}\, Y_{l_1}^{m_1}(\hat{r})\, Y_{l_2}^{m_2}(\hat{r}). \tag{1.23}$$

Equation (1.23) is an explicit statement of eqn (1.19) for the rotation group. It embodies what is usually called the addition of angular momenta l_1 and l_2.

Actually, these equations can be generalized to cover the product of any two multiplets of the rotation group. For example, instead of being multiplets relating to two distinct particles, M_{l_1} and M_{l_2} may as well be multiplets corresponding to orbital angular momentum and spin respectively. In this case, one of the spherical harmonics in the product $Y_{l_1}^{m_1} Y_{l_2}^{m_2}$ is replaced by a spin eigenfunction and Y_L^M by a higher order spinor with two components, i.e. an eigenfunction of the total angular momentum j and its projection μ on the z-axis. We can write this as

$$|j\mu\rangle = \sum_{\substack{m \\ m+\tau=\mu \\ \Delta(l\frac{1}{2}j)}} \sum_{\tau} C^{l\frac{1}{2}j}_{m\tau\mu} Y^m_l(\theta, \varphi)\, \chi^\tau_{1/2}, \tag{1.24}$$

where $\tau = \pm\frac{1}{2}$ and $\chi^\tau_{1/2}$ are spin eigenfunctions.

Chapter Two
Theory of the Auger Process

1. Introduction

When a neutral atom is ionized in an inner shell S, the system consisting of the ionized atom and the ejected electron at rest at infinity is characterized by a positive energy E_S depending upon the level from which the latter has been ejected. Such a system tends to reorganize itself so as to attain a state of lower energy through either of two processes:

1. The inner shell vacancy is filled by an electron from an outer shell S', the excess energy appearing as radiation (X-ray) of frequency ν given by

$$h\nu = E_S - E_{S'}. \tag{2.1}$$

2. Alternatively, the inner shell vacancy is still filled by an electron from the shell S', but the excess energy is imparted to another electron of the same atom belonging to a shell R which is ejected, so that the atom in its final state is left doubly ionized. The ejected electron leaves the atom with the kinetic energy

$$T_A = E_S - E_{S'} - E'_R, \tag{2.2}$$

where E'_R is the ionization potential for the shell R with respect to an atom which is already singly ionized in an inner shell. Obviously, the process is energetically allowed only if $E_S - E_{S'} > E'_R$.

The second alternative, namely, the process of radiationless reorganization of an atom just described, is known as the Auger effect. Continuation of this process may result in an atom with multiple vacancies in its outer shells, i.e. a highly charged ion. The object here, however, is to try to understand the basic process described above. Before this is gone into, it may be convenient to illustrate the classification of Auger transitions by means of a simple example.

Consider a primary vacancy in the K-shell. This vacancy is filled by an electron from some outer shell, say the L_2 shell. In case this induces an Auger transition, the energy so released is transferred to another bound electron which is ejected. If the latter happens to come from the L_3 shell, we have a KL_2L_3 Auger transition. Similarly, if an electron from the L_3 shell jumps in to fill a primary vacancy in the K shell and the balance of energy results in the ejection of another electron from the M_1 shell, we have a KL_3M_1 Auger transition. All Auger electrons, i.e. ejected electrons, resulting from a transition of the type KX_pY_q are called K Auger electrons. By lumping together processes involving final vacancies in different subshells of the L or M shells, as the case may be, one talks about the KLL or KLM Auger spectrum and so on. In a similar vein, one talks about L, M, etc. Auger electrons (corresponding to primary vacancies in the L, M, etc. shells) and about the LMM, LMN, MNN or MNO Auger spectra, etc.

A special type of Auger transition, called the Coster–Kronig transition, involves the filling of a primary vacancy by an electron from the same major atomic shell (i.e. belonging to the same principal quantum number N) but a different subshell. For example, an $L_1L_{2,3}M_{4,5}$ Coster–Kronig transition is said to occur when a primary vacancy in the L_1 subshell is filled by an electron from the L_2 or L_3 subshell and an electron from the M_4 or M_5 subshell is expelled to the continuum as a result of this rearrangement. Such transitions are energetically possible only in limited regions of the periodic table, because the energy difference $E(L_1) - E(L_{2,3})$ must in the first place be sufficient to effect ionization in the $M_{4,5}$ subshell.

A word of caution is now necessary. Since the electrons are indistinguishable, one cannot say which electron jumps into the inner shell vacancy and which is ejected. Thus in a KL_3M_1 transition, one can equally well say that the M_1 electron jumps in and the L_3 electron is ejected. Both processes enter symmetrically in a quantum-mechanical description of the transition. They give rise to the direct and exchange terms.

2. Mechanism of the Auger process. Real and virtual photons

From the description given so far, certain questions begin to emerge about the nature of these transitions.

1. From the nomenclature given above and also from its historic association with radiative transitions, the impression may be gained that the Auger process is an inner photo-effect of X-rays. For example, the notion of a fluorescence yield ω hinges on the experimental observation that an inner shell vacancy does not always lead to the manifest emission of a physical photon, the other possibility being the ejection of an outer-shell electron. The number ω_i for any given shell is defined by

$$\omega_i = \frac{P_{x_i}}{P_{a_i} + P_{x_i}}, \tag{2.3}$$

P_{a_i} and P_{x_i} being the Auger and X-ray transition probabilities for this shell. One is thus tempted to hypothesize that, part of the time, the X-ray photon resulting from the filling of the inner-shell vacancy gets absorbed by an electron which, in its turn, is ejected. The question naturally arises: is this process mediated by a real, physical photon? For example, in the KL_1L_2 transition, does the transition $L_1 \rightarrow K$ give rise to a physical photon which ejects an L_2 electron?

2. Another aspect of the process is its obvious relationship to the nuclear phenomenon known as internal conversion. Internal conversion, in the usual sense of nuclear de-excitation resulting in the ejection of an atomic electron, is to be distinguished from the Auger effect so far as questions of internal structure are concerned. There is indeed a difference of scale involved which is best illustrated by the different orders of magnitude of the energy transfer. However, in both types of conversion this transfer occurs via the electromagnetic field, the properties of which are well understood. For one thing, this makes it possible to apply the standard techniques of perturbation theory to both phenomena. In one case the basic electromagnetic interaction is between an atomic electron and the nucleus, in the other between two atomic electrons. In either case the matrix element for the transition factors out into two parts depending on the two constituents of the interaction. For nuclear internal conversion, the dynamical description of one of these constituents, i.e. the nucleus, depends intimately upon the nature of nuclear forces, which is less well-known than the electromagnetic interaction. Nevertheless, one can make some progress by writing the nuclear part of the matrix element in a general form, a detailed study of the latter being relegated to nuclear structure theory. In the limit of the point nucleus, the problem becomes simpler, the nucleus being regarded only as a source of electromagnetic field with definite energy, angular momentum and parity. Details of nuclear structure are suppressed in this limiting model. For atomic internal conversion, i.e. the Auger process, the Hamiltonian for the two participating electrons can be written explicitly.

The underlying similarity between the two processes is particularly interesting, because at least on heuristic grounds one would expect that the mechanism of energy transfer by the electromagnetic field should be similar in these two cases. In other words, the question raised under item (1) above should not have different answers for the atomic and nuclear processes. If in a nuclear internal conversion the nucleus transfers energy to an atomic electron via an actual, physical photon, then it would seem likely that a very similar thing happens when an atomic electron imparts its excess energy to an Auger electron. The electromagnetic field should not be able to distinguish between a nucleus and an electron so long as they act only as sources of quanta having definite energy, angular momentum and parity.

A definitive answer to the question of the mechanism of nuclear internal conversion was given by Taylor and Mott (1933). For simplicity, they considered the gamma rays to be emitted by an excited particle of discrete charge in the nucleus, so that the analogy with the atomic process is clearly brought out. As this particle returns to its normal state, an inner shell electron is ejected to the continuum. The transition probability for this process is calculated using first order perturbation theory, with the interaction corresponding to a dipole or quadrupole transition. The probability per unit time that an electron is ejected from the K shell is denoted by b. As we shall see,

$$b = \frac{4\pi^2}{h^2} \left| \int \Psi_f{}^{\dagger} \{-e\varphi - e(\boldsymbol{\alpha} \cdot \mathbf{A})\} \Psi_i \, \mathrm{d}\tau \right|^2, \tag{2.4}$$

where Ψ_i is the initial wave function of the K electron and Ψ_f its wave function in the final free state.‡ φ and $\mathbf{A}$ are the scalar and vector potentials representing the electromagnetic field of the nucleus and $\boldsymbol{\alpha}$ the Dirac velocity vector. Taylor and Mott also calculated the radiative transition rate p corresponding to these field potentials. For dipole transitions

$$p = 16\pi^2 B^2 \nu / 3hc, \tag{2.5}$$

where ν is the frequency of the gamma quantum for the above transition and B is an undetermined constant. In eqn (2.5), p represents the radiative rate in the absence of the K shell electron, i.e. it is a purely nuclear quantity. Now the question which they ask is: "Does p represent the rate of emission of gamma quanta by the nucleus even in the presence of the K electron?" If one regards the ejection of the K electron purely as an internal conversion of physical gamma quanta, then, of course, the book-keeping would proceed as follows:

1. Total rate of emission of gamma quanta per unit time from the nucleus $= p$.
2. Rate of ejection of conversion electrons $= b$.
3. Rate of emission of gamma quanta actually observed $\equiv g =$ difference of (1) and (2) $= p - b$.

Since the internal conversion coefficient is defined by

$$\alpha \equiv \frac{\text{rate of ejection of conversion electrons}}{\text{rate of emission of gamma quanta actually observed}},$$

it follows from the above considerations that

$$\alpha = \frac{b}{g} = \frac{b}{p-b}. \tag{2.6}$$

Unfortunately, this simple state of affairs does not hold. What happens is that the K electron perturbs the nucleus, and the total probability that the nucleus would make a radiative transition becomes greater than it would be if the K electron were not there.

‡ Here Ψ_f is normalized to a flux of one electron/unit time (see eqn 2.37b).

Actually, the enhancement is nearly equal to b, the rate of ejection of conversion electrons. Thus the total decay rate of the nucleus in the presence of K-shell electrons is nearly equal to $p+b$. Thus $g \approx (p+b)-b=p$. Hence

$$\alpha = \frac{b}{g} \approx \frac{b}{p}. \tag{2.7}$$

Thus the rate of emission of gamma quanta actually observed seems to be effectively equal to p, i.e. the rate is almost unaffected by the presence of the K electrons. In fact this conclusion is supported fairly well by experiments.

Equation (2.7) has some interesting implications so far as the mechanism of the process is concerned. It is quite possible to conceive of a nucleus, of low atomic number Z and radius larger than that of most radioactive nuclei, where $b > g$, g being the emission rate of actually observed photons. But because $g \approx p$, this implies $b > p$, a situation where the number of "converted electrons" is greater than the total number of quanta emitted by the bare nucleus in any given time interval. This is an eventuality which cannot simply be understood if the ejection of the electron is thought of as due to the absorption of a physical gamma ray quantum. The inevitable conclusion is that the ejection of electrons is nearly all due to direct interaction between the nucleus and the electrons. The term "internal conversion" loses its meaning in the sense that not many physical gammas are actually "converted" into continuum electrons. Taylor and Mott showed, however, that a small fraction (of the order of $e^2/\hbar c$) is responsible for the ejection of such electrons. Strictly speaking the two effects cannot be separated, because the atomic inner shell electrons do perturb the nucleus, so that there is a small probability ($O(e^2/\hbar c)$) of the nucleus making a transition to the ground state (or a lower excited state) through the emission of a physical photon which ejects an electron. But the ejection process is heavily weighted in favour of direct interaction between the nucleus and the electron.

The fact that the Auger process has also to be explained in terms of direct interaction between atomic electrons was clearly pointed out by Wentzel (1927), who gave the first perturbation-theoretic derivation of the transition probability for the non-relativistic situation. In fact, by writing the energy denominator in the expression for the transition amplitude

$$\sum_p \int_0^\infty \frac{\langle f|H_{\text{int}}|p\rangle\langle p|H_{\text{int}}|i\rangle}{E_i - E_f}\,\mathrm{d}E$$

in the form

$$E_i - E_f = T_A - E = (E_S - E_{S'}) - |E_R'| - E,$$

(cf. eqn 2.2), where E is the integration variable corresponding to the continuous energy spectrum of the ejected electron and H_{int} the perturbation, he showed that the amplitude peaks at

$$E = (E_S - E_{S'}) - |E'_R|. \tag{2.8}$$

Formally, it looks exactly as if a photon of energy $(E_S - E_{S'})$ has knocked out an electron with binding energy E'_R. However, Wentzel emphasized the point that despite this formal similarity with a radiative transition, the Auger process is radiationless in the strictest sense of the term.

Physically, the perturbation causing the transition arises from the Coulomb interaction between neighbouring electrons. In a way, this is a bound-state analogue of inelastic electron scattering by atoms.

Now, it is well known that non-relativistic quantum mechanics does not provide a complete framework for the description of radiative transitions. For a radiationless transition involving the interaction of two atomic electrons, non-relativistic perturbation theory such as that employed by Wentzel would be adequate so long as (a) retardation of the interaction can be neglected and (b) the velocities of the electrons are non-relativistic. Broadly speaking, these conditions seem to hold for the lighter elements. However, it should be made clear that, actually, there is only one correct and satisfactory theory, namely, the Lorentz-covariant theory of quantum electrodynamics. Viewed from the standpoint of this theory, the Auger effect is caused by the retarded electromagnetic interaction between two bound-state Dirac electrons. The electromagnetic interaction consists of (a) the charge–charge interaction or Coulomb interaction (i.e. the time-like part of the four-current interaction) and (b) the current–current or magnetic interaction (the space-like part). In the non-relativistic limit $v/c \to 0$, the space-like part drops out and we have a pure Coulomb-type interaction. That is why Wentzel's theory gives correct results in this limit.

Once we begin to think in terms of a retarded electromagnetic interaction between two atomic electrons, the question arises as to the nature of the "signal" exchanged between these electrons. Even in classical electrodynamics, this signal is taken to consist of so-called virtual photons travelling with the velocity of light. Quantum electrodynamics makes things more formal and says that the interaction is *mediated* by the exchange of virtual quanta of various frequencies. The spectrum of virtual photons for any given process is determined by the physical conditions governing the interaction. In contrast to real photons, virtual photons are basically a conceptual aid. Mathematically, the justification for replacing the effect of the electromagnetic field of a charged particle by means of a virtual radiation spectrum rests on the Fourier theorem—one analyses the perturbing field of the particle into pure harmonic waves and identifies them as the components of such a spectrum. The fact that these components have an ephemeral or virtual character can be seen from the fact that such Fourier decomposition allows negative as well as positive frequencies.‡ Physically, the procedure is understood by remembering that quantum mechanics makes no *a priori* qualitative distinction between the field of a charged particle and a radiation field. In any given physical process, a system may be regarded as passing from an initial physical state to a final physical state via some virtual intermediate states through the emission or absorption of virtual quanta. These virtual states are not constrained to conserve energy or momentum, because they are not experimentally observed (Rose, 1955).

We must point out here another reason for considering the effect to be a radiationless transition rather than a conversion process. Consider, for example, the KL_1L_2 transition. Let us assume that there are two successive steps: (a) there is a radiative transition $L_1 \rightarrow K$ giving rise to a physical photon and (b) the latter now ejects an L_2 electron into the continuum. Now, the transition $L_1 \rightarrow K$ is forbidden by the dipole selection rule $\Delta l = \pm 1$ for radiative transitions. We have here an Auger "conversion" for which there is no radiative counterpart. The effect must clearly be regarded as arising from a direct interaction between electrons.

The concept of virtual quanta outlined above was originally introduced by Fermi (1924) in the course of his treatment of atomic excitation by alpha particle bombardment. The method was later extended to relativistic problems by von Weizsäcker (1934) and Williams (1935). Williams tried to provide a phenomenological justification for "replacing" the electromagnetic field of a charged particle by an "equivalent" radiation pulse. In doing so, he ran into some rather stringent conditions for the existence of such equivalence. Present-day quantum field theory does not encounter this difficulty, because here the Fourier decomposition of the interaction is a purely formal procedure. The concept of virtual quanta now provides the framework for any relativistic treatment of interactions, electromagnetic or otherwise.

Let us now recall eqn (2.8) and Wentzel's remarks in this connection. The Auger transition which corresponds to an initial vacancy in the shell S and to final vacancies in the shells S' and R exhibits a peaking of the transition amplitude at the energy

$$E = (E_S - E_{S'}) - |E'_R|.$$

This is the kinetic energy which the ejected electron is most likely to have. Formally, it looks as if a photon of energy $\hbar\omega = E_S - E_{S'}$ has caused the ejection of an electron from a shell R with binding energy E'_R. Using the virtual photon description, we can say that the virtual photon spectrum peaks sharply at the frequency $\omega = (E_S - E_{S'})/\hbar$ (obviously, ejection can take place only if $(E_S - E_{S'}) \geqslant |E'_R|$). In other words, the peak of the virtual photon spectrum is

‡ Negative frequencies imply negative energies, i.e. absorption rather than emission.

taken to occur at the energy which just brings the Auger electron to the peak of its observed energy spectrum. Physically, this is just a statement of the conservation of energy for states which are experimentally observed.‡ But in calculations involving the virtual spectrum, the sharp peaking behaviour ensures that in practice we have to deal with a narrow band of frequencies and not the entire range from $-\infty$ to $+\infty$.

3. The Møller formula

We can now try to write down Møller's relativistic formula for the probability amplitude of radiationless transitions. Møller (1931) treated this as a time-independent two-electron problem within the framework of relativistic single-particle theory. One of these electrons acts as the source of virtual photons with which the second electron interacts. Making a Fourier expansion of the charge and current densities into harmonic components, we have

$$\rho(\mathbf{r}, t) = \int e^{i\omega t} \rho_\omega(\mathbf{r})\, d\omega + \text{c.c.} \tag{2.9}$$

and

$$\mathbf{j}(\mathbf{r}, t) = \int e^{i\omega t} \mathbf{j}_\omega(\mathbf{r})\, d\omega + \text{c.c.} \tag{2.10}$$

With these densities, the scalar and vector potentials at the position of the second electron due to the first are obtained by employing the usual Green's function,

$$\varphi(\mathbf{r}_2, t) = \int \rho(\mathbf{r}_1, t) \frac{e^{ikr_{12}}}{4\pi r_{12}} d^3\mathbf{r}_1 \tag{2.11}$$

and

$$\mathbf{A}(\mathbf{r}_2, t) = \int \mathbf{j}(\mathbf{r}_1, t) \frac{e^{ikr_{12}}}{4\pi r_{12}} d^3\mathbf{r}_1, \tag{2.12}$$

with

$$r_{12} = |\mathbf{r}_1 - \mathbf{r}_2|.$$

Thus the interaction energy between the two harmonic charge and current systems is given by

$$H_{\text{int}} = \iint d^3 r_1\, d^3 r_2 \frac{e^{ikr_{12}}}{4\pi r_{12}} [\rho_{1,\omega}(\mathbf{r}_1)\, \rho_{2,\omega}(\mathbf{r}_2) + \mathbf{j}_{1,\omega}(\mathbf{r}_1) \cdot \mathbf{j}_{2,\omega}(\mathbf{r}_2)]. \tag{2.13}$$

‡ The dispersion around the peak is a consequence of the uncertainty principle.

The retardation in the interaction is expressed through the scalar Green's function $e^{ikr_{12}}/(4\pi r_{12})$, where $k = \omega/c$, the wave number of the virtual photon with frequency ω. Remembering the peaking of ω at $E_S - E_{S'}/\hbar$, we now formally write this frequency as $\omega_{if} = (E_i - E_f)/\hbar$, where E_i and E_f are the initial and final stationary energy values involved in the virtual photon transition (i.e. $E_i = E_S$ and $E_f = E_{S'}$). Then

$$\rho_\omega(\mathbf{r}) \sim \rho_{if}\,\delta(\omega - \omega_{if}), \tag{2.14}$$

and

$$\mathbf{j}_\omega(\mathbf{r}) \sim \mathbf{j}_{if}\,\delta(\omega - \omega_{if}), \tag{2.15}$$

so that equations (2.9) and (2.10) reduce to

$$\rho(\mathbf{r}, t) = \rho_{if}\,e^{i\omega_{if}t}, \tag{2.16}$$

and

$$\mathbf{j}(\mathbf{r}, t) = \mathbf{j}_{if}\,e^{i\omega_{if}t}. \tag{2.17}$$

In his derivation of the probability amplitude, Møller used eqns (2.16) and (2.17) to interpret the atom as a classical oscillating distribution of charge and current densities. At the same time, for the densities themselves he used the expressions

$$\rho_{if} = -e\psi_f^+\psi_i \tag{2.18}$$

and

$$\mathbf{j}_{if} = -e\psi_f^+\boldsymbol{\alpha}\psi_i, \tag{2.19}$$

where ψ_i and ψ_f are the time-independent parts of the initial and final four-component Dirac wave functions of an electron in a Coulomb field and $\boldsymbol{\alpha}$ is the Dirac velocity vector. Thus ρ_{if} and $\mathbf{j}_{if}$ are obviously the transition charge and current densities for such an electron. This embodies a semiclassical approach which may be avoided by interpreting ω_{if} not as the classical frequency of oscillation of the atomic charge–current system, but as the frequency of the virtual photons mediating the radiationless transition.

Using the preceding equations, one finally obtains the probability amplitude

$$\langle f|H_{\text{int}}|i\rangle = e^2 \iint \psi_f^+(2)\,\psi_f^+(1)\,(1 - \boldsymbol{\alpha}_1\cdot\boldsymbol{\alpha}_2)\frac{e^{ikr_{12}}}{4\pi r_{12}}\,\psi_i(2)\,\psi_i(1)\,d^3r_1 d^3r_2. \tag{2.20}$$

The expression on the right-hand side may be regarded as the matrix element of the interaction operator

$$\mathcal{O} = e^2(1 - \boldsymbol{\alpha}_1\cdot\boldsymbol{\alpha}_2)\frac{e^{ikr_{12}}}{r_{12}}, \tag{2.21}$$

which was used in the relativistic theory of the Auger effect (Massey and Burhop, 1936; Rose, 1955). It was first obtained by Hulme (1936). The first term in this operator represents the Coulomb repulsion between the two electrons and the second is the relativistic current–current interaction. Equation (2.21) may be put into a multipole form if we use an identity for the Green's function

$$\frac{e^{ik|\mathbf{r}_1-\mathbf{r}_2|}}{|\mathbf{r}_1-\mathbf{r}_2|}=4\pi ik\sum_{l=0}^{\infty}\sum_{m=-l}^{l}j_l(kr_<)\,h_l(kr_>)\,Y_l^{m*}(\theta_1,\varphi_1)Y_l^m(\theta_2,\varphi_2), \tag{2.22}$$

where $r_<$ denotes the lesser and $r_>$ the greater of r_1 and r_2. $j_l(kr)$ and $h_l(kr)$ are the spherical Bessel functions. Equation (2.22) exhibits the superposition of multipoles in the virtual radiation field.

Going over to the non-relativistic limit, $|\boldsymbol{\alpha}|\to 0$ and neglecting the retardation, one obtains from eqn (2.20) Wentzel's non-relativistic expression for the probability amplitude

$$\langle f|H_{\text{int}}^{\text{NR}}|i\rangle=e^2\int\!\!\int\psi_f^\dagger(2)\psi_f^\dagger(1)\frac{1}{R}\psi_i(1)\psi_i(2)\,\mathrm{d}^3r_1\,\mathrm{d}^3r_2, \tag{2.23}$$

which is just the matrix element of the static Coulomb interaction between the two electrons.

4. Generalized derivation of the Møller formula

From the discussion so far, it is clear that the Auger effect may be treated as a second order process in quantum electro-dynamics,‡ the complete interaction being visualized in terms of the retarded exchange of a virtual photon between the participating electrons. It can be shown that for light charged particles such as electrons, the energy transfer may be described in terms of the exchange of a single photon (Biedenharn and Brussaard, 1965). The fact that the perturbing energy, i.e. the interaction energy of the two electrons, is small is also significant in another sense. It ensures that the probability is small for the transition to take place in a time equal to the period of the atom, so that it is meaningful in the first place to regard the electrons as being initially in definite stationary states (Mott and Massey, 1965).

Consider the second order scattering operator $S^{(2)}$ given by

$$S^{(2)}(t,t_0)=-\tfrac{1}{2}\int_{t_0}^{t}\mathrm{d}t_1\int_{t_0}^{t}\mathrm{d}t_2\,P[V(t_1)V(t_2)], \tag{2.24}$$

‡ i.e. a process for which the S-matrix element is represented by a second order Feynman graph with two vertex points.

where t_0 and t refer to the initial and final instants of time for the scattering process and P is the Dyson chronological operator (Dyson, 1949a). $V(t_1)$ and $V(t_2)$ are the values of the interaction Hamiltonian at times t_1 and t_2, $V(t)$ being given by

$$V(t) = -\int j_\mu(x)\, A_\mu(x)\, \mathrm{d}^3\mathbf{r}, \tag{2.25}$$

where A_μ and j_μ are the potential and current density four vectors. The latter is given by

$$j_\mu(x) = ie\mathrm{N}(\bar{\psi}(x)\,\gamma_\mu\,\psi(x)), \tag{2.26}$$

where ψ and $\bar{\psi}$ are the electron and positron field operators and N the normal-ordering operator.

Using eqn (2.25) in eqn (2.24), we get

$$S^{(2)} = -\tfrac{1}{2}\int P[j_\mu(x) j_\nu(y)]\, P[A_\mu(x)\, A_\nu(y)]\, \mathrm{d}^4x\, \mathrm{d}^4y, \tag{2.27}$$

where x and y are two space-time points.

We now specialize to the matrix element ${}^eS^{(2)}$ between states of the system which contain no photon. Replacing the operator $P[A_\mu(x) A_\nu(y)]$ by its vacuum expectation value

$$\langle P[A_\mu(x)\, A_\nu(y)]\rangle_0 = \tfrac{1}{2}\delta_{\mu\nu}\, D^C(x-y), \tag{2.28}$$

where

$$D^C(x-y) = \frac{-2i}{(2\pi)^4}\int \frac{e^{ik(x-y)}}{k^2}\, \mathrm{d}^4k, \tag{2.29}$$

we get

$${}^eS^{(2)} = -\tfrac{1}{2}\iint j_\mu(y)\, D^C(y-x)\, j_\mu(x)\, \mathrm{d}^4y\, \mathrm{d}^4x. \tag{2.30}$$

D^C is the well-known photon propagator.

We now make the usual expansion of the fermion field operators in terms of the appropriate Dirac basis functions. On using the anti-commutation properties of the fermion creation and annihilation operators $a^\dagger$ and a which appear in this expansion, the only non-vanishing terms in eqn (2.30) turn out to be

$$\begin{aligned}
& a_C^\dagger a_A a_D^\dagger a_B [\bar{\psi}_C(x)\gamma_\mu\psi_A(x)]\,[\bar{\psi}_D(y)\gamma_\mu\psi_B(y)] \\
&+ a_D^\dagger a_B a_C^\dagger a_A [\bar{\psi}_C(y)\gamma_\mu\psi_A(y)]\,[\bar{\psi}_D(x)\gamma_\mu\psi_B(x)] \\
&+ a_C^\dagger a_B a_D^\dagger a_A [\bar{\psi}_C(x)\gamma_\mu\psi_B(x)]\,[\bar{\psi}_D(y)\gamma_\mu\psi_A(y)] \\
&+ a_D^\dagger a_A a_C^\dagger a_B [\bar{\psi}_C(y)\gamma_\mu\psi_B(y)]\,[\bar{\psi}_D(x)\gamma_\mu\psi_A(x)].
\end{aligned} \tag{2.31}$$

Here ψ and $\bar{\psi}$ are the electron wave function and its adjoint respectively. The labels A and B stand for initial electron states while C and D stand for final electron states. A little algebraic manipulation yields the matrix element

$$^{e}S^{(2)}_{i\to f} = e^2 \iint \{\bar{\psi}_D(y)\gamma_\mu \psi_B(y)\, D^C(y-x)\,\bar{\psi}_C(x)\gamma_\mu\psi_A(x) - \bar{\psi}_C(y)\gamma_\mu\psi_B(y)\, D^C(y-x)\,\bar{\psi}_D(x)\gamma_\mu\psi_A(x)\}\,\mathrm{d}^4 y\,\mathrm{d}^4 x. \quad (2.32)$$

In eqn (2.32) the two terms which differ only in the interchange of the final state indices represent the direct and exchange terms respectively. Figures 2.1(a) and (b) are the corresponding Feynman graphs. We may write

$$^{e}S^{(2)}_{i\to f} = S^{(2)}_{AB;CD} - S^{(2)}_{AB;DC}, \quad (2.33)$$

where

$$S^{(2)}_{AB;CD} = e^2 \iint \bar{\psi}_D(y)\gamma_\mu\psi_B(y)\, D^C(y-x)\,\bar{\psi}_C(x)\gamma_\mu\psi_A(x)\,\mathrm{d}^4 y\,\mathrm{d}^4 x. \quad (2.34)$$

Let us write out explicitly the time-dependence of the wave functions

$$\psi_A(x) = \psi_A(\mathbf{r})\,\mathrm{e}^{-i\omega_A t_1}, \text{ etc.}$$

We also note that

$$D^C(y-x) = -\frac{i}{(2\pi)^4}\int \frac{\mathrm{e}^{i\mathbf{k}\cdot(\mathbf{r}_1-\mathbf{r}_2)-i\omega(t_1-t_2)}}{k^2-\omega^2}\,\mathrm{d}^4 k. \quad (2.35)$$

Using

$$\int \mathrm{e}^{i(\omega_D-\omega_B+\omega)t_2}\,\mathrm{d}t_2 = 2\pi\delta(\omega_D-\omega_B+\omega),$$

$$\int \frac{\mathrm{e}^{i\mathbf{k}\cdot(\mathbf{r}_1-\mathbf{r}_2)}}{k^2-\omega^2}\mathrm{d}^3 k = \frac{4\pi}{|\mathbf{r}_1-\mathbf{r}_2|}\int_0^\infty \frac{\sin k|\mathbf{r}_1-\mathbf{r}_2|}{k^2-\omega^2}\,k\,\mathrm{d}k,$$

and

$$\int_0^\infty \frac{\sin k|\mathbf{r}_1-\mathbf{r}_2|}{k^2-\omega^2}\,k\,\mathrm{d}k = \frac{\pi}{2}\mathrm{e}^{i\omega|\mathbf{r}_1-\mathbf{r}_2|/c},$$

we finally get, after integrating over t,

$$\begin{aligned} S^{(2)}_{AB;CD} &= -\frac{i}{2}\delta(\omega_{CA}+\omega_{BD})\int \frac{\mathrm{e}^{i\omega_{AC}|\mathbf{r}_1-\mathbf{r}_2|/c}}{|\mathbf{r}_1-\mathbf{r}_2|}(\rho_{CA}\rho_{DB}-\mathbf{j}_{CA}\cdot\mathbf{j}_{DB})\,\mathrm{d}^3 r_1\,\mathrm{d}^3 r_2 \\ &= \delta(\omega_{CA}+\omega_{BD})\,U_{AB;CD}, \end{aligned} \quad (2.36)$$

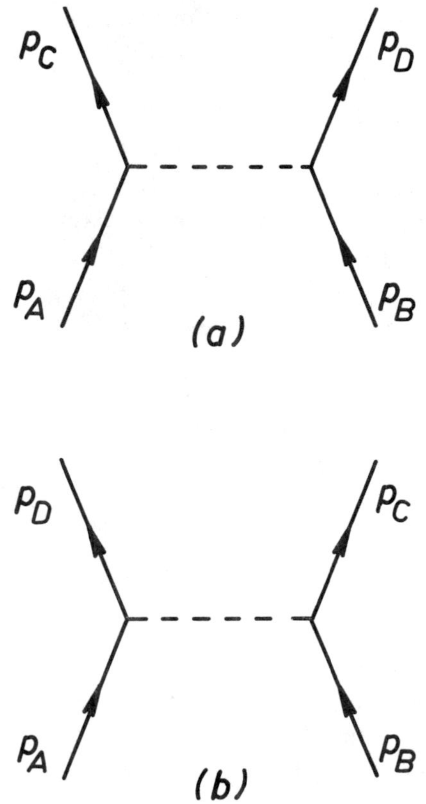

Fig. 2.1. (a) and (b) Feynman diagrams for the direct and exchange terms in the two-electron interaction.

where $U_{AB;CD}$ stands for the factors other than the δ-function. Here $\omega_{AC} = \omega_A - \omega_C$ and the other symbols have their usual meaning. As before, the ρ's and **j**'s represent transition charge and current densities. The δ-function implies conservation of energy. We have now obtained the Møller formula from quite general electrodynamic considerations, with the direct and exchange terms occurring in a natural way. Our derivation explicitly shows the Lorentz invariant character of this formula.

The transition probability per unit time, including both direct and exchange terms, is

$$P = (2\pi/\hbar)|U_{AB;CD} - U_{AB;DC}|^2 \,\delta(\omega_{CA} + \omega_{BD})\,\rho(E_f), \qquad (2.37)$$

where $\rho(E_f)$ is the density of final states. Equation (2.37) gives the general expression for the transition rate. In making calculations, one must specify the normalization of the electron ejected into the continuum. Wentzel normalized the continuum orbital to a flux of one electron per unit energy interval, which gives $\rho(E_f) = 1$. Equation (2.37) then assumes the form

$$P = \frac{2\pi}{\hbar}\,|U_{AB;CD} - U_{AB;DC}|^2\,\delta(\omega_{CA} + \omega_{BD}). \tag{2.37a}$$

An alternative procedure is to take a flux of one ejected electron per unit time (Bambynek *et al.*, 1972; Oppenheimer, 1929; Gaunt, 1930), so that $\rho(E_f) = 1/h$ and the transition rate becomes

$$\frac{1}{\hbar^2}\,|U_{AB;CD} - U_{AB;DC}|^2\,\delta(\omega_{CA} + \omega_{BD}). \tag{2.37b}$$

We must caution the reader about the different types of normalization occurring in the literature. A commonly used unit of energy is the Hartree (1 Hartree $= e^2/a_0 = 27.2$ eV), which is used in conjunction with the atomic unit of time $\tau = \hbar a_0/e^2 = \hbar^3/me^4 = 2.42 \times 10^{-17}$ s and the Bohr radius a_0 as the unit of length (McGuire, 1974a). Using these units, the transition rate becomes, in the non-relativistic limit,

$$\frac{2\pi}{\tau}\left|\left\langle i\left|\frac{1}{|\mathbf{r}_1 - \mathbf{r}_2|}\right|f\right\rangle\right|^2, \tag{2.37c}$$

where i and f denote the initial and final two-electron states. A further simplification occurs if the continuum orbital is normalized to 2π per Hartree, when the factor $2\pi/\tau$ drops out, the transition rate being measured per atomic time unit.

We must of course sum the expression (2.37) over the quantum numbers describing the transitions under consideration. For example, in *jj* coupling the *KLL* group of Auger electrons resolves into the six lines $KL_aL_b(a, b = 1, 2, 3)$. Each of the six transition probabilities P_{ab} is given by

$$P_{ab} = \frac{2\pi}{\hbar}\sum_{m_{L_a}}\sum_{m_{L_b}}\sum_{\kappa_\infty}|U_{ab} - U_{ba}|^2\,\delta(E_{L_a} + E_{L_b} - E_K - E_\infty), \tag{2.38}$$

where we have rewritten the δ-function in a more explicit form. Here κ is the Dirac quantum number for the single electron, related to the orbital angular momentum l and the total angular momentum j by

$$\kappa = (l - j)(2j + 1). \tag{2.39}$$

κ_∞ is the value of κ in the continuum state and is well defined by the conservation of angular momentum and parity. m_{L_a} and m_{L_b} are the z-components of the total angular momentum of the L shell electrons. Summation and averaging over the two states of the K shell electron gives a factor of unity.

The expressions derived in this section form the starting point of any relativistic calculation of the Auger transition probability. To proceed further, one must begin by specifying the single-particle wave functions entering into the Møller formula and then get on with the task of performing the integration over the spatial coordinates of the two electrons. In the non-relativistic limit, one neglects the current–current interaction and calculates the matrix element of only the electrostatic part. In either case, this is the stage where one begins to worry about coupling schemes corresponding to various limits of the interaction, the question of screening and of integrating the appropriate differential equations to obtain the radial parts of the single-electron wave functions. We shall presently examine the state of the art in these matters of technique.

5. Some related concepts and processes

Although the primary object of this work is to present the theoretical aspects of the type of radiationless transitions known as the Auger process, we shall briefly introduce here some related concepts and processes which are essential to this study.

The *fluorescence yield* of an atomic shell or subshell is defined as the probability that a vacancy in that shell or subshell is filled through a radiative transition. Historically, the Auger effect was discovered in connection with the study of such radiative transitions when it was found that the emission of monoenergetic electrons constituted an alternative competing process through which de-excitation of atoms can occur. We shall not, however, be concerned with the original definition of fluorescence yield given by Barkla (1918) in connection with this early work.

Consider an atom with a vacancy in an inner shell. Obviously, such an atom is in an excited state and decays through the channels available to it. These channels are:

1. radiative transitions
2. Auger transitions
3. Coster–Kronig transitions (where allowed).

We have already seen what a Coster–Kronig transition is. The total width Γ of such a state is therefore given by

$$\begin{aligned}\Gamma &= \hbar\,(\text{total transition probability } P \text{ per unit time})\\ &= \hbar\,(P_R + P_A + P_{CK}),\end{aligned} \tag{2.40}$$

where P_R, P_A and P_{CK} are the transition probabilities per unit time for radiative, Auger and Coster–Kronig transitions respectively. The fluorescence yield ω is given by

$$\omega = \frac{P_R}{P} = \frac{P_R}{P_R + P_A + P_{CK}}. \tag{2.41}$$

Correspondingly, the Auger yield a is defined by

$$a = \frac{P_A}{P}. \tag{2.42}$$

If we neglect Coster–Kronig transitions, then

$$a = \frac{P_A}{P_R + P_A} = 1 - \omega. \tag{2.43}$$

Now coming back to item (3), when a vacancy in an inner shell of an atom is filled by another electron of the same shell but different subshell with the ejection of an electron from an outer shell with lower energy, the resulting radiationless transition is a *Coster–Kronig transition.* In effect we have here a transfer of vacancies between the two subshells. This provides the third mode of de-excitation referred to above. The energy released by this vacancy shifting is transferred to an electron in a higher shell via Coulomb interaction. Take, for example, the $L_1L_{2,3}M_{4,5}$ transition. Here an initial vacancy in the L_1 subshell is filled by an L_2 or L_3 electron and the balance of energy results in the ejection of an electron from the M_4 or M_5 subshell. Coster–Kronig transitions find an important application in the interpretation of L-series line intensities. For many elements, $L\beta_3(L_1 \to M_3)$ and $L\beta_4(L_1 \to M_2)$ lines are found to be either absent or abnormally weak. On the other hand, lines originating in L_2 and L_3 levels appear to have considerable intensity. This can be accounted for by the occurrence of Coster–Kronig transitions with an initial vacancy in the L_1 subshell. They enhance the level widths in the L_1 state and diminish the relative intensity of the $L\beta_4$ or $L\beta_3$ line relative to lines originating in the L_2 or L_3 state. Since after a Coster–Kronig transition the atom is left in a state of double ionization, these transitions have also been found responsible for the enhancement of certain L-series satellite line intensities. A transition in which both the final vacancies occur in the same shell as the initial vacancy—but in a different subshell or two different subshells—is called a super Coster–Kronig transition. For example, a vacancy in the N_4 or N_5 subshell may decay through a transition of the type $N_{4,5}N_{6,7}N_{6,7}$.

Because the energy necessary for the ejection of an outer-shell electron is released in the first place by the shifting of vacancies between two different subshells of the same shell (i.e. it is the difference between the binding energies of these subshells), Coster–Kronig transitions are energetically possible only for certain regions of Z values. The transition rates exhibit sharp cut-off at critical Z values.

An accurate knowledge of atomic fluorescence yields becomes necessary in many practical fields (Bambynek *et al.*, 1972). One must then devise some operational method for applying the definition of ω given above. For a sample of many atoms, the probabilistic definition may be interpreted in terms of the number of primary vacancies present and the number of photons emitted when these vacancies are filled. Take, for example, the *K*-shell for which this is straightforward. This shell has only one l value, $l = 0$, and contains two $S_{1/2}$ electrons. For a given sample, the fluorescence yield of the *K*-shell may be interpreted as

$$\omega_K = \frac{I_K}{n_K}, \tag{2.44}$$

where I_K is the total number of characteristic *K* X-ray photons emitted from the sample and n_K is the number of primary *K*-shell vacancies.

The definition of fluorescence yields becomes more complicated for higher shells because of two factors:

1. Shells other than the *K*-shell consist of more than one subshells, with l values ranging from 0 to N-1, N being the principal quantum number. Different methods of ionization cause different distributions of primary vacancies in these subshells, so the average fluorescence yield for any given shell depends on how it was ionized, i.e. on the history of the system.
2. Coster–Kronig transitions make it possible for a primary vacancy created in one of the subshells to shift to a higher subshell before the vacancy is filled by another transition, radiative or otherwise. This effectively causes a mix-up in the counting of primary vacancies.

Great care must, therefore, be taken in formulating proper definitions of the quantities that are measured and in interpreting experimental results in a manner which is consistent with these definitions. For details the reader is referred to Bambynek *et al.* (1972) and Fink *et al.* (1966).

Chapter Three

The Coulomb Field and Coulomb Wave Functions

1. Introduction

The Auger effect involves energy transfer by the electromagnetic field. In the non-relativistic limit with no retardation, this field may be represented by the static Coulomb potential $\alpha Z/r$,‡ α being the fine-structure constant. To a first approximation, the two electrons participating in the process are described by the well-known single particle bound state and continuum wave functions for a point Coulomb potential. Physically, such a potential is not attainable in a real many-electron atom and one must consider the field due to the other electrons and the nucleus. However, given a certain faith in the pedagogical importance of simple models, one observes that these wave functions approximately describe a single electron outside a closed atomic shell, or a hole in an otherwise closed shell, in so far as the field can be taken to be effectively central and the nuclear charge Z is replaced by an effective charge $Z_{\text{eff}} = Z - \sigma$, σ being a screening constant. Such a central field has the well-known virtue that its solutions resolve into separate radial and spin-angular parts. Neglecting spin for the present, we can write

$$\Psi_{Nlm}(r, \theta, \varphi) = R_{Nl}(r)\, Y_l^m(\theta, \varphi).$$

Henceforth, we shall call these the screened analytic hydrogenic wave functions, the screening being fixed by the physical requirements of the problem under consideration. We should only like to remark here that for small values of the principal quantum number N, i.e. for the inner orbits, the wave function is vanishingly small outside a small range of the radial distance r. For an electron in such a state, the radial dependence $\sim r^{-1}$ of the hydrogenic potential should not differ too much from the actual potential experienced by the electron in a real atom, particularly if Z is large (which makes the average radius r decrease by a corresponding scale factor). For larger values of N, the hydrogenic

‡ Using units with $\hbar = c = 1$. Much of the discussion in this chapter is made more transparent if we express the Coulomb potential in terms of the fine-structure constant α. Generally, however, we shall prefer to use absolute units, because this will enable us to show the normalization explicitly.

approximation differs substantially from more realistic wave functions (Bambynek *et al.*, 1972).

Having pointed out the limitations of the screened Coulomb model, we must emphasize its usefulness as a basis for a quantitative treatment of atomic structure. Although physically approximate, its strength lies in its conceptual simplicity and, for that very reason, the mathematical tractability of its solutions. Non-relativistic Coulomb wave functions, which can be simply expressed in terms of confluent hypergeometric functions for the continuum (and associated Laguerre polynomials for bound states) are among the simplest of the higher transcendental functions. Their properties are well known and matrix elements involving them can be easily calculated. As we shall presently see, this simplicity derives from the well-known symmetry of the Coulomb field. Even where these wave functions have only limited physical validity, they can be used as a framework for discussing more sophisticated models. For example, relativity destroys the "accidental degeneracy" (or l-degeneracy) of the Coulomb field. The corresponding changes in the wave function may be described by a parameter γ ($\equiv \sqrt{\kappa^2 - \alpha^2 Z^2}$, κ being the Dirac quantum number) which is just a measure of this symmetry-breaking. In fact, it can be shown that the resulting relativistic Coulomb wave function can be obtained from the non-relativistic solution by means of a simple mapping depending on γ. As for more sophisticated calculational techniques, such as Hartree–Fock theory, the screened Coulomb model serves as a point of departure for these methods.

We shall, therefore, lean heavily on the Coulomb field and the Coulomb wave functions in our discussion of Auger calculations. We shall do this keeping fully in mind the limitations of this model for a many-electron atom. These many-electron aspects will be discussed in a later chapter. Our object in this chapter will be to review the symmetry properties of the Coulomb field and to see how they enter into a calculation of Auger transition probabilities.

One important aspect of this calculation is that it involves only one continuum (electron) wave function, the other single-electron states entering into the matrix element being bound states. This is in contrast to the more widely discussed scattering problem, which involves continuum states only. This makes for a relative simplicity of the Auger matrix elements. For non-relativistic (NR) Coulomb wave functions, we shall see that the radial integrals occurring in these matrix elements are just the Gaussian hypergeometric functions in the limit of no retardation. Even for more realistic wave functions, these integrals are simpler than those occurring in collision problems. Again this simplicity is not spurious but rooted in the symmetry of the Coulomb field. So far as NR Coulomb wave functions are concerned, this symmetry implies a concise and complete definition of the wave functions by means of their well-known recursion relations. This applies equally to discrete (i.e. bound) and continuum states. One might hope, therefore, that there would be corresponding recursion relations for the Auger radial integrals. This indeed is the case, and these relations turn out to be the well-known recursion relations for the ${}_2F_1$ functions. Quite apart from their theoretical interest, they result in considerable simplifications in the computation of Auger transition rates. In fact Callan (1961) used them to generate the ${}_2F_1$ functions by a simple matrix method on the computer, i.e. without recourse to the tedious and much less accurate method of series summation.

2. Symmetry properties of the Coulomb field

The symmetry of the Coulomb field is quite deep, having its origins in classical mechanics. Consider the non-relativistic (NR) Hamiltonian for the motion of a projectile with charge $-e$ in the Coulomb field of another particle having charge Ze

$$H_{\mathrm{NR}} = \frac{p^2}{2m} - \frac{Ze^2}{r}. \tag{3.1}$$

It is well known that the classical orbital angular momentum $\mathbf{L}_{\mathrm{cl}} = \mathbf{r} \times \mathbf{p}$ is a vector constant of the motion. Pauli (1926) pointed out that the vector $\mathbf{A}_{\mathrm{cl}}$ defined by

$$\mathbf{A}_{\mathrm{cl}} = |2mH_{\mathrm{NR}}|^{-1/2}\{Ze^2 m\hat{r} + \mathbf{L}_{\mathrm{cl}} \times \mathbf{p}\} \tag{3.2}$$

is also a constant of the motion, as can be verified by computing the Poisson brackets of its scalar components with H_{NR}.

We have here two vector invariants, $\mathbf{L}_{\mathrm{cl}}$ and $\mathbf{A}_{\mathrm{cl}}$, knowledge of which constitutes a complete integration of the non-relativistic Kepler problem (Biedenharn, 1965). Since the equation of notion is a second order differential equation, it cannot have more than two independent constants of integration, so that the Hamiltonian itself must be a function of $\mathbf{L}_{\mathrm{cl}}$ and $\mathbf{A}_{\mathrm{cl}}$. Actually, by defining the number η_{cl} so that

$$\eta_{\mathrm{cl}}^2 = \mathbf{A}_{\mathrm{cl}}^2 - \mathbf{L}_{\mathrm{cl}}^2 \quad \text{or} \quad \eta_{\mathrm{cl}} = -Ze^2 m|2mH_{\mathrm{NR}}|^{-1/2}, \tag{3.3}$$

it may be shown that

$$H_{\mathrm{NR}} = (Ze^2 m)^2/2m\eta_{\mathrm{cl}}^2. \tag{3.4}$$

While the vector $\mathbf{L}_{\mathrm{cl}}$ defines the plane of the orbit, $\mathbf{A}_{\mathrm{cl}}$ lies in this plane and is orthogonal to both $\mathbf{L}_{\mathrm{cl}}$ and $\mathbf{p}$. Thus $\mathbf{A}_{\mathrm{cl}}$ and $\mathbf{L}_{\mathrm{cl}}$ together form a convenient orthogonal basis for describing the orbits. Specifically, $\mathbf{A}_{\mathrm{cl}}$ defines the instantaneous direction of motion of the projectile, being perpendicular to it in the plane of the orbit. The parameter η_{cl} defines the energy, and, therefore, the eccentricity of the orbit.

Going over now to quantum mechanics, we define the vectors

$$\mathbf{L} = \mathbf{r} \times \mathbf{p} \tag{3.5}$$

and

$$\mathbf{A} = |2mH_{\text{NR}}|^{-1/2}\left\{Ze^2 m\hat{r} + \frac{\hbar}{2}(\mathbf{L} \times \mathbf{p} - \mathbf{p} \times \mathbf{L})\right\}, \tag{3.6}$$

where we have introduced Planck's quantum of action and symmetrized the cross-product of **L** and **p**. **L** and **A** are again the two vector invariants which completely define the problem.

The commutation properties of these vectors are

$$[L_i, L_j] = i\varepsilon_{ijk} L_k, \tag{3.7}$$

$$[L_i, A_j] = i\varepsilon_{ijk} A_k \tag{3.8}$$

and

$$[A_i, A_j] = \pm i\varepsilon_{ijk} L_k, \tag{3.9}$$

where ε_{ijk} is the unit antisymmetric tensor and summation is implied over k. In eqn (3.9), the plus sign applies only to the bound state problem and the minus sign to positive energy states. For bound states, the discrete eigenvalues of energy are given by

$$E_{\text{NR}} = -\tfrac{1}{2}m\left(\frac{Ze^2}{N}\right)^2, \tag{3.10}$$

where N is an integer. For continuum states, the eigenvalues E_{NR}^c are obtained by the mapping $N \to i\eta$,

$$E_{\text{NR}}^c = \tfrac{1}{2}m\left(\frac{Ze^2}{\eta}\right)^2 = \frac{(Ze^2 m)^2}{2m\eta^2}, \tag{3.11}$$

η now being a continuous parameter. It is called Sommerfeld's parameter, being the quantity which determines the eccentricity of the corresponding hyperbolic orbits in the Bohr–Sommerfeld theory.

The vector **A** defined in eqn (3.6) is known as the Runge–Lenz vector. It was first pointed out by Pauli (1926) that eqns (3.7) through (3.9) are the generator relations for the O_4 group (or the $O_{3,1}$ group, if one uses the minus sign in eqn (3.9)). The Casimir operators for this group are $\mathbf{L}^2 + \mathbf{A}^2$, $\mathbf{L}\cdot\mathbf{A}$ and $\mathbf{A}\cdot\mathbf{L}$, but it may be shown that $\mathbf{L}\cdot\mathbf{A} = \mathbf{A}\cdot\mathbf{L} = 0$. For positive energy states one makes the correspondence

$$\mathbf{A}^2 + \mathbf{L}^2 + 1 \to \eta^2 = \frac{(Ze^2 m)^2}{2mH_{\text{NR}}}, \tag{3.12}$$

so that the surviving Casimir operator is associated with the energy as expected.

The generator relations enumerated above form a perfectly good formal statement of the symmetry properties of the Coulomb field. In fact, by using the algebra of these generators along with the property of spin, one can go on to deduce the multiplet structure for this system, i.e. the basis states which form a complete quantum mechanical description of the system (Biedenharn, 1965). In our present discussion, we are actually more interested in these basis states, i.e. the eigenfunctions, and some properties of these eigenfunctions which lead to a considerable simplification of Auger effect calculations. These properties take the form of recursion relations between the neighbouring members of a multiplet, i.e. eigenstates labelled by a common eigenvalue of the Casimir operator, but by different neighbouring eigenvalues of some other invariant operator such as **L**. Understandably, these recursion relations are just an alternative statement of the symmetry properties of the system. As such, they completely define the eigenfunctions of the system. And the Lie group generated by the infinitesimal differential operators occurring in these recursion relations happens to be the symmetry group O_4 or $O_{3,1}$ of the system. In other words, starting from a knowledge of the "ladder structure" of the eigenfunctions, one may proceed to exhibit the symmetry in action. An advantage of this approach is that the associated Lie algebra (of the infinitesimal generators) provides a framework for the discussion of mathematical properties of the eigenfunctions such as expansions and addition theorems without recourse to analysis. The method has some generality, because recursion relations form a complete definition of many of the special functions (Kaufman, 1966). Interesting symmetries may thus be revealed. For example, the Lie algebra associated with the recursion relations for the Gegenbauer polynomials is that of O_3. We shall, however, be concerned only with the Coulomb wave functions for the bound and continuum states (Chattarji, 1967).

A. Continuum states

The non-relativistic radial Coulomb wave function for continuum states is completely defined by the recursion relations

$$\left[\frac{l^2}{\rho}+\eta+l\frac{\mathrm{d}}{\mathrm{d}\rho}\right]F_{\eta l}(\rho)=(l^2+\eta^2)^{1/2}F_{\eta,l-1}(\rho), \tag{3.13a}$$

$$\left[\frac{(l+1)^2}{\rho}+\eta-(l+1)\frac{\mathrm{d}}{\mathrm{d}\rho}\right]F_{\eta l}(\rho)=[(l+1)^2+\eta^2]^{1/2}F_{\eta,l+1}(\rho), \tag{3.13b}$$

where the quantum number l labels the eigenvalues of the angular momentum operator $\mathbf{L}$, $\mathbf{L}^2\to l(l+1)$, and $\rho=kr$, k being the wave number. The Sommerfeld parameter $\eta = Ze^2 m/(k\hbar^2)$. From eqns (3.13) one obtains the radial Schrodinger equation for the Coulomb field,

$$\left[\frac{d^2}{d\rho^2} + \left\{1 - \frac{2\eta}{\rho} - \frac{l(l+1)}{\rho^2}\right\}\right] F_{\eta l}(\rho) = 0, \tag{3.14}$$

($\rho > 0$, $-\infty < \eta < \infty$, l a non-negative integer). Equations (3.13b) and (3.13a) give the raising and lowering operators R and L with respect to the angular momentum. In order to obtain R and L in a form which does not refer to l and η, we introduce the spurious variables z and t and define the function

$$u_l(z, t, \rho) \equiv \exp(lz + \eta t) F_{\eta l}(\rho). \tag{3.15}$$

The recursion relations for u_l are

$$\mathscr{R} u_l(z, t, \rho) = [(l+1)^2 + \eta^2]^{1/2} u_{l+1}(z, t, \rho) \tag{3.16a}$$

and

$$\mathscr{L} u_l(z, t, \rho) = [l^2 + \eta^2]^{1/2} u_{l-1}(z, t, \rho), \tag{3.16b}$$

where the raising and lowering operators are given by

$$\mathscr{R} = \left[\frac{1}{\rho}\frac{\partial^2}{\partial z^2} + \frac{\partial}{\partial t} - \frac{\partial}{\partial \rho}\frac{\partial}{\partial z}\right] e^z, \tag{3.17a}$$

$$\mathscr{L} = e^{-z}\left[\frac{1}{\rho}\frac{\partial^2}{\partial z^2} + \frac{\partial}{\partial t} + \frac{\partial}{\partial \rho}\frac{\partial}{\partial z}\right]. \tag{3.17b}$$

The commutation relations for these operators are

$$\left.\begin{aligned} &[\mathscr{L}, \mathscr{R}] \equiv \mathscr{M} = 2\frac{\partial}{\partial z} + 1, \\ &[\mathscr{R}, \mathscr{M}] = -2\mathscr{R},\ [\mathscr{L}, \mathscr{M}] = 2\mathscr{L}. \end{aligned}\right\} \tag{3.18}$$

Remembering that the argument of the Coulomb function is actually $2i\rho$, which is complex, and that $u_l(z, t, \rho) \to F_{\eta l}(\rho)$ in the limit $z, t \to 0$, it may be shown that commutation relations (3.18) imply the group $O_{2,1}$.

Let us consider the complete Coulomb wave function

$$F_{\eta l}(\rho)\, Y_l^m(\theta, \varphi). \tag{3.19}$$

It is well known that the spherical harmonics Y_l^m form the basis states for the group O_3. It may be shown by the above method that this symmetry corresponds to the infinitesimal operators which raise and lower m. If these are adjoined to the l-shifting operations on the radial functions discussed above, the algebra generated is that of $O_{3,1}$, the three-dimensional Lorentz group.

B. Bound states

For bound states, the Sommerfeld number η is replaced by the principal quantum number N the mapping being $N \rightleftarrows i\eta$. Instead of the wave number k, we introduce a, the first Bohr radius of the atom, by

$$a = \frac{\hbar^2}{Ze^2 m}. \tag{3.20}$$

It may be seen that Na is the discrete state analogue of $1/k$ and, therefore, a natural unit of distance for the bound state. Defining the radial wave function $R_{Nl}(r)$ by

$$R_{Nl}(r) \equiv \frac{Na}{r} F_{Nl}(k_{\text{NR}} r), \qquad (k_{\text{NR}} = 1/Na), \tag{3.21}$$

we get the recursion relations

$$\left[1 + la\left(\frac{\mathrm{d}}{\mathrm{d}r} + \frac{l+1}{r}\right)\right] F_{Nl}(k_{\text{NR}} r)/r = \left(1 - \frac{l^2}{N^2}\right)^{1/2} F_{N,l-1}(k_{\text{NR}} r)/r,$$

$$\left[1 - (l+1)\,a\left(\frac{\mathrm{d}}{\mathrm{d}r} - \frac{l}{r}\right)\right] F_{Nl}(k_{\text{NR}} r)/r = \left[1 - \frac{(l+1)^2}{N^2}\right]^{1/2} F_{N,l+1}(k_{\text{NR}} r)/r, \tag{3.22}$$

which is the discrete analogue of eqns (3.13). Thus the Lie algebra follows exactly as before, except that we now have the symmetry $O_4 \simeq O_3 \times O_3$, the group O_4 being locally isomorphic to the direct product of two O_3 groups.‡

Our discussion of the symmetry properties of the Coulomb potential must seem rather informal, because our primary objective is to exploit them for purposes of computation. Actually, our arguments have their foundation in the formal work done by Pauli (1926) as well as by Fock (1935) and Bargmann (1936). Fock established this symmetry by transforming the Schrodinger eqn into the three-dimensional momentum space and taking the stereographic projection of this space onto the unit hypersphere in four dimensions. Bargmann related this global method of Fock to the infinitesimal method of Pauli.

3. The non-relativistic spin-$\frac{1}{2}$ (NRS) particle in a Coulomb field

So far we have not introduced spin into our discussion of the Coulomb field. However, it is an interesting fact that the motion of a Pauli particle (non-

‡ The same result follows from the Lie algebra if we define the auxiliary vectors as $\mathbf{M} = \frac{1}{2}(\mathbf{L} + \mathbf{A})$ and $\mathbf{N} = \frac{1}{2}(\mathbf{L} - \mathbf{A})$. They satisfy the commutation relations $[M_j, M_k] = i\varepsilon_{jkl} M_l$, $[N_j, N_k] = i\varepsilon_{jkl} N_l$, $[M_j, N_k] = 0$ (Chacon *et al.*, 1971).

relativistic spin-$\frac{1}{2}$ particle with dynamically independent spin) in such a field is easier to discuss than that of a spinless particle (Biedenharn and Swamy, 1964; Biedenharn, 1965). Actually, the full significance of O_4 symmetry is best brought out by the introduction of spin. The spin-angular part of the resulting wave functions are easily handled by the well-known methods of Wigner algebra. This part is common to both non-relativistic and relativistic wave functions, provided of course that we ignore for the present the question of the dynamical origin of spin. A discussion of the Pauli particle in a Coulomb field can, therefore, form a kind of preamble for the more general relativistic approach, i.e. the study of a Dirac particle in a Coulomb field.

In this Section, we shall introduce spin only in so far as it is relevant to the understanding of the basic properties of the Kepler problem. In particular, we shall see that the recursion relations for the radial Coulomb functions given in Section 2 can be systematized by adjoining spin to the orbital angular momentum. The discussion of coupling schemes between the spin and the angular momentum of different electrons will be deferred to a later chapter.

The coupled spin-angle functions are defined by

$$\chi_\kappa^\mu = \sum_\tau C_{\mu-\tau\,\tau\,\mu}^{l(1/2)j} Y_l^{\mu-\tau}(\theta, \varphi)\, \chi_{1/2}^\tau, \tag{3.23}$$

where $\chi_{1/2}^\tau$ is the Pauli spinor ($\tau = \pm\frac{1}{2}$) for which we shall use the representation

$$\chi_{1/2}^{1/2} = \begin{pmatrix} 1 \\ 0 \end{pmatrix}, \qquad \chi_{1/2}^{-1/2} = \begin{pmatrix} 0 \\ 1 \end{pmatrix}. \tag{3.24}$$

The coupled spinors χ_κ^μ, which are designated as "Pauli central-field spinors" are the basis functions of the group $O_3 \otimes SU_2$. They obey the eigenvalue relations

$$\begin{aligned} &\mathbf{J}^2 \equiv (\mathbf{L} + \tfrac{1}{2}\boldsymbol{\sigma})^2 \to j(j+1);\ \mathbf{L}^2 \to l(l+1); \\ &\mathbf{S}^2 \equiv \tfrac{1}{4}\boldsymbol{\sigma}^2 \to \tfrac{1}{2}(\tfrac{1}{2}+1) \qquad \text{and} \qquad J_z \to \mu. \end{aligned} \tag{3.25}$$

It may be noted here that we have introduced the Dirac quantum number κ, which is a compact way of simultaneously designating the total angular momentum and parity of the system. κ is defined by the eigenvalue relation

$$(\boldsymbol{\sigma}\cdot\mathbf{L} + 1)\, \chi_\kappa^\mu = -\kappa \chi_\kappa^\mu. \tag{3.26}$$

Both $\boldsymbol{\sigma}$ and $\mathbf{L}$ are pseudovectors, so $\boldsymbol{\sigma}\cdot\mathbf{L} + 1 \equiv K$ is a scalar operator in the direct product space of spin and orbital angular momentum. From eqns (3.25) and (3.26) we can see that $K^2 = (\boldsymbol{\sigma}\cdot\mathbf{L} + 1)^2 = \mathbf{J}^2 + \frac{1}{4}$ and $K^2 + K = \mathbf{L}^2$. Thus it follows that $\kappa^2 = (j + \frac{1}{2})^2$ and $\kappa(\kappa + 1) = l(l+1)$. We can thus make the association

$$j(\kappa) = |\kappa| - \tfrac{1}{2} \qquad \text{and} \qquad l(\kappa) = |\kappa| + \tfrac{1}{2}\{\operatorname{sgn}(\kappa) - 1\}. \tag{3.27}$$

Hence κ simultaneously determines the spin and parity of the system.

Another pseudo-scalar operator in this product space is $\boldsymbol{\sigma}\cdot\hat{r}$. It leaves j unchanged, but, being pseudoscalar, reverses sign under space reflection, so that parity is changed, i.e. l changes by one unit. Thus $\boldsymbol{\sigma}\cdot\hat{r}$ reverses the sign of κ. We write

$$\boldsymbol{\sigma}\cdot\hat{r}\,\chi_{\kappa}^{\mu} = -\chi_{-\kappa}^{\mu}. \tag{3.28}$$

Equation (3.28) follows from the above reasoning and the fact that $(\boldsymbol{\sigma}\cdot\hat{r})^2 = 1$. We choose the phase so as to have the minus sign.

Consider now the wave functions for a particle of spin $\frac{1}{2}$ in a Coulomb field. The complete continuum wave function assumes the form

$$|\eta\kappa\mu\rangle = (kr)^{-1}\,F_{\eta l(\kappa)}(kr)\,\chi_{\kappa}^{\mu}(\theta,\varphi). \tag{3.29}$$

One may now define an additional pseudo-scalar operator $\boldsymbol{\sigma}\cdot\mathbf{A}$, where $\mathbf{A}$ is the Runge–Lenz vector. It may be shown that (Biedenharn, 1965)

$$\boldsymbol{\sigma}\cdot\mathbf{A}/\eta = \boldsymbol{\sigma}\cdot\hat{r}\left[1 + \frac{\hbar K}{Ze^2 m}\{i\hat{r}\cdot\mathbf{p} + \hbar r^{-1}(1+K)\}\right]. \tag{3.30}$$

The most important property of the operator $\boldsymbol{\sigma}\cdot\mathbf{A}$ is that it anticommutes with the Dirac operator K, so it connects states whose κ values differ only in sign:

$$\boldsymbol{\sigma}\cdot\mathbf{A}|\eta\kappa\mu\rangle = (\text{const})|\eta - \kappa\mu\rangle. \tag{3.31}$$

From eqn (3.31), letting $\boldsymbol{\sigma}\cdot\mathbf{A}$ operate on the function $|\eta\kappa\mu\rangle$ one may directly obtain the radial equation

$$\left[1 + \frac{\kappa\hbar^2}{Ze^2 m}\left(\frac{\mathrm{d}}{\mathrm{d}r} + \frac{\kappa+1}{r}\right)\right]\frac{F_{\eta,\,l(\kappa)}(kr)}{r} = +\left(1 + \frac{\kappa^2}{\eta^2}\right)^{1/2}\frac{F_{\eta,l(-\kappa)}(kr)}{r}. \tag{3.32}$$

Equation (3.32) contains in a compact form both the recursion relations (3.13). We can see that the raising and lowering operators both change the sign of κ, one from positive to negative and the other from negative to positive. Spin thus helps us to see the recursion relations for the radial functions from a unified viewpoint. Actually $\boldsymbol{\sigma}\cdot\mathbf{A}$ is the "ladder operator" for the complete Coulomb wave function in this approximation. This is what we mean when we say that introduction of spin makes the symmetry more transparent.

Similar remarks apply to the bound state wave functions. Following our earlier notation, the wave functions for the spinless particle may be written

$$|Nlm\rangle = (r/aN)^{-1}\,F_{Nl}(k_{\mathrm{NR}}r)\,Y_l^m(\theta,\varphi), \tag{3.33}$$

where

$$F_{Nl}(k_{\text{NR}}\,r) = C_{N,l}\left(\frac{r}{aN}\right)^{l+1} e^{-r/aN}\,{}_1F_1\left(l+1-N, 2l+2; \frac{2r}{aN}\right) \tag{3.34}$$

and

$$C_{N,l} = (-)^{N-l-1}\frac{2^{l+1}}{\Gamma(2l+2)}\left\{\frac{\Gamma(N+l+1)}{a^3 N^4 \Gamma(N-l)}\right\}. \tag{3.35}$$

The wave function for spin-$\frac{1}{2}$ particles is

$$|N\kappa\mu\rangle = \left(\frac{r}{aN}\right)^{-1} F_{Nl}(k_{\text{NR}}\,r)\,\chi_\kappa^\mu(\theta, \varphi). \tag{3.36}$$

4. Some Coulomb field calculations of Auger transitions

To see how these properties of the Coulomb wave function and the symmetry underlying them lead to simplifications in the calculation of Auger matrix elements, two particular cases will be considered in this section. These are

1. the Pauli (NRS) system, where the full symmetry of the Coulomb field is at work, and
2. the relativistic Dirac–Coulomb problem where the symmetry is broken and the manner of its breaking provides some insight into the dynamics of the system.

We shall be mainly concerned with the radial part of the matrix element where these simplifications are clearly seen. Discussion of the spin-angular part, along with the various coupling schemes, is an important subject in itself and will be taken up in detail in the next chapter.

A. The Pauli (NRS) system

The general formula for the transition probability is

$$P = \frac{2\pi}{\hbar}\left|\iint \frac{1}{|\mathbf{r}_1 - \mathbf{r}_2|}\exp[2\pi\nu_{fi}|\mathbf{r}_1 - \mathbf{r}_2|/c](\rho_1\rho_2 - \mathbf{j}_1\cdot\mathbf{j}_2)\,d\tau_1\,d\tau_2\right|^2. \tag{3.37}$$

For the NRS system, the current–current interaction is negligible, so we have to consider only the charge-density part of the integral.

$$\rho_1 = -e\langle N_1'\,\kappa_1'\,\mu_1'|N_1\,\kappa_1\mu_1\rangle, \tag{3.38a}$$

$$\rho_2 = -e\langle \eta\kappa_2'\,\mu_2'|N_2\,\kappa_2\,\mu_2\rangle. \tag{3.38b}$$

Here $|N\kappa\mu\rangle$ are the complete Pauli wave functions. Primes denote final states and unprimed quantum numbers or parameters denote initial states. The subscripts 1 and 2 refer to the first and second electrons respectively.

We write, for the bound state,

$$|N\kappa\mu\rangle = F_{Nl(\kappa)}(k_{\mathrm{NR}}r)\,\chi_\kappa^\mu. \tag{3.39}$$

The radial function $F_{Nl}(k_{\mathrm{NR}}r)$ may be written in the form

$$F_{Nl}(k_{\mathrm{NR}}r) = C_{Nl}\exp\left(-\frac{Zr}{Na_0}\right)r^l\sum_{s=0}^{N-l-1}B_{Nls}r^s, \tag{3.40}$$

where the sum is just the associated Laguerre polynomial $L_{N+l}^{2l+1}(2Zr/Na_0)$, so that

$$B_{Nls} = (-)^{s+1}\{[(N+l)!]^2/(2l+1+s)!\,s!(N-l-1-s)!\}\left(\frac{2Z}{Na_0}\right)^s \tag{3.41}$$

and

$$C_{Nl} = -\{(N-l-1)!/2N[(N+l)!]^3\}^{1/2}\left(\frac{2Z}{Na_0}\right)^{l+3/2}. \tag{3.42}$$

Here, we have used $a_0 = \hbar^2/me^2$, the first Bohr radius of hydrogen. For the continuum wave function we write

$$|\eta\kappa\mu\rangle = F_{\eta l(\kappa)}(kr)\,\chi_\kappa^\mu, \tag{3.43}$$

where $F_{\eta l(\kappa)}(kr)$ is obtained from $F_{Nl}(r)$ by the usual mapping $N \to i\eta$,

$$F_{\eta l(\kappa)}(kr) = C_l(\eta)(kr)^l\,\mathrm{e}^{-ikr}F(l+1-i\eta,\,2l+2;\,2ikr) \tag{3.44}$$

and

$$C_l(\eta) = \frac{2^l\,\mathrm{e}^{-\pi\eta/2}|\Gamma(l+1+i\eta)|}{\Gamma(2l+2)}. \tag{3.45}$$

Eventually, the continuum wave function given here may be normalized to represent one electron ejected per unit energy interval by multiplying (3.44) by the factor $2(2\pi km)^{1/2}$, i.e. by letting $C_l(\eta) \to 2(2\pi km)^{1/2}C_l(\eta)$. Considering the limit of no retardation, the interaction term may be expanded in a series of Legendre polynomials

$$\frac{1}{|\mathbf{r}_1-\mathbf{r}_2|} = \sum_{L=0}^{\infty}\frac{r_<^L}{r_>^{L+1}}P_L(\cos\theta_{12}). \tag{3.46}$$

Using the addition theorem for spherical harmonics, we have

$$P_L(\cos\theta_{12}) = \frac{4\pi}{2L+1} \sum_{M=-L}^{L} \overset{*}{Y}{}_L^M(\theta_1, \varphi_1)\, Y_L^M(\theta_2, \varphi_2). \tag{3.47}$$

The matrix element now assumes the form (Kostroun *et al.*, 1971)

$$\begin{aligned} M = e^2 \sum_{L=0}^{\infty} \sum_{M=-L}^{L} & C_{N_1 l_1} C_{N_2 l_2} C_{N_1' l_1'} C_{l_2'}(\eta) \int_0^\infty r_2^2\, dr_2 \left\{ \int_0^\infty r_1^2\, dr_1 \frac{r_<^L}{r_>^{L+1}} \cdot \right. \\ & \cdot \sum_{s_2=0}^{N_2-l_2-1} \sum_{s_1'=0}^{N_1'-l_1'-1} \sum_{s_1=0}^{N_1-l_1-1} B_{N_1' l_1' s_1'} B_{N_1 l_1 s_1} B_{N_2 l_2 s_2} r_1^{l_1+s_1} r_1^{l_1'+s_1'} r_2^{l_2+s_2}. \\ & \cdot \exp\left(-\frac{Z_1 r_1}{N_1 a_0}\right) \exp\left(-\frac{Z_1' r_1}{N_1' a_0}\right) \exp\left(-\frac{Z_2 r_2}{N_2 a_0}\right). \\ & \cdot (kr_2)^{l_2'} e^{ikr_2}\, {}_1F_1(l_2' + 4i\eta_2, 2l_2' + 2; -2ikr_2). \\ & \cdot \frac{4\pi}{2L+1} \int \overset{\dagger}{\chi}{}_{\kappa_1'}^{\mu_1'} Y_L^M(\theta_1, \varphi_1)\, \chi_{\kappa_1}^{\mu_1}\, d\Omega_1 \int \overset{\dagger}{\chi}{}_{\kappa_2'}^{\mu_2'} Y_L^M(\theta_2, \varphi_2)\, \chi_{\kappa_2}^{\mu_2}\, d\Omega \end{aligned} \tag{3.48}$$

The exchange term is obtained by interchanging the labels 1 and 2 in the final state.

The spin-angular part of the matrix element is evaluated by using the Wigner–Eckart theorem, e.g.

$$\begin{aligned} \int \overset{\dagger}{\chi}{}_{\kappa_1'}^{\mu_1'} \overset{*}{Y}{}_L^M(\theta_1, \varphi_1)\, \chi_{\kappa_1}^{\mu_1}\, d\Omega_1 = (-)^{j_1'+L+\frac{1}{2}+M} \left[\frac{(2l_1+1)(2l_1'+1)(2j+1)}{4\pi} \right]^{1/2}. \\ \cdot C_{000}^{l_1 l_1' L}\, C_{\mu_1 - M \mu_1'}^{j_1 L j_1'}\, W(j_1' l_1' j_1 l_1; \tfrac{1}{2} L). \end{aligned} \tag{3.49}$$

In eqn (3.49), the Clebsch–Gordan coefficient $C_{0\,0\,0}^{l_1 l_1' L}$ vanishes unless $l_1 + l_1' + L$ is an even integer, so it contains the parity selection rule. The coefficient $C_{\mu_1 - M \mu_1'}^{j_1 L j_1'}$ contains the conservation of angular momentum, i.e. the triangle rule for j_1, j_1' and L, and the relation $\mu_1 - M = \mu_1'$. W is the Racah coefficient. In the limit we are considering here, the total angular momentum for each electron is conserved, i.e. we have *jj* coupling. The sum over L in eqn (3.48) is reduced to a finite number of terms consistent with the triangle rule and the parity selection rule.

With L values subject to this constraint, we can write the inner integration, i.e. the integration over r_1, in the form

$$I_1 = \int_0^\infty \frac{r_<^L}{r_>^{L+1}} e^{-\left(\frac{Z_1}{N_1} + \frac{Z_1'}{N_1'}\right)\frac{r_1}{a_0}} r_1^{l_1+l_1'+s_1+s_1'} r_1^2\, dr_1. \tag{3.51}$$

We now break this up into two integrals

$$\int_0^\infty \mathrm{d}r_1 \cdots = \int_0^{r_2} \mathrm{d}r_1 \cdots + \int_{r_2}^\infty \mathrm{d}r_1 \cdots. \tag{3.52}$$

Using the results

$$\int_0^R r^p \mathrm{e}^{-qr}\,\mathrm{d}r = \frac{\Gamma(p+1)}{q^{p+1}}\left[1 - \mathrm{e}^{-qR}\sum_{i=1}^{p+1}\frac{(qR)^{p+1-i}}{(p+1-i)!}\right], \tag{3.53}$$

and

$$\int_R^\infty r^p \mathrm{e}^{-qr}\,\mathrm{d}r = \frac{\Gamma(p+1)}{q^{p+1}}\mathrm{e}^{-qr}\sum_{i=1}^{p+1}\frac{(qR)^{p+1-i}}{(p+1-i)!}, \tag{3.54}$$

we get

$$\begin{aligned} I_1 &= \frac{1}{r_2^{L+1}}\int_0^{r_2} \mathrm{e}^{-qr_1} r_1^p\,\mathrm{d}r_1 + r_2^L \int_{r_2}^\infty \mathrm{e}^{-qr_1} r_1^{p-2L-1}\,\mathrm{d}r_1 = \\ &= \frac{1}{r_2^{L+1}}\frac{\Gamma(p+1)}{q^{p+1}}\left[1 - \mathrm{e}^{-qr_2}\sum_{i=1}^{p+1}\frac{(qr_2)^{p+1-i}}{(p+1-i)!}\right] + \\ &\quad + r_2^L\frac{\Gamma(p-2L)}{q^{p-2L}}\left[\mathrm{e}^{-qr_2}\sum_{i=1}^{p-2L}\frac{(qr_2)^{p-2L-i}}{(p-2L-i)!}\right], \qquad \left(q = \frac{1}{a_0}\left(\frac{Z_1}{N_1} + \frac{Z_1'}{N_1'}\right)\right). \end{aligned} \tag{3.55}$$

The integral over r_2 now takes the form

$$I_2 = \int_0^\infty r_2^2\,\mathrm{d}r_2\, r_2^{l_2+s_2}(kr_2)^{l_2'}\mathrm{e}^{ikr_2}\mathrm{e}^{-Z_2r_2/N_2a_0}\, I_1\, {}_1F_1(l_2'+1+i\eta, 2l_2'+2; -2ikr_2). \tag{3.56}$$

Using Kummer's transformation,

$${}_1F_1(a, b; z) = \mathrm{e}^z\,{}_1F_1(b-a, b; -z), \tag{3.57}$$

one finally gets

$$\begin{aligned} I_2 &= \frac{\Gamma(p+1)}{q^{p+1}}k^{l_2'}A(l_2, l_2', L, k, \eta, s_2) - \frac{\Gamma(p+1)}{q^{p+1}}k^{l_2'}\sum_{i=1}^{p+1}\frac{q^{p+1-i}}{(p+1-i)!}\cdot \\ &\quad \cdot B(l_2, l_2', L, k, \eta, s_2, i) + \\ &\quad + \frac{\Gamma(p-2L)}{q^{p-2L}}k^{l_2'}\sum_{i=1}^{p-2L}\frac{q^{p-2L-i}}{(p-2L-i)!}B(l_2, l_2', L, k, \eta, s_2, i), \end{aligned} \tag{3.58}$$

where the radial integrals A and B are given by

$$A(l_2, l_2', L, k, \eta, s_2) = \int_0^\infty dr_2\, r_2^{\,l_2+l_2'-L+s_2+1}\, e^{-(q'+ik)r_2}\, {}_1F_1(l_2' 1 - i\eta, 2l_2' + 2;\, 2ikr_2) \tag{3.59a}$$

and

$$B(l_2, l_2', L, k, \eta, s_2, i) = \int_0^\infty dr_2\, r_2^{\,l_1+l_1'+l_2+l_2'+s_1+s_1'+s_2+4-i}\, e^{-(q+q'+ik)r_2} \times \times\, {}_1F_1(l_2' + 1 - i\eta, 2l_2' + 2;\, 2ikr_2). \tag{3.59b}$$

with $q' = Z_2/N_2 a_0$. Integrals A and B are both of the form

$$\int_0^\infty dr r^P e^{-Qr}\, {}_1F_1(a, b;\, cr) = \frac{\Gamma(P+1)}{Q^{P+1}}\, {}_2F_1\left(a, P+1, b;\, \frac{c}{Q}\right), \tag{3.60}$$

where P is an integer and Q a complex number.

We can see that the different integrals belonging to any one category, i.e. category A or category B, differ only in the power P of r contained in the integrand, this power always being an integer. Correspondingly, the calculation of the matrix element involves Gaussian hypergeometric functions with different integral values of the parameter P. The problem thus reduces to the rather simple task of generating these functions by making use of the well-known three-term recurrence relations for contiguous hypergeometric functions (Erdelyi *et al.*, 1953). This was first pointed out by Burhop (1937), who used these relations in a non-relativistic calculation of the KL_1L_3 transition. He showed that the most convenient way of calculating the relevant hypergeometric functions was to first evaluate

$${}_2F_1(a, c, c;\, z) = (1 - z)^{-a}. \tag{3.61}$$

Writing the other functions in the generic form ${}_2F_1(a, b, c;\, z)$, for $b > c$ the other required functions can be obtained by using the recurrence relation

$$b(1-z)\,{}_2F_1(a, b+1, c;\, z) = \{(b-a)(1-z) - (c-b-a)\}\,{}_2F_1(a, b, c; z) + + (c-b)\,{}_2F_1(a, b-1, c; z). \tag{3.62}$$

For $b < c$, one computes numerically ${}_2F_1(a, c-2, c;\, z)$ and then uses eqn (3.62). Using eqns (3.61) and (3.62) and the fact that

$${}_2F_1(a, 0, c;\, z) = 1, \tag{3.63}$$

Callan *et al.* (1964) developed an elegant matrix method for simultaneously computing a series of hypergeometric functions differing in b, the b being positive integers with some maximum value m depending on the particular transition. The method consists of writing out eqn (3.62) for successively increasing values of b in the form of a matrix equation. With the notation

$$\begin{aligned} b(1-z) &\equiv \alpha_b,\ (a-b)(1-z)+(c-b-a) \equiv \beta_b, \\ b-c &\equiv \gamma_b \quad \text{and} \quad {}_2F_1(a,b,c;z) \equiv F(b), \end{aligned} \tag{3.64}$$

the matrix equation is

$$[A]\cdot|F\rangle = |B\rangle, \tag{3.65}$$

where $[A]$ is an $(m+1)\times(m+1)$ square matrix

$$[A] = \begin{bmatrix} \gamma_1 & \beta_1 & \alpha_1 & 0 & \cdots\cdots & 0 & 0 & 0 \\ 0 & \gamma_2 & \beta_2 & \alpha_2 & \cdots\cdots & 0 & 0 & 0 \\ \cdots & \cdots & \cdots & \cdots & \cdots & \cdots & \cdots & \cdots \\ \cdots & \cdots & \cdots & \cdots & \cdots & \cdots & \cdots & \cdots \\ \cdots & \cdots & \cdots & \cdots & \cdots & \cdots & \cdots & \cdots \\ 0 & 0 & 0 & 0 & \cdots\cdots & \gamma_{m-1} & \beta_{m-1} & \alpha_{m-1} \\ 1 & 0 & 0 & 0 & \cdots\cdots & 0 & 0 & 0 \\ 0 & 0 & 0 & 1 & \cdots\cdots & 0 & 0 & 0 \end{bmatrix}, \tag{3.66}$$

and $|F\rangle$ and $|B\rangle$ are column vectors with $m+1$ elements each,

$$|F\rangle = \begin{bmatrix} F(0) \\ F(1) \\ F(2) \\ \vdots \\ F(m-1) \\ F(m) \end{bmatrix}, \quad |B\rangle = \begin{bmatrix} 0 \\ 0 \\ 0 \\ \vdots \\ 1 \\ (1-z)^{-a} \end{bmatrix}. \tag{3.67}$$

The unit element in the last row of $[A]$ is in position A_{mb}. The system of simultaneous eqns (3.65) can be solved directly on a computer by Gaussian elimination methods, which ensure a high degree of accuracy by avoiding the truncation errors involved in series summation methods.

The fact that such simple techniques are available for the calculation of matrix elements may directly be ascribed to the symmetry properties of the Coulomb field. The recursion relations of the Coulomb wave functions are a consequence of this symmetry. And it may be shown that the radial integrals for the Auger transition, i.e. the Gaussian hypergeometric functions described above, are amenable to a simple recursive treatment precisely because they are derived from these wave functions.

B. The Dirac–Coulomb problem (The relativistic Kepler problem)

Earlier in this work, we have emphasized the point that a complete picture of the Auger effect can be given only by means of a relativistic treatment. Calculations in the non-relativistic approximation are fairly accurate for low atomic numbers. For medium values of the atomic number Z there is disagreement in the order of 30–50% with experiment and it reaches a factor of 3–4 in the case of heavier elements ($Z \geqslant 80$). The fact that a major source of this disagreement lies in the neglect of relativity has been demonstrated by subsequent calculations (Burhop, 1935; Massey and Burhop, 1936; Asaad, 1959; Listengarten, 1960; Bhalla and Ramsdale, 1970a, b; Bhalla *et al.*, 1970; Bhalla, 1970). Neglect of relativity implies not only the use of non-relativistic wave functions for the charge–charge interaction, but also the neglect of the current–current interaction altogether. It also means that one is overlooking the mode of interaction between the participating electrons, namely, the retarded exchange of virtual quanta. For, the concept of a retarded interaction is inherently a relativistic concept. Furthermore, electron spin is a dynamic attribute of relativistic theory, so that Pauli theory can only be an incomplete description of the problem.

Our treatment of relativistic effects will be confined here to the limits of one-particle Dirac theory. As Burhop pointed out in his pioneering calculation of this kind (Burhop, 1937), use of Møller's formula provides a test, not only of Dirac's one-electron theory, but also of its extension, in the first approximation, to two electrons. In the following paragraphs, we shall use the approach introduced by Biedenharn (1962).

The introduction of relativity spoils the O_4 (or $O_{3,1}$, as the case may be) symmetry of the Coulomb field discussed in the last section. The degeneracy of states with the same principal quantum number N but different values of the orbital angular momentum l is removed. This symmetry-breaking is caused by the fine-structure interaction which makes the potential depart from its strict $1/r$ dependence. To use the language of representations, O_4 being a larger symmetry than O_3, states labelled by different l-values (i.e. corresponding to different eigenvalues of the invariant of O_3) can belong to the same irreducible representation of O_4. With the loss of O_4 symmetry, this is no more the case.

To understand what this symmetry-breaking does to the dynamics of the system, we note that in all relativistic calculations involving the Coulomb potential, there appears a non-integer number $\gamma(l) = (l^2 - \alpha^2 Z^2)^{1/2}$. This is the number which parametrizes relativistic Coulomb wave functions instead of l, which appears in non-relativistic wave functions. As we shall presently see, the decrement $\alpha^2 Z^2$ in l^2 implies a corresponding decrement in the centrifugal potential experienced by the orbiting particle, which arises from the coupling of the particle to the Coulomb field. Thus, in course of its motion in the orbit, the particle experiences a continuously changing centrifugal potential corresponding to the changing value of the effective nuclear charge Z_{eff} "seen" by it. This is the classical meaning of the famous "rosette" motion or precession of the orbit of an electron introduced by Sommerfeld. Because $\gamma(l)$ is not an integer, we have an "open" orbit. Interestingly, the strength of the coupling which causes this effect is determined by the product αZ. The coupling vanishes for (a) $Z \to 0$, i.e. if there is no Coulomb field or (b) if α is regarded as so small that αZ may be neglected for a given Z. These two limits correspond to the Dirac plane wave and the non-relativistic Coulomb field problem.

Assuming that a Coulomb field is present, that αZ should be a measure of relativistic effects can be seen if we remember that $\alpha Z/N = \langle v \rangle/c$, where $\langle v \rangle$ is the expectation value of the velocity. Non-relativistic theory takes $\langle v \rangle/c$ and by implication αZ to be small.

To see the meaning of these remarks, we start out by writing the Dirac plane wave equation in the form (Biedenharn, 1962),

$$\mathcal{O}_+ \Psi \equiv \left(\rho_2 \boldsymbol{\sigma}\cdot\nabla - \rho_3 \frac{E}{\hbar c} + \frac{mc}{\hbar} \right) \Psi = 0. \tag{3.68}$$

Here the ρ's are Dirac matrices, E is the total energy and the operator $\mathcal{O}_+$ has the dimensions of a wave number. We now iterate this to obtain the second order equation

$$\mathcal{O}_- \mathcal{O}_+ \Phi = \left(\rho_2 \boldsymbol{\sigma}\cdot\nabla - \rho_3 \frac{E}{\hbar c} - \frac{mc}{\hbar} \right) \left(\rho_2 \boldsymbol{\sigma}\cdot\nabla - \rho_3 \frac{E}{\hbar c} + \frac{mc}{\hbar} \right) \Phi =$$

$$= \left(\frac{1}{r^2} \frac{\partial}{\partial r} r^2 \frac{\partial}{\partial r} - \frac{\mathbf{L}^2}{r^2} + k^2 \right) \Phi = 0, \tag{3.69}$$

with $k^2 = (E^2 - m^2 c^4)/\hbar^2 c^2$. Every solution of eqn (3.68) is a solution also of eqn (3.69), but not conversely. We distinguish the solution of (3.69) by the symbol Φ.

The Legendre operator $\mathbf{L}^2$ may be written in the spinor form

$$\mathbf{L}^2 = (\boldsymbol{\sigma}\cdot\mathbf{L} + 1)^2 - (\boldsymbol{\sigma}\cdot\mathbf{L} + 1) = K^2 - \rho_3 K, \tag{3.70}$$

so that we have

$$l(l+1) = \kappa^2 + \kappa = \kappa(\kappa + 1).$$

The solutions of eqn (3.69) are of the form

$$\Phi_+ \equiv \Phi_{\rho_3 = +} = \begin{pmatrix} j_{l(-\kappa)}(kr)\, \chi^{\mu}_{-\kappa} \\ 0 \end{pmatrix} \tag{3.71}$$

and

$$\Phi_- \equiv \Phi_{\rho_3=-} = \begin{pmatrix} 0 \\ j_{l(\kappa)}(kr)\,\chi_\kappa^\mu \end{pmatrix}. \tag{3.72}$$

We can now go over to the Dirac–Coulomb wave equation

$$\mathcal{O}_+^c\,\Psi = \left[\rho_2\,\boldsymbol{\sigma}\cdot\boldsymbol{\nabla} - \rho_3\left(\frac{E}{\hbar c} + \frac{\alpha Z}{r}\right) + \frac{mc}{\hbar}\right]\Psi = 0. \tag{3.73}$$

Iterating as before

$$\mathcal{O}_-^c\,\mathcal{O}_+^c\,\Phi = \left[\frac{1}{r^2}\frac{\partial}{\partial r}r^2\frac{\partial}{\partial r} - \frac{K^2-\alpha^2Z^2}{r^2} - \frac{2\alpha ZE}{\hbar cr} + \frac{1}{r^2}(\rho_3 K + i\alpha Z\rho_1\,\boldsymbol{\sigma}\cdot\hat{r}) + k^2\right]\Phi = 0, \tag{3.74}$$

we now define the operator

$$\Gamma = \rho_3 K + i\alpha Z\rho_1\,\boldsymbol{\sigma}\cdot r. \tag{3.75}$$

Since $\rho_3 K$ and $\rho_1\boldsymbol{\sigma}\cdot\hat{r}$ anticommute, it follows that

$$\Gamma^2 = K^2 - \alpha^2 Z^2. \tag{3.76}$$

Thus eqn (3.74) may be written in a form exactly analogous to eqn (3.69),

$$\left(\frac{1}{r^2}\frac{\partial}{\partial r}r^2\frac{\partial}{\partial r} - \frac{\Gamma(\Gamma-1)}{r^2} - \frac{2k\eta}{r} + k^2\right)\Phi = 0. \tag{3.77}$$

Here η is the relativistic analogue of Sommerfeld's parameter,

$$\eta = \frac{\alpha ZE}{\hbar ck}. \tag{3.78}$$

Remembering that $\mathbf{L}^2 = (\boldsymbol{\sigma}\cdot\mathbf{L}+1)^2 - (\boldsymbol{\sigma}\cdot\mathbf{L}+1)$, we can at once define a non-integer orbital angular momentum $l(\gamma)$ by the relation

$$\Gamma^2 - \Gamma = l(l+1), \qquad \text{where} \qquad \Gamma \to \gamma = \pm\left|(\kappa^2 - \alpha^2 Z^2)^{1/2}\right|.$$

Obviously, Γ is the Coulomb analogue of the operator $\rho_3 K = \boldsymbol{\sigma}\cdot\mathbf{L}+1$ of the plane wave case. It may be regarded as a generalized angular momentum which incorporates Coulomb-field effects in addition to spin.

One might ask at this stage, "is it possible to go over to a new coordinate system, analogous to the rotating frame introduced by Sommerfeld, so that the fine structure interaction $i\alpha Z\rho_1\boldsymbol{\sigma}\cdot\hat{r}$ occurring in eqn (3.75), is turned off and we have closed orbits corresponding to a formally nonrelativistic motion in this new frame?" A transformation which accomplishes this was indeed given by Biedenharn and Swamy (1964). The resulting symmetric Hamiltonian has two vector invariants **J** and **K** ($|\mathbf{K}| = K$), which are analogues of **L** and **A** of the spinless non-relativistic problem. It has the form

$$H_{\text{sym}} = c\rho_1\,\boldsymbol{\sigma}\cdot\mathbf{p} + \rho_3 mc^2 - \frac{Ze^2}{r} + c\rho_2\frac{\boldsymbol{\sigma}\cdot\hat{r}}{r}K\left[\sqrt{\left(1+\left(\frac{\alpha Z}{K}\right)^2\right)} - 1\right], \quad (3.79)$$

i.e. it differs from the exact Dirac Hamiltonian with a Coulomb potential by the fine structure term

$$H_{\text{fs}} = c\rho_2\frac{\boldsymbol{\sigma}\cdot\hat{r}}{r}K\left[\sqrt{\left(1+\left(\frac{\alpha Z}{K}\right)^2\right)} - 1\right]. \quad (3.80)$$

Because of the formal "closing" of orbits implied in this Hamiltonian, its solutions have integer angular momenta and may be used to calculate Auger transition probabilities in a manner closely analogous to the Pauli system. The difference is that here one is taking relativistic effects into account to within the above approximation (Chattarji and Talukdar, 1968; Talukdar and Chattarji, 1970).

Let us now return to eqn (3.73) and try to write down its solutions for bound and continuum states. We shall do this by exploiting the second-order eqn (3.77). Note that Γ is a non-integer orbital angular momentum which is not sharp in the usual Dirac representation. Nevertheless, we can still write down the solution Φ of this second-order equation just by exploiting its formal similarity with the non-relativistic Coulomb case. In other words, for $l(\kappa)$ we can write $l(\gamma)$, remembering that this is not sharp. We associate positive energy solutions Φ_-, i.e. electrons, with $\rho_3 = -1$ in Dirac space and negative energy solutions Φ_+, i.e. positrons, with $\rho_3 = 1$. Then

$$\Phi_- = \begin{pmatrix} 0 \\ |\eta\gamma\mu\rangle \end{pmatrix} = \begin{pmatrix} 0 \\ F_{\eta l(\gamma)}(kr)\,\chi_\kappa^\mu \end{pmatrix} \quad (3.81)$$

and

$$\Phi_+ = \begin{pmatrix} |\eta - \gamma\mu\rangle \\ 0 \end{pmatrix}, \quad (3.82)$$

where

$$F_{\eta l(\gamma)}(kr) = C_l(\eta)(kr)^{l(\gamma)}\,e^{-ikr}\,{}_1F_1(l+1-i\eta, 2l+2; 2ikr),$$
$$C_l(\eta) = [2^l\,e^{-\pi\eta/2}|\Gamma(l+1+i\eta)|/\Gamma(2l+2)]. \quad (3.83)$$

To get the solution of the first order equation, we note that the operators $\mathcal{O}_+^c$ and $\mathcal{O}_-^c$ commute, so that

$$\mathcal{O}_-^c\,\mathcal{O}_+^c\,\Phi = \mathcal{O}_+^c(\mathcal{O}_-^c\,\Phi) = 0,$$

i.e. the function $\mathcal{O}_-^c \Phi$ is a solution of the first order eqn (3.73). Using this and remembering that the operator ρ_2 is an odd operator in Dirac space,‡ we can see that the solution Ψ contains mixtures of $l(\gamma)$ and $l(-\gamma)$ in the "big" and "little" spinor components. Actually,

$$\mathcal{O}_-^c = \begin{pmatrix} -\left(E/\hbar c + \dfrac{\alpha Z}{r} + \dfrac{mc}{\hbar}\right) & -i\boldsymbol{\sigma}\cdot\hat{r}\left(\dfrac{\partial}{\partial r} + \dfrac{1+\gamma}{r}\right) \\ i\boldsymbol{\sigma}\cdot r\left(\dfrac{\partial}{\partial r} + \dfrac{1-\gamma}{r}\right) & \left(E/\hbar c + \dfrac{\alpha Z}{r} - \dfrac{mc}{\hbar}\right) \end{pmatrix}. \tag{3.84}$$

Analogous to eqn (3.32) we have the operator identity for the radial functions,

$$\left[\frac{k\eta}{\gamma} + \frac{\mathrm{d}}{\mathrm{d}r} + \frac{1+\gamma}{r}\right] F_{\eta l(\gamma)}(kr) = k\,\mathrm{sgn}\,(\gamma)\left|\left(1 + \frac{\eta^2}{\gamma^2}\right)^{1/2}\right| F_{\eta l(-\gamma)}(kr). \tag{3.85}$$

It is evident that the off-diagonal elements of $\mathcal{O}_-^c$ behave as lowering and raising operators for these functions. This is what causes the mixing.

We thus obtain, after normalization in the energy scale, $\int d\mathbf{r}\Psi_{E'}^{\dagger}\Psi_E = \delta(E - E')$,

$$\Psi = \begin{pmatrix} if_\kappa(r)\,\chi_{-\kappa}^{\mu} \\ g_\kappa(r)\,\chi_\kappa^{\mu} \end{pmatrix}, \tag{3.86}$$

where

$$rf_\kappa(r) = \frac{i(E-1)^{1/2}(2kr)^{\gamma}\,e^{\pi\eta/2}|\Gamma(\gamma + i\eta)|}{2(\pi k)^{1/2}\Gamma(2\gamma+1)} \cdot \{e^{-ikr+i\theta}(\gamma + i\eta)\,{}_1F_1(\gamma + 1 + i\eta, 2\gamma + 1; 2ikr) - \text{c.c.}\}, \tag{3.87}$$

$$rg_\kappa(r) = \frac{(E+1)^{1/2}(2kr)^{\gamma}\,e^{\pi\eta/2}|\Gamma(\gamma + i\eta)|}{2(\pi k)^{1/2}\Gamma(2\gamma+1)} \cdot \{e^{-ikr+i\theta}(\gamma + i\eta)\,{}_1F_1(\gamma + 1 + i\eta, 2\gamma + 1; 2ikr) + \text{c.c.}\}. \tag{3.88}$$

Here θ is given by

$$e^{2i\theta} = -\frac{\kappa - i\eta/E}{\gamma + i\eta}.$$

In eqn (3.87), we have used the Kummer transformation

$${}_1F_1(a, c; x) = e^{x}\,{}_1F_1(c - a, c; -x)$$

‡ I.e. it has non-zero off-diagonal elements.

to write the expression within braces in a symmetrical form. The bound state solutions may be written out in an entirely analogous manner, keeping in view the mapping $N \rightleftarrows i\eta$.

Using the same spinor form as in eqn (3.86), the normalized radial eigenfunctions for the bound state are:

$$f_\kappa(r) = -\frac{\sqrt{(\Gamma(2\gamma+n'+1))}}{\Gamma(2\gamma+1)\sqrt{n'!}} \sqrt{\left(\frac{(1-\varepsilon)}{4N'(N'-\kappa)}\right)} \left(\frac{2Z}{N' a_0}\right)^{3/2} e^{-Zr/N'a_0} \left(\frac{2Zr}{N' a_0}\right)^{\gamma-1} \cdot$$

$$\cdot \left[n' \, {}_1F_1\left(-n'+1, 2\gamma+1; \frac{2Zr}{N' a_0}\right) + (N'-\kappa) {}_1F_1\left(-n', 2\gamma+1; \frac{2Zr}{N' a_0}\right)\right], \tag{3.89}$$

$$g_\kappa(r) = -\frac{\sqrt{(\Gamma(2\gamma+n'+1))}}{\Gamma(2\gamma+1)\sqrt{n'!}} \sqrt{\left(\frac{1+\varepsilon}{4N'(N'-\kappa)}\right)} \left(\frac{2Z}{N' a_0}\right)^{3/2} e^{-Zr/N'a_0} \left(\frac{2Zr}{N' a_0}\right)^{\gamma-1} \cdot$$

$$\cdot \left[-n' \, {}_1F_1\left(-n'+1, 2\gamma+1; \frac{2Zr}{N' a_0}\right) + \right.$$

$$\left. + (N'-\kappa) {}_1F_1\left(-n', 2\gamma+1; \frac{2Zr}{N' a_0}\right)\right]. \tag{3.90}$$

Here we have used

$$\varepsilon = E/mc^2 = \frac{1}{[1+\alpha^2 Z^2/(n'+\gamma)^2]^{1/2}},$$

$$n' = \frac{\alpha Z \varepsilon}{\sqrt{(1-\varepsilon^2)}} - \gamma = \frac{\alpha Z \varepsilon}{\sqrt{(1-\varepsilon^2)}} - \sqrt{(\kappa^2 - \alpha^2 Z^2)}, \tag{3.91}$$

N' = apparent principal quantum number

$$= \sqrt{(N^2 - 2n'(|\kappa| - \sqrt{\kappa^2 - \alpha^2 Z^2}))}.$$

Quite obviously, introduction of relativity has led to the non-integer "principal quantum number" N' and the non-integer γ in the expressions. This makes it difficult to obtain results in closed form, numerical computation being often the only way to get results. The first order Dirac–Coulomb wave equation can be written in the form

$$[E - c\rho_1 \boldsymbol{\sigma} \cdot \mathbf{p} - \rho_3 mc^2] \Psi = -\frac{Ze^2}{r} \Psi. \tag{3.92}$$

Using the operator identity (Dirac, 1958)

$$\boldsymbol{\sigma}\cdot\mathbf{p} = -i\hbar\boldsymbol{\sigma}\cdot\hat{r}\left(\frac{\partial}{\partial r}+\frac{1}{r}\right) + i\hbar\frac{\boldsymbol{\sigma}\cdot\hat{r}}{r}\rho_3 K, \tag{3.93}$$

eqn (3.92) may be written

$$\left[E + i\rho_1 \hbar c\boldsymbol{\sigma}\cdot\hat{r}\left\{\frac{\partial}{\partial r}+\frac{1}{r}-\frac{\rho_3 K}{r}\right\} - \rho_3 mc^2 + \frac{Ze^2}{r}\right]\Psi = 0,$$

or

$$\left[\frac{E+\alpha Z/r}{\hbar c} + i\rho_1\boldsymbol{\sigma}\cdot\hat{r}\left\{\frac{\partial}{\partial r}+\frac{1}{r}-\frac{\boldsymbol{\sigma}\cdot\mathbf{L}+1}{r}\right\} - \frac{\rho_3 mc}{\hbar}\right]\Psi = 0. \tag{3.94}$$

Substituting from eqn (3.86) and using the properties of the spinor operators, we get the coupled radial equations (cf. Rose, 1961)

$$\left(\frac{E+\alpha Z/r}{\hbar c}+\frac{mc}{\hbar}\right)f_\kappa + \frac{\mathrm{d}g_\kappa}{\mathrm{d}r} + \frac{\kappa+1}{r}g_\kappa = 0,$$

$$\left(\frac{E+\alpha Z/r}{\hbar c}-\frac{mc}{\hbar}\right)g_\kappa + \frac{\mathrm{d}f_\kappa}{\mathrm{d}r} - \frac{\kappa-1}{r}f_\kappa = 0. \tag{3.95}$$

It may be noted that these two equations completely define the functions $f_\kappa(r)$ and $g_\kappa(r)$. In any numerical computation, these are the basic equations for generating the radial part of the solution. Formally, they are immediately related to the recursive relation (3.85).

To illustrate the use of these solutions in Auger calculations, we consider the earliest work in this field, namely, that of Massey and Burhop (1936) on the transition rates for the KL_1L_1, KL_1L_2 and KL_1L_3 lines for gold ($Z = 79$). These authors treated the Auger process as a two-electron problem and used screened hydrogen-like wave functions to compute the charge–charge and current–current interactions in Møller's formula. The retardation term was included where it was significant. Their work was a direct application of the approach described in the earlier sections of this chapter, namely, the screened Coulomb field, to an exact relativistic calculation of the Auger process.

Since our object is to present the technique of their calculation, we shall only be concerned with the KL_1L_1 transition.

Massey and Burhop used the solutions of the Dirac–Coulomb problem in the form given by Darwin (1928). To quote the latter, his approach was based on "making use of the ordinary methods of differential equations instead of the non-commutative algebra used by him (Dirac)". In the present work, we have been inclined to use this non-commutative algebra, whenever possible, in order to exploit the physical symmetries of the problem. In contrast,

Darwin's approach (Darwin, 1928) was to clearly exhibit all four spinor components of the solution Ψ‡ and to use the usual techniques of differential equations on these components. The components happen to have a simple form for the wave functions we shall be concerned with and, because all four components are explicitly written, spherical harmonics of the type $Y_l^m(\theta,\varphi)$, and not the Pauli central field spinors $\chi_\kappa^\mu(\theta,\varphi)$, appear in the calculation.

For the K-shell, these authors use ;§

$$\begin{aligned}\Psi_1 &= -iN_K f(r)\cos\theta,\\ \Psi_2 &= -iN_K f(r)\sin\theta\, e^{i\varphi},\\ \Psi_3 &= N_K g(r),\\ \Psi_4 &= 0,\end{aligned} \qquad (3.96)$$

where

$$f(r) = \frac{\alpha Z_K}{1+\sqrt{(1-\alpha^2 Z_K^2)}}\, e^{-Z_K r/a_0}\, r^{\sqrt{(1-\alpha^2 Z_K^2)}-1},$$

$$g(r) = e^{-Z_K r/a_0}\, r^{\sqrt{(1-\alpha^2 Z_K^2)}-1},$$

(Z_K = effective nuclear charge for the K-shell)

and N_K represents a normalizing factor:

$$N_K = \frac{1}{\sqrt{(4\pi)}}\left[\left(\frac{2Z_K}{a_0}\right)^{1+2\sqrt{(1-\alpha^2 Z_K^2)}} \frac{(1+\sqrt{(1-\alpha^2 Z_K^2)})}{2\Gamma(1+2\sqrt{(1-\alpha^2 Z_K^2)})}\right]^{1/2}.$$

It may be noted that the four components written out in eqn (3.96) are of the form

$$\Psi_1 = -if_{l(\kappa)}(r)\, Y_{l+1}^m(\theta,\varphi), \quad \Psi_2 = -if_{l(\kappa)}(r)\, Y_{l+1}^{m+1}(\theta,\varphi),$$
$$\Psi_3 = (l+m+1)\, g_{l(\kappa)}(r) Y_l^m(\theta,\varphi), \quad \Psi_4 = (-l+m) g_{l(\kappa)}(r) Y_l^{m+1}(\theta,\varphi), \quad (3.97)$$

for the particular case $N=1$, $l=0$, $\kappa=-1$, $m=0$.

Equation (3.97) gives the general expressions for the spinor components, written out in Darwin's explicit form, for the case $j = l+\frac{1}{2}(\kappa<0)$.

For the case $j = l-\frac{1}{2}(\kappa>0)$, the corresponding expressions are

$$\Psi_1 = -i(l+m) f_{l(-\kappa)}(r)\, Y_{l-1}^m(\theta,\varphi), \quad \Psi_2 = -i(l-m+1) f_{l(-\kappa)}(r) Y_{l-1}^{m+1}(\theta,\varphi),$$
$$\Psi_3 = g_{l(-\kappa)}(r) Y_l^m(\theta,\varphi), \quad \Psi_4 = g_{l(-\kappa)}(r) Y_l^{m+1}(\theta,\varphi), \qquad (3.98)$$

the changed subscripts for g and f denoting the fact that different radial functions (corresponding to $l \to -l-1$) are involved. Using eqn (3.97), we

‡ Two in Dirac space for positive and negative energy solutions and two in Pauli space for the two values of spin.

§ The authors used Darwin's phases, namely,

$$\Psi = \begin{pmatrix} -if(r)\chi_{-\kappa}^\mu \\ g(r)\chi_\kappa^\mu \end{pmatrix}.$$

can write out the two wave functions for the L_1 subshell ($N=2$, $l=0$, $\kappa=-1$) corresponding to the two allowed values of m, i.e. $m=0$ and $m=-1$.

$$L_1^0(m=0): \left.\begin{aligned} \Psi_1 &= -iN_{L_1}F_0(r)\cos\theta \\ \Psi_2 &= -iN_{L_1}F_0(r)\sin\theta\, e^{i\varphi} \\ \Psi_3 &= N_{L_1}G_0(r) \\ \Psi_4 &= 0 \end{aligned}\right\}, \tag{3.99}$$

and

$$L_1^{-1}(m=-1): \left.\begin{aligned} \Psi_1 &= iN_{L_1}F_0(r)\sin\theta\, e^{-i\varphi} \\ \Psi_2 &= -iN_{L_1}F_0(r)\cos\theta \\ \Psi_3 &= 0 \\ \Psi_4 &= -N_{L_1}G_0(r) \end{aligned}\right\}, \tag{3.100}$$

where

$$F_0(r) = \frac{\alpha Z_L}{\sqrt{(1-\alpha^2 Z_L^2)} + N' + 1}\, e^{-Z_L r/a_0 N'_L} \left\{ r^{\sqrt{(1-\alpha^2 Z_L^2)}}(N'+1) - \right.$$

$$\left. - r^{\sqrt{(1-\alpha^2 Z_L^2)}-1} \frac{a_0}{Z_L} \frac{N'(N'+2)(1+2\sqrt{(1-\alpha^2 Z_L^2)})}{2} \right\},$$

$$G_0(r) = e^{-Z_L r/N'_L a_0} \left\{ r^{\sqrt{(1-\alpha^2 Z_L^2)}}(N'+1) - r^{\sqrt{(1-\alpha^2 Z_L^2)}} \frac{a_0}{Z_L} N'^2 \frac{(1+2\sqrt{(1-\alpha^2 Z_L^2)})}{2} \right\},$$

$$N' = \sqrt{(2 + 2\sqrt{(1-\alpha^2 Z_L^2)})},$$

and

$$N_{L_1} = \frac{1}{4\pi}\left(\frac{2Z_L}{a_0 N'}\right)^{(3+2\sqrt{(1-\alpha^2 Z_L^2)})/2} \left[\Gamma(2+2\sqrt{(1-\alpha^2 Z_L^2)})\{N'(N'+2) + \right.$$

$$\left. + 2(1+\sqrt{(1-\alpha^2 Z_L^2)})\}\left\{1 + \frac{\alpha^2 Z_L^2}{1 + N' + \sqrt{(1-\alpha^2 Z_L^2)}}\right\}\right]^{-1/2}.$$

For the continuum also there are two types of states corresponding to (A) $\kappa<0$ ($j=l+\frac{1}{2}$) and (B) $\kappa>0$ ($j=l-\frac{1}{2}$), respectively. In our case, i.e. for transitions of the type $L_1L_1 \to K\infty$, the only allowed transition is

$$L_1^0 L_1^{-1} \to K^0\infty \quad \text{(type A).}$$

For this type of continuum state, the components are

$$\begin{aligned} \Psi_1 &= -iN_{\infty A}\mathscr{H}_l(r)\, Y_{l+1}^m(\theta,\varphi), \\ \Psi_2 &= -iN_{\infty A}\mathscr{H}_l(r)\, Y_{l+1}^{m+1}(\theta,\varphi), \\ \Psi_3 &= N_{\infty A}(l+m+1)\mathscr{G}_l(r)\, Y_l^m(\theta,\varphi), \\ \Psi_4 &= N_{\infty A}(-l+m)\mathscr{G}_l(r)\, Y_l^{m+1}(\theta,\varphi). \end{aligned} \tag{3.101}$$

Here the radial functions $\mathscr{H}_l(r)$ and $\mathscr{G}_l(r)$ are given by

$$\mathscr{H}_l(r) = \frac{r^s e^{-ikr}(2k)^{2s+2} e^{-\pi\eta}}{2A\,\Gamma(2s+3)} [\{(l-s) - i(\eta+d)\}\Gamma(s+2-i\eta)\Gamma(s+1+i\eta) \cdot$$
$$\cdot\ {}_1F_1(s+1+i\eta, 2s+3; 2ikr) + \{(l-s)+i(\eta+d)\}\Gamma(s+1-i\eta) \cdot$$
$$\cdot\ \Gamma(s+2+i\eta)\,{}_1F_1(s+2+i\eta, 2s+3; 2ikr)], \qquad (3.102)$$

and

$$\mathscr{G}_l(r) = \frac{-ir^s e^{-ikr}(2k)^{2s+2} e^{-\pi\eta}}{2B\,\Gamma(2s+3)} [\{(l-s) - i(\eta+d)\}\Gamma(s+2+i\eta) \cdot$$
$$\cdot\ \Gamma(s+1+i\eta)\,{}_1F_1(s+1+i\eta, 2s+3; 2ikr) - \{(l-s)+i(\eta+d)\} \cdot$$
$$\cdot\ \Gamma(s+2+i\eta)\Gamma(s+1-i\eta)\,{}_1F_1(s+2+i\eta, 2s+3; 2ikr)], \qquad (3.103)$$

with

$$A^2 = \frac{mc^2+E}{\hbar c}, \quad B^2 = \frac{mc^2-E}{\hbar c}, \quad s = \sqrt{(l+1)^2 - \alpha^2 Z_L^2} - 1,$$

$$\eta = \frac{\alpha ZE}{\hbar ck}, \quad d = \frac{\eta mc^2}{E}, \quad N_{\infty A} = \frac{1}{[4\pi(l+m+1)!(_l - m)!]^{1/2}} \cdot$$
$$\cdot \left(\frac{2\pi E}{\hbar c^2 k}\right)^{1/2} \frac{A|B|}{\{A^2+|B|^2\}^{1/2}} C^{-1},$$

$$C = \tfrac{1}{2}[(l-s)^2 + (\eta+d)^2]^{1/2} |\Gamma(s+1+i\eta)|\, e^{-3\pi\eta/2}(2k)^{s+1}.$$

To evaluate the matrix element, we note that the usual definition of transition charge and current densities,

$$\rho = -e\Psi_f^\dagger \Psi_i$$

and

$$\mathbf{j} = ce\Psi_f^\dagger \boldsymbol{\alpha}\Psi_i,$$

yields, on multiplication, the following expressions in terms of the components:

$$\rho = -e[(\Psi_f)_1^*(\Psi_i)_1 + (\Psi_f)_2^*(\Psi_i)_2 + (\Psi_f)_3^*(\Psi_i)_3 + (\Psi_f)_4^*(\Psi_i)_4],$$
$$j_x = ce[(\Psi_f)_4^*(\Psi_i)_1 + (\Psi_f)_3^*(\Psi_i)_2 + (\Psi_f)_2^*(\Psi_i)_3 + (\Psi_f)_1^*(\Psi_i)_4],$$
$$j\ = ice[(\Psi_f)_4^*(D_i)_1 - (\Psi_f)_3^*(\Psi_i)_2 + (\Psi_f)_2^*(\Psi_i)_3 - (\Psi_f)_1^*(\Psi_i)_4],$$
$$j_z = ce[(\Psi_f)_3^*(\Psi_i)_1 - (\Psi_f)_4^*(\Psi_i)_2 + (\Psi_f)_1^*(\Psi_i)_3 - (\Psi_f)_2^*(\Psi_i)_4]. \qquad (3.104)$$

For the KL_1L_1 transition, we can write,

1. for the first electron $\Psi_i = \Psi(L_1^0)$, $\Psi_f = \Psi(K^0)$,
2. for the second electron $\Psi_i = \Psi(L_1^{-1})$, $\Psi_f = \Psi(\infty A, l=0, m=-1)$.

The exchange integral here is the same as the direct integral. To illustrate the kind of expressions one gets, the charge and current densities for (1) are of the form

$$\begin{aligned}
\rho &= \frac{1}{4\pi}(f^* F_0 + g^* G_0)\, N_{L_1} N_K, \\
j_x &= \frac{i}{4\pi}(-F_0 g^* + G_0 f^* \mathrm{e}^{-2i\varphi}) \sin\theta\, \mathrm{e}^{i\varphi} N_{L_1} N_K, \\
j_y &= -\frac{1}{4\pi}(F_0 g^* + f^* G_0 \mathrm{e}^{-2i\varphi}) \sin\theta\, \mathrm{e}^{i\varphi} N_{L_1} N_K, \\
j_z &= -\frac{i}{4\pi}(F_0 g^* - f^* G_0) \cos\theta N_{L_1} N_K.
\end{aligned} \tag{3.105}$$

For (2) one gets similar expressions with $f, g \to F_0, G_0, F_0, G_0 \to H_0, G_0$ and $N_K, N_{L_1} \to N_{L_1}, N_{\infty A}$ $(l=0, m=-1)$.

Consider now the retardation term

$$\frac{1}{|\mathbf{r}_1 - \mathbf{r}_2|} \mathrm{e}^{iq|\mathbf{r}_1 - \mathbf{r}_2|} = 4\pi i q \sum_{LM} \Phi_L(r_1, r_2)\, Y_L^M(\theta_1, \varphi_1)\, Y_L^{M*}(\theta_2, \varphi_2), \tag{3.106}$$

where $q = 2\pi v_{KL}/c$ and the subscripts 1 and 2 stand for the two electrons as before.

$$\Phi_L(r_1, r_2) = \begin{cases} h_L(qr_1)\, j_L(qr_2) & \text{for} \quad r_1 > r_2, \\ j_L(qr_1)\, h_L(qr_2) & \text{for} \quad r_2 > r_1. \end{cases} \tag{3.107}$$

Note that this is a generalization of eqns (3.46) and (3.47), because the product of the spherical Bessel and spherical Hankel functions is a series expansion with terms like $r_<^L / r_>^{L+1}$. Obviously, the angular integration proceeds as before with the help of the Wigner–Eckart theorem.

The more interesting question relates to the radial integration. The integrals involved here are of the general form (Listengarten, 1961)

$$\begin{aligned}
&\int_0^\infty \{R_\infty(r_2)\, R_{L_1}(r_2)\, h_L(qr_2) \int_0^{r_2} j_L(qr_1)\, R_K(r_1)\, R_{L_1}(r_1)\, r_1^2 \,\mathrm{d}r_1 + \\
&\quad + R_K(r_2)\, R_{L_1}(r_2)\, h_L(qr_2) \int_0^{r_2} j_L(qr_1)\, R_\infty(r_1)\, R_{L_1}(r_1)\, r_1^2 \,\mathrm{d}r_1\}\, r_2^2 \,\mathrm{d}r_2,
\end{aligned} \tag{3.108}$$

where $R_\infty(r)$ are the functions $\mathscr{H}_0$ and $\mathscr{G}_0$, $R_K(r)$ the functions f and g, and $R_{L_1}(r)$ the functions F_0 and G_0. From the angular integration, we know that $L=0$ for the charge–charge terms and $L=1$ for the current–current terms. This means we have to evaluate only a small number of integrals. The real question, however, is how far are these integrals amenable to the analytic methods described in the earlier sections of this chapter?

We can see that even the discrete integral

$$\int_0^{r_2} j_L(qr_1)\, R_K(r_1)\, R_L(r_1)\, r_1^2\, \mathrm{d}r_1 \tag{3.109}$$

is more complicated than the corresponding integrals occurring in non-relativistic theory. The bound state radial functions R_K and R_{L_1} contain non-integer powers of r. Thus the expression (3.109) resolves into primary integrals of the form

$$\int_0^{r} r'^s J_{L+1/2}(qr')\, \mathrm{e}^{-\mu r'}\, \mathrm{d}r',$$

where s is not an integer. Massey and Burhop evaluated these integrals from series expansions for small values of r' ($r' < 4a_0/Z_L$) and used asymptotic expansions for larger values of r'. It is to be noted that the non-integer power or r' comes from the mapping $\kappa \to \gamma = \sqrt{\kappa^2 - \alpha^2 Z^2}$. The analogous non-relativistic integral without retardation,

$$\int_0^{r} r'^p \mathrm{e}^{-\mu r'}\mathrm{d}r' = \frac{\Gamma(p+1)}{\mu^{p+1}}\left[1 - \mathrm{e}^{-\mu r}\sum_{i=1}^{p+1}\frac{(\mu r)^{p+1-i}}{(p+1-i)!}\right] \quad (p \text{ integer}),$$

can be evaluated in closed form for any finite p.

5. Screening parameters

An important point about Coulomb-type wave functions is the choice of a screening parameter. Because the choice is crucial to all calculations involving such wave functions, we shall try to outline here the philosophy of the screening approach.

The basic assumption underlying the use of a screened hydrogen-like wave function to describe a many-electron atom is that for a single electron outside a closed shell, or a hole in an otherwise filled shell, the field due to the nucleus and the other electrons taken together can be assumed to behave as $1/r$. Once this assumption is made, the effect of the other electrons is accommodated by replacing the nuclear charge Z by an appropriate effective charge $Z_{\text{eff}} = Z - \sigma$, σ being a number characteristic of the N and l values of the electronic shell.‡ For incomplete shells with more than a single electron (or a single vacancy), description of these electrons (or vacancies) by screened hydrogen-like functions becomes even more of an approximation. However, the real

‡ Historically, this approach evolved from the observed similarities between alkali and hydrogen spectra.

advantage of this approach is that it greatly simplifies calculations and therefore serves as a kind of prototype for other more sophisticated methods.

Now the actual choice of σ would depend on what criterion one uses to compare a physical many-electron atom with its idealized hydrogenic counterpart. In other words, different atomic properties are differently affected by the presence of the other electrons and one has to decide what property one would use to define σ, assuming that the total effect on this property can be represented by an apparent decrease in the nuclear charge. Hartree (1957) suggested that a suitable candidate could be found in a typical length R, which might be either the mean radius

$$\bar{r}_{Nl} = \int_0^\infty r P_{Nl}^2(r)\,\mathrm{d}r, \tag{3.110}$$

$P_{Nl}(r)$ being the self-consistent-field (SCF) wave function for the shell (Nl), or the median radius r_m such that

$$\int_0^{r_m} P_{Nl}^2(r)\,\mathrm{d}r = \tfrac{1}{2}. \tag{3.111}$$

Equation (3.110) represents the most popularly used parameter in such calculations. Having decided on this parameter, one makes another vital assumption. One assumes that for a given (Nl), the normalized radial wave functions for atoms of different atomic number Z are geometrically similar, i.e. they are replicas of the same function on different linear scales. One thus takes the entire effect of "screening" to be represented by a scaling of this length R by a factor equal to the screened nuclear charge, this scaling being measured with respect to the hydrogen atom. Thus

$$R = R_{\mathrm{H}}/(Z-\sigma), \tag{3.112}$$

where R_{H} is the value of R for the (Nl) wave function of the hydrogen atom. In other words,

$$\sigma = Z - \frac{R_{\mathrm{H}}}{R}. \tag{3.113}$$

Hartree's definition of the screening parameter has the obvious advantage of simplicity, a mean radius being one of the simplest parameters one could associate with an atom. However, one must remember that most realistic wave functions show considerable departures from scaled hydrogenic wave functions, so that there can be no standard choice of σ. Hartree's prescription simply represents a plausible choice.

The parameter R, e.g. the mean radius $\bar{r}_{Nl}$ given by eqn (3.110) must be calculated by using some SCF wave functions. Results differ between different calculations. Froese (1966) has given a method which can be used to obtain screening parameters for any particular group (Nl) for a series of Z up to 46. The method consists in interpolating σ by means of the relation

$$\sigma = \bar{r}_{Nl}\frac{d\sigma}{d\bar{r}_{Nl}} + \sigma_0, \tag{3.114}$$

where σ_0 is the ultimate screening parameter for an atom with atomic number Z as the mean radius $\bar{r}_{Nl} \to 0$. With $\bar{r}_{Nl}$ and $d\sigma/d\bar{r}_{Nl}$ known for a number of configurations, eqn (3.114) may be used to calculate σ. Callan (1961) has given the following relations which are useful in finding r_{2s} and r_{2p} for any given Z.

$$2s\colon Z = 2.4528 + 6.3405/\bar{r}, \tag{3.115}$$

$$2p\colon Z = 3.6911 + 5.3005/\bar{r}. \tag{3.116}$$

A very different choice of σ depends on using the binding energy as the physical property for comparing a many-electron atom with its hydrogenic prototype. Here, σ is chosen so that a point nucleus of charge $Z_{\text{eff}} = Z - \sigma$ binds a single hydrogenic electron in the state (Nl) with the same binding energy as is observed for an electron in the corresponding state in the actual atom of nuclear charge Z. This is the genre to which the earliest prescription of Slater (1930) belongs, although, by ignoring features like nodes in the wave function and by following a simple empirical fitting procedure, Slater obtained numbers which are much smaller than are obtained from more realistic calculations (O'Connell and Carroll, 1966). In a relativistic wave function, there is a mixing of $l(\kappa)$ and $l(-\kappa)$, which makes the choice of screening a non-trivial question. The following is a possible option (Chattarji and Talukdar, 1968) and it relates to a KL_1L_1 transition. The K-shell corresponds to a pure $1s_{1/2}$ state and the screening constant σ_K for this shell shows little variation with Z. The authors assigned to it the Hartree value of 0.3 for all atoms. For the L_1 subshell‡ they defined the effective nuclear charge

$$Z_L = |A|^2 Z_{2s} + |B|^2 Z_{2p}, \tag{3.117}$$

where A and B are the dimensionless weight factors multiplying the "big" and "little" spinors $|N=2,\ \kappa=1,\ \mu\rangle$ and $|N=2,\ \kappa=-1,\ \mu\rangle$ occurring in the relativistic eigenfunction $\Psi_{N\kappa\mu}$. Z_{2s} and Z_{2p} are screened nuclear charges evaluated by Froese's method.

6. The hydrogenic atom—some dividends

In spite of its remarkable simplicity, the hydrogenic model, i.e. the approximation that a many-electron atom may be described by a screened (or, to be even

‡ Strictly speaking, one can no more talk about the L_1 subshell, because of the mixing of $l(\kappa)$ and $l(-\kappa)$. However, we do so by keeping in mind that the "big" spinor is the one associated with χ^{μ}_{κ}, which is the component that survives in the NRS limit.

more naive, unscreened) Coulomb field, does yield some interesting results. Consider, for example, the Auger yield a_W and the fluorescence yield ω_W, resulting from a inner shell vacancy W, as functions of the atomic number Z. If we look at a particular Auger transition WXY, then it was shown by Wentzel (1927) that the transition rate is practically independent of Z. He used the asymptotic form of the wave function for the Auger electron. Even without this simplification, this is a kind of result which is more easily seen in the hydrogenic model. For example, it was shown by Burhop (1935) that the amplitude for the KL_1L_3 transition in NR theory is given by

$$M = \frac{\sqrt{6}}{27}\left(\frac{\pi}{h}\right)^{3/2}\frac{m^{1/2}}{a_0^{11/2}} Z_L^4 Z_K^{3/2} k^{3/2} e^{\pi\eta/2} |\Gamma(2+i\eta)|\, D, \tag{3.118}$$

where

$$\begin{aligned} D = &-\frac{3Z_L}{a_0\gamma^2}\frac{5!}{\alpha^6}\,{}_2F_1\left(2+i\eta, 6, 4; \frac{2ik}{\alpha}\right) + \left(\frac{6}{\gamma^2} - \frac{12Z_L}{a_0\gamma^3}\right)\frac{4!}{\alpha^5}\,{}_2F_1\left(2+i\eta, 5, 4; \frac{2ik}{\alpha}\right) + \\ &+ \left(\frac{24}{\gamma^3} - \frac{24Z_L}{a_0\gamma^4}\right)\frac{3!}{\alpha^4}\,{}_2F_1\left(2+i\eta, 4, 4; \frac{2ik}{\alpha}\right) + \left(\frac{48}{\gamma^4} - \frac{24Z_L}{a_0\gamma^5}\right)\cdot \\ &\cdot \frac{2!}{\alpha^3}\,{}_2F_1\left(2+i\eta, 3, 4; \frac{2ik}{\alpha}\right) + \frac{48}{\gamma^5\alpha^2}\,{}_2F_1\left(2+i\eta, 2, 4; \frac{2ik}{\alpha}\right) + \frac{24Z_L}{a_0\gamma^5}\cdot \\ &\cdot \frac{2!}{\beta^3}\,{}_2F_1\left(2+i\eta, 3, 4; \frac{2ik}{\beta}\right) - \frac{48}{\gamma^5}\frac{1}{\beta^2}\,{}_2F_1\left(2+i\eta, 2, 4; \frac{2ik}{\beta}\right). \end{aligned} \tag{3.119}$$

$$\alpha = \frac{1}{a_0}(Z_L + Z_K) + ik,\ \beta = \frac{Z_L}{2a_0} + ik,\ \gamma = \frac{1}{a_0}\left(\frac{Z_L}{2} + Z_K\right). \tag{3.120}$$

Equations (3.118) and (3.119) follow directly from eqn (3.48). It can be at once seen that in the unscreened limit, i.e. with $Z_L \approx Z_K = Z$, the dominant behaviour of D is like Z^{-8} and this is exactly compensated by the Z-dependence of the other factors in eqn (3.118). We remember that $k = Z_L/a_0\eta$ in this case. Even with screening, individual Auger yields are not really sensitive to Z except for the lighter elements. This was shown very clearly by Callan (1961) who carried out calculations for forty elements in the LS coupling (Fig. 3.1). He used screening parameters derived from Hartree–Fock calculations and used eqn (3.114) to get values for different subshells. The only marked deviation comes in the $KL_{2,3}L_{2,3}(^1D)$ and $KL_1L_{2,3}(^1P)$ cases, particularly in the former, where there is a steep rise for $Z < 40$. The levelling off of individual KL_xL_y rates as one goes to high Z values was explained by Callan in terms of the average radius $\bar{r}$, which varies as $1/Z$. For high Z, $\bar{r} \to 0$, and the screening parameter $\sigma \to \sigma_0$. Thus the wave functions for heavy atoms correspond to

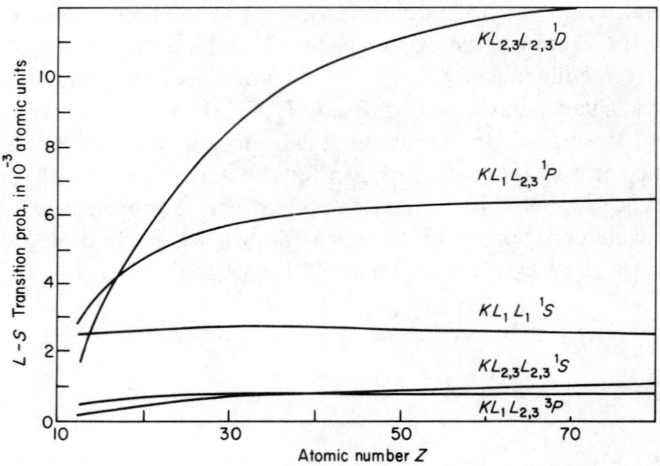

Fig. 3.1. Variation of *KLL* Auger transition probabilities with atomic number Z in Russell–Saunders (*LS*) coupling, using nonrelativistic screened hydrogenic wave functions (from Callan, 1961, courtesy of the American Institute of Physics).

this ultimate screening parameter, and the Auger transition probability settles down to a limiting value. The limit $\bar{r} \to 0$ may be regarded as the limiting "hydrogenic" case, i.e. the strong Coulomb field of the nucleus has led to a kind of "collapse" of the atom to a small region near it.

One must, however, bring in relativity at this stage because of the high velocity of the electrons.

Total *KLL* transition rates obtained by Callan are shown in Fig. 3.2. They increase monotonically at first up to $Z = 50$ and then settle down to a saturation value. Because the total rate is independent of the coupling schemes which we shall discuss in the next chapter, Fig. 3.2 may be regarded as representative of the screened hydrogenic model.

The radiative dipole transition rate corresponding to an initial state Ψ_i and a final state Ψ_f is given by

$$P_R = \frac{16\pi^4 \nu^3 e^2}{3hc^3} \left| \int \Psi_f^* \mathbf{r} \Psi_i \, d\tau \right|^2, \tag{3.121}$$

where $\nu = (E_i - E_f)/h = RZ^2[(1/n_f^2) - (1/n_i^2)]$, R being the Rydberg energy. For hydrogenic wave functions, this varies as Z^4. For example, in the $L_3 \to K$ transition, we have

$$\Psi_i = \Psi(2p_{3/2}) = \frac{1}{8\sqrt{\pi}} \left(\frac{Z_L}{a_0} \right)^{5/2} r\, e^{-Z_L r/2a_0} \sin\theta\, e^{\pm i\varphi},$$

$$\Psi_f = \Psi(1s_{1/2}) = \frac{1}{\sqrt{\pi}} \left(\frac{Z_K}{a_0} \right)^{3/2} e^{-Z_K r/a_0}.$$

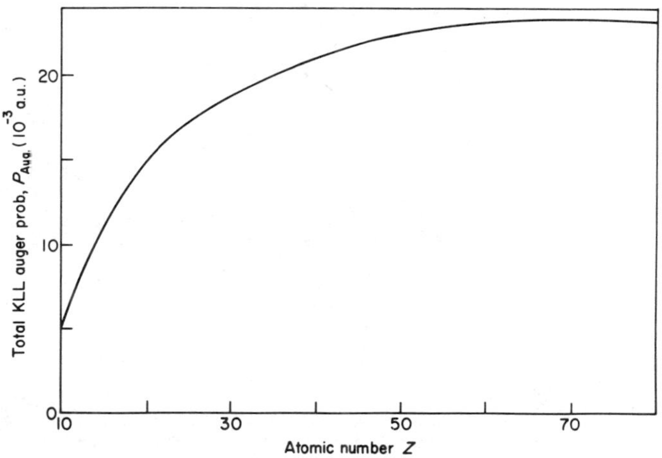

Fig. 3.2. Total *KLL* Auger transition probability as a function of *Z*, with nonrelativistic screened hydrogenic wave functions (from Callan, 1961, courtesy of the American Institute of Physics).

Setting $Z_K \approx Z_L = Z$, and using the result $\int_0^\infty e^{-ar} r^n dr = n!/a^{n+1}$, it may be shown that $P_R \propto Z^4$. If we now set, for the *K*-shell,

$$P_R = aZ^4, \quad P_A = b,$$

then

$$\omega_K = (1 + \alpha Z^{-4})^{-1} \quad (\alpha = b/a). \tag{3.122}$$

Burhop (1952) estimated the constant α to be about 1.12×10^6. Subsequently, he gave a modified formula

$$\left(\frac{\omega_K}{1 - \omega_K}\right)^{1/4} = -A + BZ - CZ^3 \tag{3.123}$$

where the constants *A* and *C* describe the effects of screening and relativity, respectively (Burhop, 1955).

A comparison of theoretical values of ω_K with experiment was made by Callan as well as by Byrne and Howarth (1970). Figure 3.3 shows the curves given by Callan (1961), where he includes the calculated values of Burhop (1955) and of Geffrion and Nadeau (1959) along with his own results. Experimental values shown in this figure were taken from Broyles *et al.* (1953). It will be seen that Callan's results are consistently higher than those obtained by others. Because $a_K = 1 - \omega_K$, the total Auger yields are too low.

Experiments show that for low atomic numbers, Auger transitions dominate. As will be seen in Chapter 8, this makes Auger electron spectroscopy a particularly suitable technique for the detection of light elements.

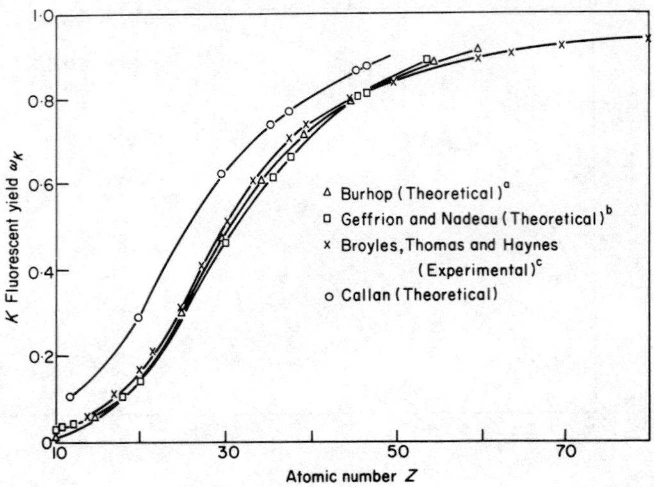

Fig. 3.3. Comparison of theoretical and experimental values of K-shell fluorescence yields (ω_K) as functions of atomic number Z. The legends represent the following references (a) Burhop (1955), (b) Geffrion and Nadeau (1959), (c) Broyles *et al.* (1953). The open circles on the upper curve are from Callan (1961; courtesy of the American Institute of Physics).

Byrne and Howarth wrote eqn (3.122) in the form

$$\ln\left(\frac{1-\omega_K}{\omega_K}\right) = -m \ln Z + \ln \alpha \tag{3.124}$$

and plotted the experimental and theoretical values of $\ln[(1-\omega_K)/\omega_K]$ against Z. The results were straight lines with slope $-m$ and intercept $\ln\alpha$ as follows:

	m	α
Experiment	3.36 ± 0.02	1.16 ± 0.07
Theory	3.41 ± 0.01	1.25 ± 0.05

The theoretical values were taken from Rubenstein and Snyder (1955), Asaad (1959), Listengarten (1962) and Callan (1962). Experimental values were taken from the compilation made by Fink *et al.* (1966). Assuming that $\omega_K \propto Z^4$, it was suggested by Byrne and Howarth that the Auger transition probability is approximately proportional to $Z^{1/2}$. They also inferred atomic shell-structure effects from the somewhat oscillating behaviour of α with increasing Z.

To summarize, we have seen in this chapter how some of the basic features of Auger yield calculations can be brought out within the limited framework of a screened Coulomb model. In the non-relativistic approximation, the model introduces considerable simplification with a matching dividend of physical insight. Unfortunately, even in this simple model, a relativistic calculation calls for numerical techniques because of the non-integer "angular momentum". This need not frighten the theoretician, but it does make the screening approach a little less worthwhile than a more precise and sophisticated many-electron approach such as the Thomas–Fermi–Dirac (TFD) model or a selfconsistent field (SCF) model such as the Hartree–Fock–Slater system. Such calculations have been made by several authors including Asaad (1959), Listengarten (1961, 1962), Bhalla and his coworkers (1970) etc. We shall discuss these calculations in Chapter 6.

Chapter Four
Symmetry-Breaking and the Classification of States

1. Introduction

In the last chapter, we discussed in detail how the symmetry properties of the Coulomb field enter into the description of the Auger process and lead to certain simplifications in the calculation of non-relativistic matrix elements. At the same time, we were careful in pointing out the approximate nature of the non-relativistic screened hydrogenic model which makes these simplifications possible. For example, the very introduction of relativity spoils the O_4 symmetry of the problem. Furthermore, even within a non-relativistic set-up, treating a many-electron atom as a screened hydrogen-like system involves an approximation. As soon as we try to view this problem in a more realistic way, we have to introduce perturbations to the pure Coulomb Hamiltonian. The most important among these perturbations are

1. electrostatic interaction between pairs of electrons in the atom, and
2. coupling between the spin and orbital motion of individual electrons.

The first perturbation is rather important in the lighter atoms, because it causes the overall electrostatic field experienced by any given electron to be non-central, the effect of the other atomic electrons being quite comparable to the nuclear Coulomb field. The second type of interaction assumes greater importance as one goes to heavier atoms, because there the magnetic interaction between the orbital magnetic moment and the intrinsic magnetic moment due to spin cannot be neglected in comparison with other forces, while the presence of many electrons causes a kind of smoothing out of their mutual electrostatic interactions into a "quasi-central" resultant. We shall presently go into the detailed nature of these interactions. For the present, however, we wish to emphasize the point that these perturbations break the O_4 symmetry of the system. To the extent that this is true, i.e. to the extent that the many-electron atom may be described by a perturbed hydrogen-like Hamiltonian, the states of this system may be classified by the broken multiplets of the symmetry group. The situation is quite analogous, for example, to that in particle physics, where one seeks to classify sets of particles by the broken super-multiplets of some symmetry group with the difference that here the basic interaction is known and the symmetry involved has a well-known geometric character. The symmetry-breaking interactions are also partly known, items (1) and (2) listed above being the most important among them. It is thus possible to write down explicit analytic expressions for the approximate states of the system. The geometric character of the symmetry is particularly useful in writing out the angular parts of the matrix elements. So far as this part is concerned, the symmetry involved is that of O_3, which depends only on the central character of the field. In other words, so long as the unperturbed Hamiltonian is of the central-field type (but not necessarily strictly Coulombic, i.e. $1/r$, in character), the angular parts of the matrix elements are characterized by the usual Wigner algebra. And even as one goes on to consider the breaking of the overall O_4 symmetry, the states of the system may nevertheless be classified in terms of this algebra. This, in fact, is one of the important advantages of the "broken symmetry" approach: one may classify the states of the perturbed system in terms of the unperturbed one, the classification becoming more exact as the perturbation gets weaker. This is the basis for using the multiplet notation ${}^{2S+1}L_J$ to describe the states of a many-electron system, where L is the total orbital angular momentum, J the total angular momentum and $2S+1$ the multiplicity of the system. Strictly speaking, these quantum numbers relate only to a central-field system. To use them to describe a many-electron system is to carry over the classification to a system with a broken symmetry.

Again, precisely because of these advantages of the central field description, even the more sophisticated calculational procedures, such as the Hartree–Fock–Slater self-consistent-field formalism, tend to retain this picture while bringing about improvements in the radial part of the wave function. The radial functions in these formulations are quite different from the Coulomb functions $F_{Nl}(r)$ or $F_{\eta l}(kr)$ introduced in the previous chapter. But the angular parts involve the familiar spherical harmonics $Y_l^m(\theta, \varphi)$, which are basis functions of O_3 and are characteristic of the central field. To do away with these would lead to considerable difficulties. For example, the substitution of tensor forces, however realistic, is not a very popular procedure. In other words, we do break the symmetry to get realistic results, but we try to do this in a gentle, "painless' manner.

Looking at the radial part of the problem, we must note that even here the correspondence with the hydrogen problem is never really lost sight of. To anticipate matters a little, consider Fock's integro-differential equation (Hartree, 1957),

$$\left[\frac{\hbar^2}{2\mu}\frac{d^2}{dr^2} + \frac{Ze^2}{r}\,Y(Nl;r) - \varepsilon_{Nl} - \frac{l(l+1)\hbar^2}{2\mu r^2}\right] P_{Nl}(r) = X(Nl;r), \qquad (4.1)$$

which defines the radial function $P_{Nl}(r)$. Here $X(Nl;r)$ on the right-hand side represents inhomogeneities or symmetry-breaking terms which go to zero in the pure hydrogenic limit. In the same limit $Y(Nl;r) \to 1$, so that we get Schrodinger's radial equation for the hydrogen problem. Coming back to our original problem, to describe the states of the perturbed many-electron system in terms of this symmetry-breaking mechanism, we must now consider, somewhat more specifically, the different interactions which can remove the degeneracy. We have already discussed the "fine-structure interaction" which removes l-degeneracy. This is the effect of relativity *per se*, which has nothing to do with the many-electron problem as such. We shall now focus our attention on two types of interaction, namely, the items (1) and (2) mentioned above.

To be specific, consider the *KLL* Auger spectrum. For simplicity, we confine our attention to an atom with an initial K-shell vacancy and an L shell completely filled, e.g. Ne^+ with the configuration $(1s)(2s)^2(2p)^6$. After a *KLL* Auger transition, the Hamiltonian of the six remaining electrons in the L-shell may be written, in the Coulomb field approximation,

$$H_0 = \sum_{i=1}^{6} \left\{ -\frac{\hbar^2}{2m}\nabla_i^2 - \frac{Ze^2}{r_i} \right\} = \sum_{i=1}^{6} H_0(i). \tag{4.2}$$

These electrons may be in one of the following three configurations:

$$(1)\ (2s)^0(2p)^6, \quad (2)\ (2s)(2p)^5 \text{ and } (3)\ (2s)^2(2p)^4.$$

The individual electronic Hamiltonian $H_0(i)$ is spherically symmetric, i.e. invariant under rotations in configuration space. Since the total Hamiltonian H_0 does not contain any coupling between particles, it must be invariant under the direct product group

$$S^{0'} = O_3(1) \otimes O_3(2) \otimes \cdots \otimes O_3(6). \tag{4.3}$$

Obviously, this is a Lie group with 18 parameters and is not a rotation group in configuration space. However, this is not the complete story. From the previous chapter, we know that the Coulomb field Ze^2/r_i has a larger symmetry group, namely, $O_4(i)$. Thus the symmetry group of H_0 is

$$S^0 = O_4(1) \otimes O_4(2) \otimes \ldots \otimes O_4(6). \tag{4.3a}$$

This is a rather large group, which contains $S^{0'}$ as a subgroup. The solutions for this system consist of direct products of the solutions for the individual electrons. Obviously, the system has a high degree of degeneracy. Specifically, it is degenerate in energy with respect to the configurations (1), (2) and (3).

Now let us switch on the perturbation (1), i.e. the mutual electrostatic interaction between the electrons

$$H_1' = \sum_{\substack{i,j \\ (i \neq j)}} \frac{e^2}{r_{ij}}. \tag{4.4}$$

Here $r_{ij} = |\mathbf{r}_i - \mathbf{r}_j|$, the inter-electronic distance, which is not invariant under independent rotations of the vectors $\mathbf{r}_i$ and $\mathbf{r}_j$. Thus even for the two interacting particles i and j, the direct product group $O_3(i)\otimes O_3(j)$ is too large. Obviously, then not even $S^{0\prime}$ is the symmetry group of the perturbed system.

We note, however, that the distance $|\mathbf{r}_i - \mathbf{r}_j|$ is invariant under simultaneous and equal rotations of $\mathbf{r}_i$ and $\mathbf{r}_j$ taken together, i.e. rotations which conserve the angle between them. Hence the symmetry group of the interaction e^2/r_{ij}, which we may write as $O_3(i,j)$ is the subgroup of $O_3(i)\otimes O_3(j)$ composed of operators of the form $\exp[i\boldsymbol{\alpha}\cdot(\mathbf{l}_i + \mathbf{l}_j)]$, with a single continuous vector parameter $\boldsymbol{\alpha}$ characterizing rotations of both the vectors $\mathbf{r}_i$ and $\mathbf{r}_j$. Obviously, these operators represent a rotation group with generators

$$\mathbf{L}_{ij} = \mathbf{l}_i + \mathbf{l}_j. \tag{4.5}$$

Generalizing this procedure, we see that the symmetry group for $H = H_0 + H_{\mathrm{I}}'$ is just the rotation group $O_3(1,2,3,4,5,6)$, which is a subgroup of S^0. Thus the symmetry group of the perturbed system is the group S_{I}' of simultaneous rotations of the coordinates of all the particles with the invariant operator

$$\mathbf{L}^2 = \sum_{i=1}^{6} \mathbf{l}_i^2. \tag{4.6}$$

Similar arguments lead to the result that $\mathbf{S}^2 = \sum_{i=1}^{6} \mathbf{s}_i^2$ is also a constant of the motion. This is the justification for the name LS coupling. From the above discussion, it should be clear that S_{I}' is a subgroup of S^0.

Now let us see what this does to the multiplets of the system, i.e. states belonging to different eigenvalues of the invariant operator. The large multiplets of S^0, called super-multiplets, are composed of the degenerate eigenstates of H_0. These supermultiplets are invariant under the subgroup S_{I}', and so must each consist of a set of multiplets of S_{I}'. But each multiplet of S_{I}', labelled by L, corresponds to a single energy level of $H(=H_0 + H_{\mathrm{I}}')$. Hence the result: if a supermultiplet of S^0 contains n multiplets of S_{I}', switching on the symmetry-breaking interaction H_{I}' will cause the corresponding energy level of H_0 to split into n distinct levels of H (McVoy, 1965)‡. As a next step, we may include spin and turn on the spin–orbit interaction

$$H_{\mathrm{II}}' = \sum_{i=1}^{6} \xi(r_i)\,\mathbf{l}_i\cdot\mathbf{s}_i, \tag{4.7}$$

where

$$\xi(r) = \frac{e}{2m^2c^2}\,\frac{1}{r}\,\frac{\partial V}{\partial r}, \tag{4.8}$$

‡ Technically, for a given configuration, states labelled by definite values of L and S are called multiplets in atomic physics.

$V(r)$ being the central field potential which now approximately describes the system apart from spin–orbit interaction. Obviously, $\mathbf{L}^2 = \sum_i \mathbf{l}_i^2$ is no more an invariant operator, but $\mathbf{J}^2 = \sum_i \mathbf{j}_i^2 = \sum_i (\mathbf{l}_i + \mathbf{s}_i)^2$ is. The symmetry group S'_{II} is now generated by the total angular momentum vectors $\mathbf{j}_i$. This causes a further splitting of the multiplets into energy levels. For heavier atoms, this splitting is quite comparable with the separation between terms caused by the electrostatic interaction H'_{I}.

We can, therefore, see that the entire classification of states of the system should depend on the relative importance of these symmetry-breaking interactions. This is not surprising, because the situation begins to be physically interesting only when a perturbation is turned on—a completely degenerate system has very few physical attributes. Once the perturbations are switched on, the broken symmetry serves as a frame of reference for classifying the states which are now physically distinguishable.

Let us see how the classification is made when one or the other of the two-symmetry-breaking perturbations, namely H'_{I} and H'_{II}, dominates. We must remember that to regard either of these as dominant implies an approximation, because for most values of Z both the electrostatic interaction and the spin–orbit coupling should be taken into account.

Another point we should like to make here is that because of the geometrical nature of the symmetry groups of the perturbations H'_{I} and H'_{II}, i.e. groups of rotations in configuration space (with certain restrictions), the generators **L** or **J** turn out to be angular momentum variables. One thus happens to talk about things like *LS* coupling or *jj* coupling when indicating the construction of these generators. In fact the coupling schemes provide a convenient means for book-keeping with regard to the states of the system. Once a coupling scheme has been decided upon, the entire apparatus of Wigner algebra becomes available for calculations involving these states. However, one must always remember that the basic criteria for choosing a particular scheme are not geometric but dynamical in character: the choice depends on which of the two dynamical perturbations, H'_{I} or H'_{II}, is dominant in a particular situation.

2. The *LS* coupling scheme (Russell–Saunders coupling)

To be specific, let us again confine our discussion to the *KLL* Auger spectrum. As we have already seen, the final state configurations are:

$$(1)\ (2s)^0(2p)^6,\quad (2)\ (2s)(2p)^5 \text{ and } (3)\ (2s)^2(2p)^4.$$

It has been shown by Condon and Shortley (1959) that the states of these nearly closed-shell configurations can be expressed in terms of those of completely closed-shell configurations together with the two-electron configurations,

$$(1)'\ (2s)^2,\ (2)'\ (2s)(2p) \text{ and } (3)'\ (2p)^2.$$

By making arguments based on particle hole symmetry and the resulting symmetry in the energetics of the system, or by pure algebra, it can be shown that the states of the two systems are equivalent, provided that the sign of the spin–orbit coupling term is reversed and certain additional terms added to the diagonal elements of the matrix of the electrostatic interaction. Since these additional terms are the same for all diagonal elements for a given configuation, they will not affect the relative positions of terms within a given configuration but only the total energy of the whole configuration, which, in the ultimate analysis, is arbitrary.

This is a very useful result, because it reduces the description of the many-electron system to the two-electron configurations indicated above so far as *KLL* Auger transitions are concerned.‡ As a result we have to worry about adding together only two orbital angular momenta, l_1 and l_2, rather than the six involved in eqn (4.6), and only two intrinsic spins $s_1 = \frac{1}{2}$ and $s_2 = \frac{1}{2}$.

For light elements with $10 \leqslant Z \leqslant 25$, the main perturbing term is the electrostatic interaction

$$H_I' = \sum_{\substack{i,j \\ i \neq j}} \frac{e^2}{r_{ij}}.$$

Since it does not contain the spin coordinates at all, it commutes with the total spin angular momentum **S**. Also, being a scalar built of space coordinates only, it commutes with **L** (i.e. it remains invariant under all equal rotations of the electronic coordinates). In other words, H_I' is a scalar both in ordinary and in spin space (De-Shalit and Talmi, 1963).

Now, considering any of the two-electron configurations (1)′, (2)′ or (3)′, H_I' assumes a diagonal form in **S** and **L**, and also in $\mathbf{J} = \mathbf{L} + \mathbf{S}$. The wave function for such a system is obtained by coupling first l_1 and l_2 to **L**, then $\mathbf{s}_1$ and $\mathbf{s}_2$ to **S**, and finally **S** and **L** to **J**. It may be written in the form of an antisymmetrized linear combination

$$\Psi_a(N_1 l_1, N_2 l_2; SLJM) = \frac{1}{\sqrt{2}} \sum_{M_S M_L} C^{S\ L\ J}_{M_S M_L M} \cdot [\varphi(N_1 l_1 N_2 l_2 L M_L) + (-1)^{l_1+l_2-L+S} \times$$
$$\times \varphi(N_2 l_2 N_1 l_1 L M_L)] \cdot \chi(s_1 s_2 S M_S), \tag{4.9}$$

‡ Generally, it implies that we can describe an Auger transition in a closed-shell atom having n electrons in the relevant closed shell in terms of *two* vacancies rather than the $n-2$ electrons actually present in the final state. This, together with the fact that all other closed shells behave formally like a vacuum with $L = 0$, $S = 0$ and $J = 0$, explains why the two-particle treatment of Chapter 2 provides a fair description of the process. Since electrons and holes are obtained by letting creation and annihilation operators $a^{\dagger}_{nlm_lm_s}$ and $a_{nlm_lm_s}$ act on this vacuum, the correspondence between configurations (1), (2), (3) and (1)′, (2)′, (3)′ is understandable so long as other interactions do not foul it up.

where

$$\varphi(N_i l_i N_j l_j L M_L) = \sum_{m_{l_i} m_{l_j}} C^{l_i\ l_j\ L}_{m_{l_i} m_{l_j} M_L} Y^{m_{l_i}}_{l_i}(\theta_1, \varphi_1)\, Y^{m_{l_j}}_{l_j}(\theta_2, \varphi_2)\, R_{N_i l_i}(r_1)\, R_{N_j l_j}(r_2), \tag{4.10}$$

and

$$\chi(s_1 s_2 S M_s) = \sum_{m_{s_1} m_{s_2}} C^{s_1\ s_2\ S}_{m_{s_1} m_{s_2} M_S} \chi^{m_{s_1}}_{s_1} \chi^{m_{s_2}}_{s_2}. \tag{4.11}$$

Here $R_{N_j l_j}(r)$, $Y^{m_{l_j}}_{l_j}(\theta, \varphi)$ and $\chi^{m_s}_m$ are the single particle radial, angular and spin wave functions with quantum numbers N_j, l_j, m_{l_j} and m_s. Equation (4.9) gives the well-known *SLJM* representation for two electrons.

In this scheme the matrix elements $\langle l_1 l_2 SLJM | H_1' | l_1 l_2 SLJM \rangle$ are diagonal in S and L. Applying eqn (4.9) to the configurations listed above, one gets the six multiplets:

$$(1)\ {}^1S; \quad (2)\ {}^1P, {}^3P \quad \text{and} \quad (3)\ {}^1S, {}^1D, {}^3P.$$

Here we have not indicated the values of J, the total angular momentum, because for given S and L, the states are degenerate in J, i.e. the energy is independent of J. This is easily understood, because

$$[H_1', \mathbf{S}] = 0 \quad \text{and} \quad [H_1', \mathbf{L}] = 0 \tag{4.12}$$

imply that the eigenvalues of H_1' must be independent of M_S and M_L and hence of the orientation of **L** and **S**. Thus states with definite S and L must have the same energy irrespective of the value of J.‡

It may be noted that the triplet ^3P belonging to the third configuration does not participate in *KLL* Auger transitions because this would violate the selection rule $\Delta L = 0$.

With the introduction of the electrostatic interaction, the complete Hamiltonian is

$$H = H_0 + H_1' = \sum_i \left(-\frac{\hbar^2}{2m} \nabla_i^2 - \frac{Ze^2}{r_i} \right) + \sum_{i \neq j} \frac{e^2}{r_{ij}}.$$

In the literature, it is customary to distinguish between single-particle operators f_i and two-particle operators g_{ij}. In this case we have

$$f_i = -\frac{\hbar^2}{2m} \nabla_i^2 - \frac{Ze^2}{r_i}, \tag{4.13}$$

‡ The complete set of operators commuting with H_1' is $\mathbf{L}^2$, $\mathbf{S}^2$, L_z, S_z and $\mathbf{J}^2$, but in the absence of a magnetic field the system is degenerate in L_z and S_z.

and

$$g_{ij}=\frac{e^2}{r_{ij}}. \tag{4.14}$$

To obtain the energies of the different multiplets belonging to a configuration, one must evaluate diagonal matrix elements of H. In the general case, one has to use many-electron wave functions, i.e. Slater determinants built up from single-particle wave functions. The procedure for this has been set out at length in the classic texts by Hartree (1957) and Slater (1960). Here we shall only introduce the standard notation for subsequent use.

1. Diagonal matrix elements of the single-particle operator $\sum_i f_i$ are obtained by summing over all the single-particle states (Bethe and Jackiw, 1968), i.e.

$$\sum_{\substack{i \\ \text{complete} \\ \text{shells}}} \langle i|f_i|i\rangle + n\langle j|f_j|j\rangle, \tag{4.15}$$

where $|i\rangle$ and $|j\rangle$ are single-particle states of the type $|N_i l_i m_{l_i} m_{s_i}\rangle$. The label j in eqn (4.15) stands for any one of the n electrons in an incomplete shell. For the single-particle matrix element $\langle i\,|\,f_i|i\rangle$ Hartree introduced the symbol $I(Nl)$, where

$$I(Nl)=\int_0^\infty\left\{\frac{\hbar^2}{2m}P_{Nl}\left[-\frac{\mathrm{d}^2P_{Nl}}{\mathrm{d}r^2}+\frac{l(l+1)}{r^2}P_{Nl}\right]-\frac{Ze^2}{r}P_{Nl}^2\right\}\mathrm{d}r, \tag{4.16}$$

The function $P_{Nl}(r)$ is defined by

$$R_{Nl}(r)=\frac{P_{Nl}(r)}{r}. \tag{4.17}$$

2. The two-particle operator $\sum_{ij} g_{ij}$ can connect two different types of many-electron states, namely, (a) states which differ only in two single-particle orbitals and (b) states which contain the same set of single-particle orbitals (Bethe and Jackiw, 1968). In either case the matrix element is a linear combination of two-particle matrix elements of the type

$$\langle ij|g_{ij}|ij\rangle - \langle ij|g_{ij}|ji\rangle,$$

the first term being called the direct term and the second the exchange term.

By making a multipole expansion of $1/r_{ij}=1/|\mathbf{r}_i-\mathbf{r}_j|$ (see eqn 3.46) a typical two-particle matrix element $\langle ab|e^2/r_{ij}|\,cd\rangle$ can be expressed as a linear combination of general Slater integrals of the type

$$R^k(ab,cd) = e^2 \int_0^\infty \int_0^\infty R_{N_a l_a}(r_1)\, R_{N_b l_b}(r_2) \frac{r_<^k}{r_>^{k+1}} R_{N_c l_c}(r_1)\, R_{N_d l_d}(r_2)\, r_1^2 r_2^2 \, \mathrm{d}r_1 \, \mathrm{d}r_2$$

$$= e^2 \int_0^\infty \int_0^\infty P_{N_a l_a}(r_1) P_{N_b l_b}(r_2) \frac{r_<^k}{r_>^{k+1}} P_{N_c l_c}(r_1) P_{N_d l_d}(r_2) \, \mathrm{d}r_1 \, \mathrm{d}r_2. \quad (4.18)$$

Corresponding to the direct and exchange matrix elements we get the direct and exchange Slater integrals

$$F^k(ij) = R^k(ij,ij)$$
$$G^k(ij) = R^k(ij,ji). \quad (4.19)$$

3. The *jj* coupling scheme

We now go over to the other limit where the spin–orbit interaction

$$H'_{\mathrm{II}} = \sum_i \xi(r_i)\, \mathbf{l}_i \cdot \mathbf{s}_i \quad (4.20)$$

is the dominant perturbation. That this limit relates to large values of Z is well known and also easy to understand provided one remembers that this is basically a relativistic effect. Once one assumes the electron to have an intrinsic angular momentum **s**, this effect may be understood even in classical terms,‡ at least in order of magnitude (Jackson, 1962). The classical equation of motion for **s** in the rest frame of the electron may be written

$$\frac{\mathrm{d}\mathbf{s}}{\mathrm{d}t} = \boldsymbol{\mu} \times \mathbf{B}', \quad (4.21)$$

where $\mathbf{B}'$ is the magnetic induction experienced by the electron in this frame and $\boldsymbol{\mu} = (e/mc)\mathbf{s}$ its magnetic moment. If **B** is the magnetic induction in the laboratory frame, we have, to order $\mathbf{v}/c$,

$$\frac{\mathrm{d}\mathbf{s}}{\mathrm{d}t} = \boldsymbol{\mu} \times \left(\mathbf{B} - \frac{\mathbf{v}}{c} \times \mathbf{E} \right). \quad (4.22)$$

The corresponding energy of interaction between $\boldsymbol{\mu}$ and **B** is

$$U' = -\boldsymbol{\mu} \cdot \left(\mathbf{B} - \frac{\mathbf{v}}{c} \times \mathbf{E} \right). \quad (4.23)$$

‡ Actually, the spin s does not occur naturally in classical mechanics.

In an atom we can approximate **E** by the negative gradient of the spherically symmetric average potential $V(r)$

$$\mathbf{E} = -\frac{\mathbf{r}}{r}\frac{\mathrm{d}V}{\mathrm{d}r}. \tag{4.24}$$

Substituting from eqn (4.24) in eqn (4.23) and using $\mathbf{l} = m(\mathbf{r} \times \mathbf{v})$, we get

$$U' = \frac{e}{mc}\mathbf{s}\cdot\mathbf{B} + \frac{e}{m^2c^2}(\mathbf{s}\cdot\mathbf{l})\frac{1}{r}\frac{\mathrm{d}V}{\mathrm{d}r}. \tag{4.25}$$

The second term on the right-hand side is the spin–orbit interaction term, except that it gets multiplied by a factor of $\frac{1}{2}$ when we consider the Thomas precession of the electron due to the acceleration experienced by it in its atomic orbit. The factor $\mathbf{v}/c$ indicates the relativistic character of this effect, which should obviously be more pronounced for inner shell electrons and for heavier atoms. For a single electron, we may write

$$\mathbf{l}\cdot\mathbf{s} = \frac{\mathbf{j}^2 - \mathbf{l}^2 - \mathbf{s}^2}{2},$$

so that the matrix elements of the operator $\xi(r)\mathbf{l}\cdot\mathbf{s}$ are given by

$$\langle Nljm|\xi(r)\mathbf{l}\cdot\mathbf{s}|Nljm\rangle = \xi_{Nl}\frac{j(j+1) - l(l+1) - \frac{3}{4}}{2}, \tag{4.26}$$

where

$$\xi_{Nl} = \hbar^2 \int_0^\infty R_{Nl}^2(r)\,\xi(r)\,r^2\,\mathrm{d}r. \tag{4.27}$$

For a two-electron configuration we note that

$$H'_{11} = \xi(r_1)(\mathbf{l}_1\cdot\mathbf{s}_1) + \xi(r_2)(\mathbf{l}_2\cdot\mathbf{s}_2) \tag{4.28}$$

commutes with $\mathbf{l}_1^2$, $\mathbf{l}_2^2$, $\mathbf{j}_1^2$, $\mathbf{j}_2^2$ and **J**. Hence its matrix elements are conveniently calculated in a scheme wherein we couple $\mathbf{s}_1$ and $\mathbf{l}_1$ to $\mathbf{j}_1$, $\mathbf{s}_2$ and $\mathbf{l}_2$ to $\mathbf{j}_2$ and then $\mathbf{j}_1$ and $\mathbf{j}_2$ to **J**. This is the *jj* coupling scheme.

The *jj* coupling wave functions are defined by

$$\Psi(N_1 l_1 j_1, N_2 l_2 j_2; JM) = \sum_{m_1 m_2} C^{j_1 j_2 J}_{m_1 m_2 M}\,\varphi(N_1 l_1 j_1 m_1)\,\varphi(N_2 l_2 j_2 m_2), \tag{4.29}$$

where

$$\varphi(N_i l_i j_i m_i) = \sum_{m_{s_i} m_{l_i}} C^{s_i l_i j_i}_{m_{s_i} m_{l_i} m_i}\, R_{N_i l_i}(r_i)\, Y_{l_i}^{m_{l_i}}(\theta_i, \varphi_i)\, \chi_{s_i}^{m_{s_i}}, \tag{4.30}$$

the symbols having their usual meaning.

Obviously, the state vector of the two-electron configuration should be essentially independent of how we couple the angular momenta, i.e. the wave functions in eqns (4.9) and (4.29) must be related by a unitary transformation:

$$\Psi(N_1 l_1 j_1, N_2 l_2 j_2; JM) = \sum_{SL} [(2S+1)(2L+1)(2j_1+1)(2j_2+1)]^{1/2} \cdot \begin{Bmatrix} \frac{1}{2} l_1 j_1 \\ \frac{1}{2} l_2 j_2 \\ SLJ \end{Bmatrix} \cdot \Psi(N_1 l_1, N_2 l_2; SLJM), \quad (4.31)$$

where the expression $\{\ldots\}$ is a 9-j symbol.

Let us see how we can classify in this scheme the states corresponding to the configurations (1), (2) and (3). The interaction H'_{II} commutes separately with the total angular momentum $\mathbf{j}_i$ of each electron, so that each individual electron has a definite j-value. Instead of talking about a $(2s)(2p)$ configuration, we have to talk about a $(2s_{1/2})(2p_{1/2})$ configuration—the degeneracy of the $2p$ subshell is removed. The usual designations are $L_1 \equiv 2s_{1/2}$, $L_2 \equiv 2p_{1/2}$ and $L_3 \equiv 2p_{3/2}$ and the final states in a KLL transition are specified by naming the two subshells in which the vacancies occur. The result is six energy levels in the final state corresponding to the vacancies L_1L_1, L_1L_2, L_1L_3, L_2L_2, L_2L_3 and L_3L_3.

The question that may now be asked is: can these energy levels be related to the six terms obtained earlier under LS coupling? Of course H'_{II}, i.e. the spin-orbit interaction, does not commute separately with L or S, so that these are no longer good quantum numbers. However, one can still trace the splitting of the terms into energy levels corresponding to different values of J (i.e. the passage from a higher to a lower symmetry). This can be done by identifying the different terms arising from a given configuration—once H'_{II} is switched on it is these terms that split into multiplets with non-degenerate energies depending on J.

A straightforward though somewhat tedious qualitative procedure for doing this was given by Condon and Shortley (1959). Table 1 shows how this

Table 1. Term structure for p^2 configuration (LS coupling)

p^2		M		
		1	0	−1
	2		(1^+1^-)	
	1	(1^+0^+)	$(1^+0^-)(1^-0^+)$	(1^-0^-)
M_L	0	(1^+-1^+)	$(1^+-1^-)(1^--1^+)(0^+0^-)$	(1^--1^-)
	−1	(0^+-1^+)	$(0^+-1^-)(0^--1^+)$	(0^--1^-)
	−2		(-1^+-1^-)	

can be done in pure *LS* coupling for the configuration p^2, i.e. for two electrons with the same principal quantum number N and both having $l = 1$. Such electrons are called equivalent electrons. This scheme would work, therefore, for the vacancy distributions $(2p_{1/2})(2p_{1/2})(L_2L_2)$, $(2p_{1/2})(2p_{3/2})(L_2L_3)$ or $(2p_{3/2})(2p_{3/2})(L_3L_3)$. In Table 1, of course, we suppress the j-values which are not relevant to *LS* coupling. The table is divided into small cells, each of which corresponds to a particular value of M_L and a particular value of M_S. These M_L and M_S values may be obtained by taking different combinations of m_l and m_s for the two individual electrons. The values of m_l and m_s are indicated within parentheses—each combination corresponding to a member of a multiplet. Values of m_l are indicated by means of positive or negative integers and values of m_s by means of the superscript + or − (corresponding to $m_s = \pm\frac{1}{2}$). Several possible combinations are ruled out by the Pauli exclusion principle; for example, the state (1^+1^+) cannot occur as this would imply identical quantum numbers for the two electrons. This means that $M_L = 1 + 1 = 2$ does not occur, so that $L = 2$ states are ruled out for $S = 1$. We can thus see that the total number of terms for two equivalent electrons is much smaller than that for two electrons with different N but the same l value. The table shows that for the p^2 configuration the terms are ^1S, ^3P, ^1D. The possible J values are such as to make these terms correspond to the multiplets $^1S_0$, $^3P_{0,1,2}$ and $^1D_2$. Going over to *jj* coupling, we can see from Table 2 how the members of these multiplets originate in this limit. Here j_1 and j_2 can take on the values $\frac{1}{2}$ and $\frac{3}{2}$, so that we have three possible combinations: (a) $j_1 = \frac{3}{2}$, $j_2 = \frac{3}{2}$, (b) $j_1 = \frac{3}{2}$, $j_2 = \frac{1}{2}$ and (c) $j_1 = \frac{1}{2}$, $j_2 = \frac{1}{2}$. In this table, the numbers in parentheses indicate the individual

Table 2. Multiplet structure for p^2 configuration (*jj* coupling)

p^2		j_1j_2		
		$\frac{3}{2},\frac{3}{2}$	$\frac{3}{2},\frac{1}{2}$	$\frac{1}{2},\frac{1}{2}$
	2	$(\frac{3}{2},\frac{1}{2})$	$(\frac{3}{2},\frac{1}{2})$	
	1	$(\frac{3}{2},-\frac{1}{2})$	$(\frac{3}{2},-\frac{1}{2})$ $(\frac{1}{2},\frac{1}{2})$	
	0	$(\frac{3}{2},-\frac{3}{2})$	$(\frac{1}{2},-\frac{1}{2})$	$(\frac{1}{2},-\frac{1}{2})$
	−1	$(-\frac{3}{2},\frac{1}{2})$	$(-\frac{3}{2},\frac{1}{2})$ $(-\frac{1}{2},-\frac{1}{2})$	
	−2	$(-\frac{3}{2},-\frac{1}{2})$	$(-\frac{3}{2},-\frac{1}{2})$	
J values		2,0	2,1	0

magnetic quantum numbers m_1 and m_2. For $j_1 = \frac{1}{2}$, $j_2 = \frac{1}{2}$, the only allowed combination of magnetic quantum numbers is $(\frac{1}{2}, -\frac{1}{2})$, because these are equivalent electrons. This gives $M = \frac{1}{2} - \frac{1}{2} = 0$. Other levels with $M = 0$ occur in other columns of the table. However, for $j_1 = \frac{1}{2}, j_2 = \frac{1}{2}$, the highest and only M value is $M = 0$. Therefore this belongs to $J = 0$. Similarly, one can identify the different members of the multiplets. The J values shown at the bottom of the table agree with those previously inferred from the term structure in LS coupling, namely ${}^1S_0$, ${}^3P_{0,1,2}$ and ${}^1D_2$. This is as it should be, because the two schemes, after all, are connected by a unitary transformation, namely the one given in eqn (4.31). The book-keeping of physical states should be unique once all the interactions are turned on.

We should like to emphasize that the formal correspondence between the classifications in the two limits is given by eqn (4.31). This is the consistent and dependable method. In some simple cases, one can make a heuristic reasoning. For example, the $J = 0$, $M = 0$ level in the $p_{1/2}p_{1/2}$ configutation is clearly a singlet, because here $m = m_s$, so that the spins are antiparallel and $S = 0$ in the LS coupling limit. Hence we make the correspondence ${}^1S_0$ L_2L_2 (i.e. $2p_{1/2}2p_{1/2}$). With the degeneracy of the multiplets completely removed, one makes the following identification of the Auger lines: $KL_1L_1({}^1S_0)$, $KL_1L_2({}^1P_1)$, $KL_1L_2({}^3P_0)$, $KL_1L_3({}^3P_1)$, $KL_1L_3({}^3P_2)$, $KL_2L_2({}^1S_0)$, $KL_2L_3({}^1D_2)$, $KL_3L_3({}^3P_0)$, $KL_3L_3({}^3P_2)$.‡

4. The intermediate coupling scheme

Throughout our discussion so far, we have been using terms like KL_1L_1, KL_2L_2, etc. to designate the different lines in the KLL Auger spectra. It should be now obvious that this nomenclature assumes jj coupling—the levels L_1L_1, L_2L_2, etc. belong to this coupling scheme. According to this scheme, there should be six energy levels in the final state resulting in six KLL Auger lines. The actual number of radiationless transitions from a K-state to a final state with two vacancies in the L shells is not six, however, but nine.§ The explanation is not far to seek. In the last section we have seen that with the complete removal of degeneracy, i.e. with both the electrostatic interaction and the spin–orbit interaction turned on, these six final states actually split up into nine levels. These are shown in Table 2. In fact, this is the most realistic description of the final states in the KLL series, the two limits of pure LS coupling and pure jj

‡ Actually, there are ten possible final states in this scheme, but transitions to the $L_2L_3({}^3P_1)$ state are forbidden. The reason for this is that even in the more general scheme of intermediate coupling, this remains a pure state and there exists no other state from which transition can occur to this state without violating the selection rule $\Delta L = 0$.

§ For very light atoms, where pure LS coupling may hold, the number is five, transition to the final ${}^3P$ state belonging to the $2s^22p^4$ configuration from the initial ${}^2S_{1/2}$ state of the $1s^1$ configuration being ruled out by the selection rule $\Delta L = 0$.

coupling being approximations where one or the other of the two symmetry-breaking interactions may be neglected. For all Z values except very low ones, one must take both these interactions into account and the picture which emerges is called, somewhat inappropriately, the intermediate coupling scheme.

The problem, therefore, is to describe the general case where the two interactions may be of any relative order of magnitude. This is best done in the representation known as the *SLJM* scheme appropriate to *LS* coupling. In this scheme the electrostatic interaction is completely diagonal and the spin–orbit interaction is diagonal in J and M. In fact we have already introduced this representation in the previous section while discussing the possible final states arising in *KLL* transitions. Obviously, the levels $^1S_0$, $^3P_1$, $^1D_2$, etc. are characterized by the quantum numbers S, L and J. In the absence of a magnetic field, the total energy is independent of M, which is retained as a label for the state. In the limit of Russell–Saunders coupling, the states are also degenerate in J, but in this description we can again retain it as a state label. The advantage in doing this is that in the general case of intermediate coupling, J becomes physically interesting, being the quantum number which determines the energy. Alternatively, the same calculation can be done in the *jj* coupling scheme which is particularly suitable for heavier atoms. Of course the two representations are related by the unitary transformation of eqn (4.31). Results pertaining to any particular state in either representation cannot be related to individual states of the other. But for a particular J (and M, when this is not a degenerate quantum number), results given by the two schemes must be identical.

Using the *SLJM* representation, let us see how we can represent the intermediate coupling wave functions. This problem was first treated in relation to *K*-Auger transitions by Asaad and Burhop (1958). The important thing to note here is that the spin–orbit interaction may be regarded as a further perturbation of the system if we take the Russell–Saunders situation to be the unperturbed one. Hence, using the language of perturbation theory, the eigenfunction Ψ of the perturbed system can be written out as a linear combination of the Russell–Saunders functions φ. In doing this, we must remember that Ψ is an eigenfunction of energy and total angular momentum, so that it must be a linear combination of φ's belonging to the same J. For example, the eigenfunctions for the triplet ^3P states of the $(2s)^2(2p)^4$ configuration may now be written

$$\begin{aligned}
&\Psi(^3P_2) = C(^3P_2)\,\varphi(^3P_2) + C(^1D_2)\,\varphi(^1D_2),\\
&\Psi(^3P_1) = \varphi(^3P_1) \text{ and}\\
&\Psi(^3P_0) = C(^3P_0)\,\varphi(^3P_0) + C(^1S_0)\,\varphi(^1S_0),
\end{aligned} \tag{4.32}$$

where the C's are the coefficients for this combination. We can see that for any given intermediate coupling wave function, we have a linear combination of all the Russell–Saunders states which belong to the same J and arise from that particular configuration. Physically the result is that transitions to the two mixed states with $J = 0$ and 2 in eqn (4.32) are now possible from the state of the initial $(1s)^1$ configuration (i.e. the K-shell vacancy).‡ One thus gets two new Auger lines from the $(2s)^2(2p)^4$. Similarly, from the $(2s)(2p)^5$ configuration there arise two new lines and the total number becomes nine, in agreement with observation (Hörnfeldt *et al.*, 1963).

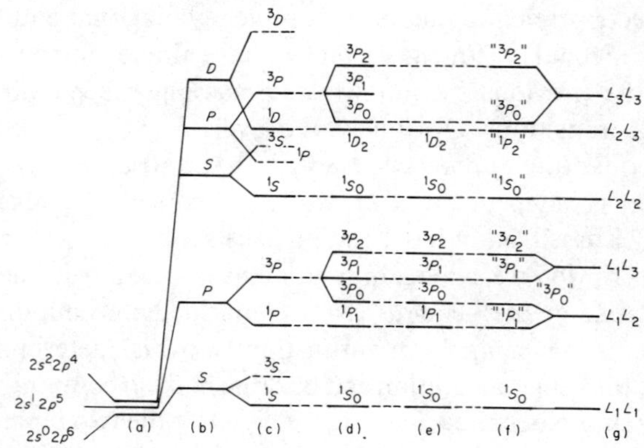

Fig. 4.1. Progressive symmetry-breaking for the final L-shell configurations in KLL transitions. (a) Completely degenerate energy with nuclear Coulomb potential (no electrostatic interaction between electrons and no spin-orbit interaction), (b) electrostatic interaction turned on between electrons, giving states of different total orbital angular momentum L, (c) splitting into triplets and singlets due to exchange interaction, (d) spin-orbit interaction turned on, (e) five allowed final states in pure LS coupling (the triplet $^3P_{0,1,2}$ belonging to the $(2s)(2p)^5$ configuration is degenerate in this limit) (f) nine allowed final states in intermediate coupling, and (g) six allowed final states in the limit of jj coupling (from Bergström and Nordling, 1965, courtesy of North-Holland Publishing Co.).

The entire sequence of symmetry-breaking starting from complete degeneracy to the general case of intermediate coupling is shown in Fig. 4.1 for the final states of the KLL Auger transitions. In Fig. 4.1(a), we have completely degenerate energies with no electrostatic or spin–orbit interaction where the electrons move in a central potential of $1/r$ shape. In Fig. 4.1(b), screening and the Coulomb interaction between electrons have been turned on, so that we have states of different total orbital angular momentum. With the introduction of exchange interaction (Fig. 4.1(c)), there is a splitting into triplet and singlet terms (among these are those excluded by the Pauli principle and shown by broken lines). Finally, spin–orbit interaction leads to a splitting of these terms into individual levels of definite angular momentum (i.e. the multiplets). This is shown in Fig. 4.1(d). In pure LS coupling, the $^3P_{0,1,2}$ multiplet of the $2p^2$ configuration (which is of course degenerate in this limit) is not an allowed

‡ Transition to the $^3P_1$ state is still not allowed because this remains a pure state.

final state for *KLL* transitions. This is shown in Fig. 4.1(e). In Fig. 4.1(f), the splitting is complete and we have nine allowed final states in the intermediate coupling. In Fig. 4.1(g), we see the other limit of *jj* coupling where there are only six allowed final states, some of the members of the nonet in Fig. 4.1(f) having merged in pairs into the single levels of Fig. 4.1.(g).

Positions of the lines of the *KLL* Auger spectrum have been calculated by Asaad and Burhop in the intermediate coupling scheme for different *Z* values. Figure 4.2, taken from Briancon (1970), shows them on the energy scale,

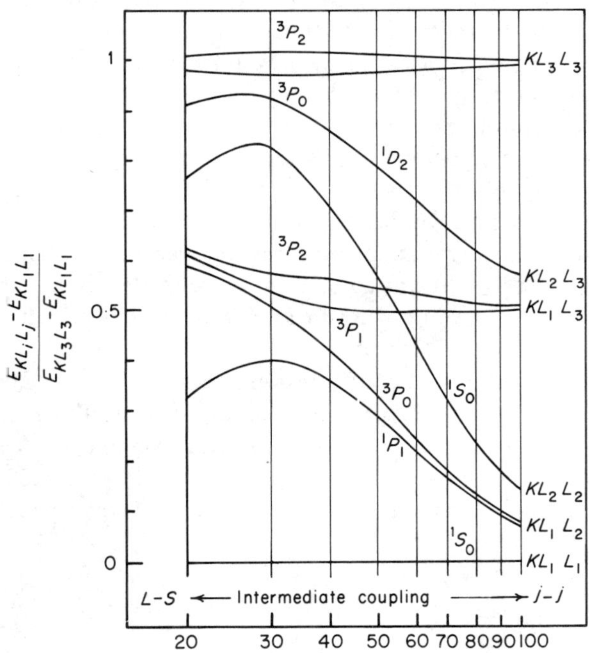

Fig. 4.2. Relative line positions in the *KLL* Auger group as a function of atomic number *Z*. Energy-difference with the KL_1L_1 line is expressed as a fraction of the difference between the KL_3L_3 and KL_1L_1 lines. The energy difference between the latter, $E_{KL_3L_3} - E_{KL_1L_1}$ ranges from 55 eV at $Z = 10$ to 17 keV at $Z = 104$ (from Briancon, 1970).

normalized to the separation between the KL_3L_3 and KL_1L_1 lines as a function of *Z*. The merging of the nine *KLL* lines into five lines in the low *Z* (*LS* coupling) limit and six lines in the high *Z* (*jj* coupling) limit is beautifully shown.

5. Configuration interaction

So far, we have neglected matrix elements of the electrostatic and spin–orbit interactions between states belonging to different configurations. In other words, we have assumed that the matrices of these interactions are diagonal between configurations. Having made this assumption, we have followed the usual practice of perturbation theory: we have written the perturbed eigenfunction, such as an intermediate coupling eigenfunction, as a linear combination of unperturbed (e.g. Russell–Saunders) eigenfunctions belonging to a definite configuration. In the language of group theory, we tacitly assumed that the eigenfunctions belonging to a given configuration form an invariant subspace of the system. In point of fact, however, all we know is that only the submatrix of the Hamiltonian defined by all states of the same configuration can be made diagonal by a suitable choice of wave functions. Since both J and π (the parity) are conserved, the wave functions which would accomplish this must be linear combinations of eigenfunctions of J^2 and π. However, the matrix of the Hamiltonian taken as a whole may have non-diagonal elements between states with the same J and π which belong to different configurations. The smaller these non-diagonal elements, the better the approximation of neglecting inter-configuration interaction will be. However, it is clear that in the general case we must consider the larger augmented matrix operating in the space defined by states belonging to different configurations with the same J and π. To do otherwise would be to work with an artificially truncated subspace of the system.

The important point is that configuration interaction needs to be taken into account even in a many-electron formulation such as the Hartree–Fock–Slater (HFS) self-consistent-field procedure. For example, if we have an N-electron system approximately described by a set of N coupled differential equations where the interaction between electrons is incorporated in a central potential $V(r)$:

$$\left(-\frac{\hbar^2}{2m}\nabla_i^2 - \frac{Ze^2}{r_i} + V(r_i)\right)\varphi_i = \varepsilon_i \varphi_i, \quad (i = 1, 2, \ldots, N),$$

the corresponding many-particle Hamiltonian is

$$H_{\mathrm{HFS}} = \sum_i \left[-\frac{\hbar^2}{2m}\nabla_i^2 - \frac{Ze^2}{r_i} + V(r_i)\right]. \tag{4.33}$$

These equations are solved for the single-particle wave functions φ_i by demanding that they yield a self-consistent potential $V(r_i)$ for each electron. However, the exact Hamiltonian of the system is $H = H_0 + H_{\mathrm{I}}' + H_{\mathrm{II}}'$. Thus the system is still under a residual perturbation (Mehlhorn, 1969) H_r, which is given by

$$\begin{aligned} H_r = H - H_{\mathrm{HFS}} &= H_{\mathrm{I}}' + H_{\mathrm{II}}' - \sum_i V(r_i) \\ &= \sum_{i,j} \frac{e^2}{r_{ij}} + \sum_i \xi(r_i)\mathbf{l}_i \cdot \mathbf{s}_i - \sum_i V(r_i). \end{aligned} \tag{4.34}$$

In the general case, H_r will be diagonal only in J and π. Thus two states belonging to different configurations but having the same values of J and π will be connected by H_r.

Physically, configuration interaction behaves as a perturbation which causes the interacting energy levels to be pushed apart. This is understandable, because the introduction of a hitherto non-existent interaction of the same sign is tantamount to augmenting their existing energy gap.

For medium atomic numbers, the intermediate coupling theory discussed in the last section gave good agreement with experiment in that the nine-line *KLL* spectrum was actually observed. However, even with the best available single-particle wave functions, discrepancies remained between experiment and non-relativistic calculations for low Z elements. Essentially they lay in the ratios of the total transition probability for each of the configurations $(2s)(2p)^5$ and $(2s)^2(2p)^4$ to that for the $(2s)^0(2p)^6$ configuration. Mehlhorn and Albridge (1963) pointed out that the theoretical values for both these ratios were too small compared with experiment. For example, with Rubinstein and Snyder's (1955) transition amplitudes, these ratios are 3.25 and 8.16 for $Z = 18$. The experimental ratios for $Z = 26$ are greater than 5 and 16 respectively. Use of numerical Hartree–Fock wave functions for the bound states (Archard, 1964) and Yost–Breit functions for continuum states did not lead to any improvement.

Since the group intensities corresponding to the configurations are independent of the coupling scheme, it appeared that something more than the latter was involved. It was inferred by Asaad that one should look for a new mechanism rather than for improved wave functions. This mechanism should be such as to greatly diminish the L_1L_1 line intensity and/or to largely enhance the total transition probability for the configurations $(2s)^1(2p)^5$ and $(2s)^2(2p)^4$. His answer was to introduce configuration interaction between the two configurations $(2s)^0(2p)^6$ and $(2s)^2(2p)^4$ for the $J = 0$ state.

The procedure is to express, as before, the unperturbed system in the *SLJM* representation in the limit of Russell–Saunders coupling. Both J and π are good quantum numbers, so the matrix of the perturbed Hamiltonian will resolve itself into a set of independent submatrices labelled by a definite pair of quantum numbers J, π. The J, π values for the three configurations involved are as follows

I. $(2s)^0(2p)^6$: $J = 0$; π even

II. $(2s)^1(2p)^5$: $J = 0, 1, 2$; π odd

III. $(2s)^2(2p)^4$: $J = 0, 1, 2$; π even.

Following the logic of our discussion, we have to consider configuration interaction between $J = 0$ states belonging to configurations I and III only, which have the same parity. Needless to say, we do not consider interaction between different J values, because our matrices are diagonal in J.

Our augmented subspace for $J = 0$ and even parity now consists of three states belonging to two configurations: the states $^3P_0$ and $^1S_0$ for configuration III and the state $^1S_0$ for configuration I (henceforth to be denoted by $^1S_0'$). The perturbed wave functions for these states completely span the $J = 0$, even parity part of the Hilbert space of the system. Let φ_1, φ_2 and φ_3 be the unperturbed functions for $^3P_0$, $^1S_0$ and $^1S_0'$, i.e. wave functions without configuration interaction. Let Ψ_1, Ψ_2 and Ψ_3 be the perturbed functions. Then we can write

$$\Psi_k = C_{kl}\,\varphi_1(^3P_0) + C_{k2}\,\varphi_2(^1S_0) + C_{k3}\,\varphi_3(^1S_0') \quad (k = 1, 2, 3). \tag{4.35}$$

The φ_k satisfy the Schrödinger equation

$$H_{\mathbf{HFS}}\,\varphi_k = \varepsilon_k\,\varphi_k \quad (k = 1, 2, 3), \tag{4.36}$$

and the perturbed wave functions satisfy

$$H\Psi_k = E_k\,\Psi_k, \tag{4.37}$$

where H is the Hamiltonian used in eqn (4.34).

Consider the case where $k = 1$,

$$H\Psi_1 = E_1\,\Psi_1. \tag{4.38}$$

Substituting for Ψ_1 and taking scalar product with φ_1, we get

$$C_{11}\langle\varphi_1|H|\varphi_1\rangle + C_{12}\langle\varphi_1|H|\varphi_2\rangle + C_{13}\langle\varphi_1|H|\varphi_3\rangle = E_1\,C_{11}.$$

Using the notation $H_{mn} = \langle\varphi_m|H|\varphi_n\rangle$, we write

$$(H_{11} - E_1)\,C_{11} + H_{12}\,C_{12} + H_{13}\,C_{13} = 0. \tag{4.39}$$

Similarly, by taking the scalar product of eqn (4.38) with φ_2 and φ_3 respectively, we get

$$H_{21}\,C_{11} + (H_{22} - E_1)\,C_{12} + H_{23}\,C_{13} = 0 \tag{4.40}$$

and

$$H_{31}\,C_{11} + H_{32}\,C_{12} + (H_{33} - E_1)\,C_{13} = 0. \tag{4.41}$$

Equations (4.39), (4.40) and (4.41) may be put in the matrix form

$$\begin{pmatrix} H_{11} - E_1 & H_{12} & H_{13} \\ H_{21} & H_{22} - E_1 & H_{23} \\ H_{31} & H_{32} & H_{33} - E_1 \end{pmatrix} \begin{pmatrix} C_{11} \\ C_{12} \\ C_{13} \end{pmatrix} = 0. \tag{4.42}$$

The corresponding secular equation

$$\begin{Vmatrix} H_{11}-E_1 & H_{12} & H_{13} \\ H_{21} & H_{22}-E_1 & H_{23} \\ H_{31} & H_{32} & H_{33}-E_1 \end{Vmatrix} = 0 \tag{4.43}$$

can be solved to yield the three eigenvalues of the system E_1, E_2, E_3. The three eigenvalues are simultaneously determined because of the Hermitian character of the Hamiltonian H. Once these eigenvalues are determined, the corresponding eigenvectors can be obtained, in principle, from eqns (4.39) to (4.41) and the orthonormality conditions

$$\sum_k c_{ik}^* c_{jk} = \delta_{ij} \quad (i, j = 1, 2, 3). \tag{4.44}$$

It was shown by Asaad (1965b) that the corresponding submatrix for the Hamiltonian is the 3 × 3 matrix

$$H_{J=0,\pi\,\mathrm{even}} = \begin{pmatrix} E_{\mathrm{III}} - \frac{3}{25}F^2(2p,2p) + \xi_{2p} & \sqrt{2}\xi_{2p} & 0 \\ \sqrt{2}\xi_{2p} & E_{\mathrm{III}} + \frac{12}{25}F^2(2p,2p) & -\frac{1}{\sqrt{3}}G^1(2s,2p) \\ 0 & -\frac{1}{\sqrt{3}}G^1(2s,2p) & E_I \end{pmatrix}, \tag{4.45}$$

where E_{III} and E_{I} are the average energies of the configurations III and I, and F, G and ξ_{Nl} have already been defined. The average energy of a configuration E_{av} is defined as the average of the energies of the multiplets belonging to it, weighted by their statistical weights.

$$E_{\mathrm{av}} = \frac{\sum_{LS} (2L+1)(2S+1)\,E(^{2S+1}L)}{\sum_{LS} (2L+1)(2S+1)}\,. \tag{4.46}$$

In this case,

$$E_{\mathrm{I}} = 2E(L_1) + F^0(2s,2s), \tag{4.45a}$$

$$E_{\mathrm{III}} = 2E(L_2) - 2\xi_{2p} + F^0(2p,2p) - \frac{2}{25}F^2(2p,2p),$$

$$\xi_{2p} = \frac{2}{3}[E(L_2) - E(L_3)].$$

$E(L_x)$ represents the binding energy of subshell L_x.

It may be at once seen that $G^1(2s, 2p)$ represents the interconfiguration term. If we put $G^1(2s, 2p) = 0$, then H resolves into a block-diagonal form with one 2×2 submatrix belonging to configuration III and one 1×1 submatrix belonging to configuration I. The two different configurations then form two different invariant subspaces of the $J = 0$, π even system. On the other hand, letting $\xi_{2p} = 0$ makes the submatrix for configuration III diagonal, but the coupling between the ${}^1S_0$ terms of the two configurations remains. Letting $\xi_{2p} = 0$ and $G^1(2s, 2p) = 0$ simultaneously implies going over to pure Russell–Saunders coupling.

Now, the inter-configuration term $G^1(2s, 2p)$ increases roughly as Z, as may be easily checked by looking at the expression for the integral.‡ On the other hand, the spin–orbit interaction ξ_{2p} increases roughly as Z^4. Thus when Z becomes sufficiently large, the matrix (4.45) effectively reduces to the block diagonal form, G^1 being negligible in comparison with ξ_{2p}. It means that for sufficiently large atomic numbers, we have intermediate coupling without configuration mixing. This explains why the discrepancy between theory and experiment appeared for low atomic numbers in the first place.

Getting down to wave functions, and restricting the discussion to the $J = 0$ subspace, the following states form the basis vectors:

$$L_3L_3({}^3P_0),\ L_2L_2({}^1S_0),\ L_1L_1({}^1S_0'). \tag{4.47}$$

Denoting the corresponding normalized eigenvectors by $\mathbf{C}_i$, $i = 1, 2, 3$, respectively, and the components of these eigenvectors by $C_i({}^{2S+1}L_j)$, we can write the wave functions Ψ_i representing the terms in (4.47) as a linear combination of Russell–Saunders eigenfunctions for the ${}^3P_0$, ${}^1S_0$ and ${}^1S_0'$ terms:

$$\begin{aligned}\Psi_1 &\equiv \Psi({}^3P_0) = C_1({}^3P_0)\,\varphi_1({}^3P_0) + C_1({}^1S_0)\,\varphi_2({}^1S_0) + C_1({}^1S_0')\,\varphi_3({}^1S_0'),\\ \Psi_2 &\equiv \Psi({}^1S_0) = C_2({}^3P_0)\,\varphi_1({}^3P_0) + C_2({}^1S_0)\,\varphi_2({}^1S_0) + C_2({}^1S_0')\,\varphi_3({}^1S_0')\\ \Psi_3 &\equiv \Psi({}^1S_0) = C_3({}^3P_0)\,\varphi_1({}^3P_0) + C_3({}^1S_0)\,\varphi_2({}^1S_0) + C_3({}^1S_0')\,\varphi_3({}^1S_0').\end{aligned} \tag{4.48}$$

Again, in eqn (4.48), Ψ_1, Ψ_2, Ψ_3 are the basis vectors for the $J = 0$, π even subspace of KLL final states with intermediate coupling and configuration mixing. The φ's are eigenfunctions for pure Russell–Saunders coupling corresponding to the terms shown in parentheses.

We must now find transition probabilities from the initial state to these final states given by eqn (4.48). The only initial state with which these states can be connected is the singlet ${}^1S_0$ term arising from the coupling between the

‡ The simplest way to do this is to substitute hydrogenic radial functions for $R_{2s}(r)$ and $R_{2p}(r)$, which gives

$$G'(2s, 2p) = \frac{16}{3}\frac{Ze^2}{a_0}\int_0^\infty \mathrm{d}x(1-x)\,\mathrm{e}^{-2x}\int_0^{2a_0x/Z}(1-y)\,\mathrm{e}^{-2y}y^3\,\mathrm{d}y.$$

For large Z, this integral behaves like Ze^2/a_0.

Table 3. Comparison of theoretical results for relative *KLL* Auger line intensities of neon obtained with and without configuration interaction and high resolution experimental data. Values without configuration interaction are based on *LS* coupling. From Bhalla (1973)

Lines	Theory, without conf. mixing (*LS* coupling) Bhalla (1973)	Theory, with conf. mixing Bhalla (1973)	Experiment Mehlhorn *et al.* (1968)	Experiment Fahlman *et al.* (1966)	Experiment Cleff and Mehlhorn (1969)
$KL_1L_1(^1S_0)$	1.0	1.0	1.0	1.0	1.0
$KL_1L_2(^1P_1)$	2.20	2.98	2.74(5)	2.87(5)	2.92(6)
$KL_1L_{2,3}(^3P_{2,1,0})$	0.73	0.99	1.00(2)	1.06(5)	1.06(5)
$KL_2L_2(^1S_0)$	0.47	0.99	1.55(3)	1.5(1)	1.5(1)
$KL_2L_3(^1D_2)$	6.2	8.3	9.31(15)	10.00(18)	10.13(24)

initial vacancy in the K shell and the hole in the continuum representing the missing electron. Other possible initial states will have either $J \neq 0$ or π odd and may, therefore, be ignored. Furthermore, the wave functions on the right-hand side of eqn (4.48) are Russell–Saunders functions, they conserve both L and S. As a result transitions from the initial ${}^1S_0$ term to the ${}^3P_0$ term of the final state are not allowed. Denoting the wave function of the initial singlet by $\Psi_c({}^1S_0)$, we get for the transition probabilities to the final states of (4.47),

$$\frac{2\pi}{\hbar}\,|\langle C_i({}^1S_0)\,\varphi_2({}^1S_0) + C_i({}^1S_0')\,\varphi_3({}^1S_0')|\frac{e^2}{r_{12}}|\Psi_c({}^1S_0)\rangle|^2 \quad (i = 1, 2, 3). \quad (4.49)$$

Note that without configuration interaction one of the ${}^1S_0$ terms would be absent from the final state. The inclusion of both terms happened to remove a large part of the discrepancies between theory and experiment. In particular, the calculated relative intensities of the sp and p^2 groups of lines, normalized to s^2, were now in better agreement with experiment.

The improvement caused in the KLL Auger spectrum of neon can be seen from Table 4, which was given by Bhalla (1973). He used the non-relativistic HFS model, with the exchange approximation of Herman, Van Dyke and Ortenburger (1972), to calculate the Auger matrix elements. The mixing coefficients were expressed in terms of the interconfiguration term G^1. The ratios of the theoretical transition probabilities of the different K–LL lines relative to the KL_1L_1 line, i.e. the relative intensities with KL_1L_1 as the base line, with and without configuration mixing are shown in columns 3 and 2 respectively. The results in column 2 are based on LS coupling. Columns 4, 5 and 6 show the high-resolution experimental values obtained by Mehlhorn *et al.* (1968), by Fahlman *et al.* (1966) and by Cleff and Mehlhorn (1969) respectively. The considerable improvement caused by the inclusion of configuration interaction is quite evident.

The story of configuration interaction is actually much longer than would appear from the above account of KLL Auger transitions. The real question is: what is the smallest invariant subspace for any problem involving a many-electron system. In most of our discussions preceding this section, the answer to this question was sought within the confines of a given configuration. For the pure Coulomb field, this is sufficient. But the pure Coulomb field is an idealization so far as a many-electron system is concerned and the practice of working within a configuration becomes inadequate as one moves away from this idealization.

6. The mixed coupling scheme

In the experimental study of L- and M-shell Auger spectra of atoms with low and medium values of Z, it may happen that the spin–orbit splitting is more clearly discernible in the initial inner-shell vacancy state than in the final two-vacancy configuration involving the outer shells. For example, Mehlhorn and Stalherm (1968) found that they could distinguish between the L_2MM and L_3MM spectra of argon ($Z = 18$), but could not separate the triplet states of the final MM configuration. For krypton ($Z = 36$), all final NN states of the M_4NN and M_5NN spectra could, however, be experimentally separated (Cleff *et al.*, 1968). This is an interesting example of the fact that, as one goes to heavier atoms, the strong intensity peaks arise from transitions involving electrons moving well within the atomic core and, therefore, the spin–orbit interaction is no more negligible. Because the initial vacancy occurs in an inner shell, the resulting splitting is greater in the initial state. One may be faced with a marginal situation, as in argon above, where the splitting in the initial state can be clearly measured, whereas that in the final state cannot. It was suggested by Asaad and Mehlhorn (1968) that in cases like this transition probabilities should be calculated assuming Russell–Saunders coupling for the outer shell holes and jj coupling for the initial holes in the inner shell and the continuum. For example, to describe the LMM series of Argon, i.e. L_1MM, L_2MM and L_3MM, they considered the following terms belonging to the final two-vacancy configurations:

$$(3s)^{-2}: {}^1S_0$$
$$(3s)^{-1}(3p)^{-1}: {}^1P_1,\ {}^3P_{0,1,2}$$
$$(3p)^{-2}: {}^1S_0,\ {}^3P_{0,1,2},\ {}^1D_2$$

For an L_2MM transition, there is an initial vacancy in the L_2 shell, and a vacancy in the continuum with a j-value which is fixed by selection rules. To be specific, let us follow Asaad and Mehlhorn (1968) and consider the term ${}^3P_2$ belonging to the $(3p)^{-2}$ configuration. Since $J = 2$, and $j_1 = \frac{1}{2}$ for the L_2 shell, the allowed values of j_2, the total angular momentum of the continuum hole, are $\frac{3}{2}$ and $\frac{5}{2}$. The allowed combinations of l_2 and j_2 are, then, $p_{3/2}$, $d_{5/2}$ and $f_{5/2}$. Parity conservation rules out the d-states. The Coulomb operator $1/r_{12}$ commutes with $\mathbf{L}$, so the f-state is ruled out because the combination $l_1 = 1$ and $l_2 = 3$ cannot give a P state. Thus the initial vacancy configuration is $(2p_{1/2})^{-1}(\varepsilon p_{3/2})^{-1}$, ε being the energy of the Auger electron. The resulting jj coupling state with $J = 2$ and π even, when transformed to the LS coupling scheme by eqn (4.31), yields the linear combination

$$\Psi_i = \frac{1}{\sqrt{6}}(\sqrt{2}\,{}^1D_2 + \sqrt{3}\,{}^3D_2 - {}^3P_2). \tag{4.50}$$

Obviously, the only non-vanishing term in the matrix-element $\langle {}^3P_2|e^2/r_{12}|\Psi_i\rangle$ is $-(1/\sqrt{6})\langle {}^3P_2|e^2/r_{12}|{}^3P_2\rangle$. On multiplying this by the appropriate statistical factors, one gets the transition amplitude $-\sqrt{\frac{5}{6}}[D(0) - D(2)]$ where

$$D(k) = e^2 \int_0^\infty \int_0^\infty R_{2p}(r_1)\, R_{\varepsilon p}(r_2) \frac{r_<^k}{r_>^{k+1}} R_{3p}(r_1)\, R_{3p}(r_2)\, r_1^2 r_2^2 \, dr_1 \, dr_2.$$

The overall result is that the amplitude is multiplied by a factor of $-1/\sqrt{6}$ as compared to the *LS* coupling case.

Subsequently, expressions for Auger transition rates in this "mixed coupling" scheme were given for M_4NN and M_5NN spectra by El Ibayari *et al.* (1972) corresponding to $(4s)^{-2}$, $(4s)^{-1}(4p)^{-1}$, $(4p)^{-2}$ and $(4d)^{-2}$ vacancy configurations. Precise measurements of these spectra are now available for the noble gases, e.g. for Kr and Xe (Werme, *et al.*, 1972, 1973; Mehlhorn *et al.*, 1972). The energy resolution is about one part in two thousand. Lines corresponding to different *j*-values for the initial *M*-hole are clearly separated. For the final state, the splitting of terms within a configuration is clearly discernible in most cases (Fig. 4.3). McGuire (1974c) has been able to predict these splittings on the

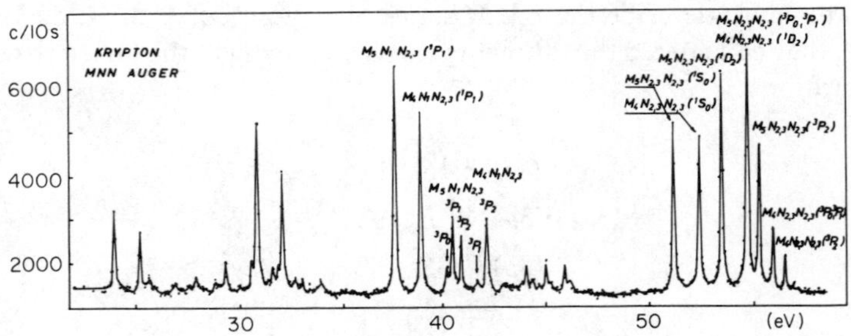

Fig. 4.3. Krypton *MNN* Auger spectrum (from Werme *et al.*, 1972, courtesy of the Royal Swedish Academy of Sciences).

basis of intermediate coupling. His calculations for the M_4NN and M_5NN spectra of krypton using a mixed coupling scheme, i.e. *jj* coupling for the initial one-hole state and intermediate coupling for the final two-hole state, have led to a correct identification of the $M_{4,5}N_1N_1$ doublet and to reasonably accurate assignment of the energies of the other prominent peaks. Recently, he has also given a generalized expression (McGuire, 1974a, b) for transition probabilities in atoms with partially filled shells which is based on mixed coupling (see Chapter 7.2). The question that naturally arises is what is the dynamical meaning of the mixed coupling scheme and where does it stand in relation to the other schemes described earlier in this chapter?

Theoretically, the difficulty in characterizing mixed coupling as a coupling scheme *per se* arises from the fact that the initial and final states in this scheme are eigenstates of different Hamiltonians. For example, in the original form proposed by Asaad and Mehlhorn (1968), the initial state Ψ_i is an eigenfunction of $H_0 + H'_{\mathrm{II}}$, and the final state Ψ_f that of $H_0 + H'_{\mathrm{I}}$. Secondly, pure *jj* coupling, i.e. a vanishingly small electrostatic interaction, represents a limiting case which is not quite attained even for $Z = 80$. The physical situation is perhaps better understood if we consider the complete many-electron wave functions for the initial and final states. Take, for example, an L-Auger transition in argon ($Z = 18$). Consider the initial and final states

$$|i\rangle = |2p^{-1}, SLJ\rangle, \; |f\rangle = |n_1 l_1^{-1} n_2 l_2^{-1}(S'L'J')\varepsilon l' SLJ\rangle.$$

Both $|i\rangle$ and $|f\rangle$ are seventeen-electron wave functions. In the final state, the two holes in the shells $n_1 l_1$ and $n_2 l_2$ couple to the angular momentum quantum numbers $S'L'J'$ through a physical interaction. The coupling scheme used to add their angular momenta must reflect the dynamics of this interaction. In argon it is predominantly electrostatic in nature, which implies LS coupling, although a small amount of spin–orbit interaction is also present. This is what suggests LS coupling to describe the residual ion core after Auger emission has taken place.

Now consider the Auger electron of energy ε and orbital angular momentum l' which is also present in the final state. We need to couple its angular momentum with that of the core $(S'L'J')$ to obtain the total final state. Here, as a first approximation, we can neglect the interaction between the core $(S'L'J')$ and the Auger electron $(\varepsilon l')$. As such we are free to choose the most convenient scheme for coupling angular momenta so as to get the same values of SLJ for the states $|i\rangle$ and $|f\rangle$. The choice may, for example, be based on the experimental situation. Interestingly, the fine-structure splitting of the initial state corresponding to the quantum numbers $2p$, $j=\frac{1}{2}$ and $2p$, $j=\frac{3}{2}$ of the initial vacancy can be experimentally detected. This suggests that we can use the quantum numbers $\varepsilon l'j'$ to describe the Auger electron and, correspondingly, jj coupling to combine its angular momentum with that of the core. Obviously, this coupling is by no means imposed by any physical assumption about the interaction between the core and the Auger electron, e.g. dominance of the spin–orbit interaction over Coulomb interaction. Actually, we are neglecting these interactions and focusing our attention on the geometrical aspect of angular momentum coupling so as to produce the physically observed multiplet states. To do this, we use jj coupling without vitiating the dynamical situation.

If we now revert to two-vacancy initial and final states

$$|i'\rangle = |2p^{-1}(\varepsilon l')^{-1}\, S'L'J'\rangle,$$

$$|f'\rangle = |(n_1 l_1)^{-1}(n_2 l_2)^{-1}\, S'L'J'\rangle,$$

we can see at once that the jj coupling between the core and the Auger electron in the many-electron picture is now transferred to the two initial-state vacancies $2p$ and $\varepsilon l'$. This happens because the physical Auger electron in the many-electron picture is now replaced by an initial-state vacancy $\varepsilon l'$ in the continuum. The final state $|f'\rangle$ is obtained by using LS coupling between the vacancies $n_1 l_1$ and $n_2 l_2$. Viewed from this angle, mixed coupling gives us a geometrical technique of combining angular momenta which helps the analysis of experimental data without conflicting with the dynamics of the transition. Obviously, it does not imply the existence of a mixed symmetry. As one goes to medium Z atoms, the final state appears to be better described by intermediate coupling, as might be expected.

Chapter Five
Some Central-Field Calculations

1. Introduction

We are now ready to take a look at the state of the art in central-field Auger calculations with the perturbations discussed in the previous chapter. A general feature of these calculations is that the electrons in the unfilled shells are considered to be moving in the effective central field of the nucleus screened by the remaining electrons in the atom. After a radiationless transition fills a single inner vacancy, the atom is left doubly ionized in the outer shells. We have seen that in such a model, the initial and final states can conveniently be represented by two-hole states in the following manner:

(a) initial state—one inner-shell vacancy in a bound state and one hole in the continuum

(b) final state—two inner-shell vacancies corresponding to bound states.

The transition probability per unit time is given by eqn (2.37a):

$$P = \frac{2\pi}{\hbar} |U_{AB;CD} - U_{AB;DC}|^2 \, \delta(\omega_{CA} + \omega_{BD}),$$

where $U_{AB;CD}$ and $U_{AB;DC}$ are the direct and exchange matrix elements.

For the non-relativistic case, we can write this as

$$P = \frac{2\pi}{\hbar} |D - E|^2 \tag{5.1}$$

where

$$D = \iint \Psi_A^\dagger(\mathbf{r}_1)\, \Psi_B^\dagger(\mathbf{r}_2)\, \frac{e^2}{|\mathbf{r}_1 - \mathbf{r}_2|}\, \Psi_C(\mathbf{r}_1)\, \Psi_D(\mathbf{r}_2)\, \mathrm{d}^3 r_1\, \mathrm{d}^3 r_2, \tag{5.2}$$

and

$$E = \iint \Psi_A^\dagger(\mathbf{r}_1^\dagger)\, \Psi_B^\dagger(\mathbf{r}_2)\, \frac{e^2}{|\mathbf{r}_1 - \mathbf{r}_2|}\, \Psi_D(\mathbf{r}_1)\, \Psi_C(\mathbf{r}_2)\, \mathrm{d}^3 r_1\, \mathrm{d}^3 r_2. \tag{5.3}$$

Here $\Psi_A^\dagger$ and $\Psi_B^\dagger$ may be regarded as the individual wave functions of the initial-state holes A and B, one of them being in an inner bound state and the other in the continuum. We shall take B to be the continuum hole. $\Psi_C^\dagger$ and $\Psi_D^\dagger$

represent the bound holes in the final state. This is the picture in terms of vacancies, which is more commonly used.

One might equally well use Ψ_A and Ψ_B as electron wave functions for the final state and Ψ_C, Ψ_D as electron wave functions for the initial state. The physical result would be the same, the direct and exchange matrix elements being the Hermitian adjoints of the expressions (5.2) and (5.3).

In this chapter, we shall outline the calculation of Auger transition probabilities and energies within the framework of the central-field picture. We shall confine our discussion to the non-relativistic case, primarily because this is the simpler situation. The advantage of the central-field model is that the angular part of the calculation can be carried through on the basis of the coupling schemes described in the last chapter, whereas the radial integration is left in a general form which may be treated freely according to the particular type of screening one wishes to use. Physically, this means that systems with diverse radial behaviour can have the same angular characteristics so long as the coupling scheme remains the same. In the next chapter we shall consider the incorporation of the many-electron aspects into the radial part of the calculation.

Rather than giving a complete survey of calculations involving the various coupling schemes, we shall try to set out in some detail the basic steps in such calculations.

2. Wigner algebra and selection rules

In a central field, the centre-of-force, namely, the nucleus, acts as a natural origin. Rotations about this origin leave the system unaltered and the generators of such rotations, i.e. the angular momentum operators, provide a convenient framework for the description of the system. For the Auger process in an atom with a single inner-shell initial hole in the ground-state configuration and otherwise complete shells, this description turns out to be particularly simple due to the following circumstances.

1. The initial and final states of such an atom are effectively described by two vacancies (or, alternatively, by two electrons) only. Hence the angular momentum eigenstates of the system are obtained by coupling the angular momenta of two objects only. In other words, the familiar techniques of Wigner algebra are adequate. Even with the inclusion of spin, this algebra remains fairly simple.

2. For the Auger process in the non-relativistic limit, the basic interaction is $e^2/|\mathbf{r}_1 - \mathbf{r}_2|$, i.e. the two-electron electrostatic interaction. It is an even operator with respect to space reflections, so it cannot change parity. It is also isotropic in configuration space and independent of spin, so it cannot change the projection quantum numbers M_L, M_S or M, as the case may be. In fact we

can at once write down the selection rules appropriate to *LS* coupling:

$$\Delta S = \Delta L = \Delta M_S = \Delta M_L = \Delta J = 0,$$

$$\pi_i = \pi_f.$$

Even in *jj* coupling, $\Delta M = 0$ and $\Delta J = 0$, and of course parity is conserved.

Another point about the interaction $e^2/|\mathbf{r}_1 - \mathbf{r}_2|$ is that it can be expanded in multipoles for all relative orientations of the vectors $\mathbf{r}_1$ and $\mathbf{r}_2$ (except when $\mathbf{r}_1 = \mathbf{r}_2$). This means we can treat the interaction as being made up of small pieces, each piece corresponding to a rather simple elementary disposition of the charges in the polar coordinates r_1, θ_1, φ_1 and r_2, θ_2, φ_2. Add to this the Wigner–Eckart theorem and we have the result that only a few selected pieces of the interaction can connect the given initial state of a participating electron to any specified final state. What these initial and final states are of course depends on the type of coupling we are working with. But for a given coupling scheme, the multipoles which can cause a transition are well defined.

3. Auger transition probabilities in *jj* coupling

To illustrate the remarks made in the last section, we shall describe a calculation made by Asaad (1963b) where he derived selection rules for Auger transitions in *jj* coupling and obtained an expression for the transition probability from any initial vacancy of the singly ionized atom to any final configuration of the doubly ionized atom.

Consider two vacancies in the states N, l, s and N', l', s', the quantum numbers having their usual meaning. We have $\mathbf{j} = \mathbf{l} + \mathbf{s}$, $\mathbf{j}' = \mathbf{l}' + \mathbf{s}'$ and $\mathbf{J} = \mathbf{j} + \mathbf{j}'$. We write the total wave function Ψ, given in eqn (4.23), in the slightly modified form

$$\Psi(Nlj, N'l'j'; JM) = \sum_m C^{j\ j'\ \ J}_{m\ M-m\ M}\, \varphi_1(Nljm)\, \varphi_2(N'l'j'\ M-m) \tag{5.4}$$

with

$$\varphi_1(Nljm) = \sum_{m_s m_l} C^{s\ \ l\ \ j}_{m_s m_l m}\, R_{Nl}(r_1)\, Y_l^{m_l}(\theta_1, \varphi_1)\, \chi_s^{m_s}(1) \tag{5.5}$$

and similarly for φ_2.

The subscripts 1 and 2 stand for the first and second vacancies respectively. Otherwise we retain the notation of the previous chapter. Also $s = s' = \frac{1}{2}$ and the χ's are Pauli spinors. We must now antisymmetrize the wave function Ψ,

$$\begin{aligned} \Psi_a = \frac{1}{\sqrt{2}} \sum_m \sum_{m_l} \sum_{m_{l'}} C^{j\ j'\ \ J}_{m\ M-m\ M}\, C^{s\ \ l\ \ j}_{m_s m_l m}\, C^{s'\ \ l'\ \ j'}_{m_{s'} m_{l'} M-m} [R_{Nl}(r_1)\, Y_l^{m_l}(\theta_1, \varphi_1)\, \cdot \\ \cdot\, \chi_s^{m_s}(1)\, R_{N'l'}(r_2)\, Y_{l'}^{m_{l'}}(\theta_2, \varphi_2)\, \chi_{s'}^{m_{s'}}(2) - \\ - R_{Nl}(r_2)\, Y_l^{m_l}(\theta_2, \varphi_2)\, \chi_s^{m_s}(2)\, R_{N'l'}(r_1)\, Y_{l'}^{m_{l'}}(\theta_1, \varphi_1)\, \chi_{s'}^{m_{s'}}(1)], \end{aligned} \tag{5.6}$$

for non-equivalent vacancies. For equivalent vacancies, with $N = N'$, $l = l'$ and $j = j'$, we have

$$\Psi_a = \tfrac{1}{2}[1-(-1)^{2j-J}] \sum_m \sum_{m_l} \sum_{m_{l'}} C^{j\ j\ J}_{m\ M-m\ M} C^{1/2\ l\ j}_{m-m_l\ m_l\ m} C^{1/2\ l\ j}_{M-m-m_{l'}\ m_{l'}\ M-m} \cdot$$
$$\cdot R_{Nl}(r_1)\, Y_l^{m_l}(\theta_1, \varphi_1)\, \chi_{1/2}^{m-m_l}(1)\, R_{Nl}(r_2)\, Y_l^{m_{l'}}(\theta_2\, \varphi_2)\, \chi_{1/2}^{M-m-m_{l'}}(2). \quad (5.7)$$

In eqn (5.7) the phase-factor $(-1)^{2j-J}$ comes from the fact that essentially $\Psi(jj'JM)$ and its exchange counterpart $\Psi(j'jJM)$ are equal, but may differ only by a phase due to exchange. For $J = j + j'$, they both equal the product wave function $\varphi(Nlj, m = j)\, \varphi(Nlj', m' = j')$. Hence the states for other values of M are at once written

$$\Psi(j'jJM) = (-1)^{j+j'-J}\, \Psi(jj'JM).$$

For two equivalent holes $j = j'$, and we have the exchange factor $(-1)^{2j-J}$. Since j is a half-integer, J must be even for Ψ_a not to vanish (Condon and Shortley, 1959).

We must now write down the matrix element of the electrostatic interaction e^2/r_{12} which causes a transition from the initial state $|i\rangle$ to the final state $|f\rangle$. Asaad (1963b) used electron states in his calculation. However, to bring his work into line with more recent calculations, we shall use vacancy states. Following the discussion of Section 1, we write

$$\Psi_A^\dagger = |N_i l_i j_i\rangle, \quad \Psi_B^\dagger = |\eta_i l_i' j_i'\rangle,$$
$$\Psi_C^\dagger = |N_f l_f j_f\rangle, \quad \Psi_D^\dagger = |N_f' l_f' j_f'\rangle.$$

In other words, $\Psi_A^\dagger$ represents the initial vacancy in the shell $(N_i l_i)$, $\Psi_B^\dagger$ the initial vacancy in the continuum and $\Psi_C^\dagger$ and $\Psi_D^\dagger$, the final state vacancies.

The Coulomb operator may be expanded:

$$\frac{1}{r_{12}} = \sum_\lambda \gamma_\lambda(r_1, r_2)\, P_\lambda(\cos\theta_{12}), \quad (5.8)$$

where

$$\gamma_\lambda(r_1, r_2) = \begin{cases} r_1^\lambda / r_2^{\lambda+1} (r_1 < r_2) \\ r_2^\lambda / r_1^{\lambda+1} (r_2 < r_1) \end{cases} \quad (5.9)$$

or, briefly,

$$\gamma_\lambda(r_1, r_2) = \frac{r_<^\lambda}{r_>^{\lambda+1}}.$$

Also

$$P_\lambda(\cos\theta_{12}) = \frac{4\pi}{2\lambda+1} \sum_\mu (-)^\mu\, Y_\lambda^\mu(\theta_1, \varphi_1)\, Y_\lambda^{-\mu}(\theta_2, \varphi_2). \quad (5.10)$$

Equation (5.10) is the well-known addition theorem for spherical harmonics.

We must now take matrix elements of this interaction between the initial state $j_i j_i' J_i M_i$ and the final state $j_f j_f' J_f M_f$. Each matrix element may be split into a direct term and an exchange term. The central-field wave function separates into radial and spin-angular parts. Using the multipole expansion of eqn (5.8), integration of these two parts proceeds separately.

Taking the angular part only, we have an integral over three spherical harmonics. We use the Wigner–Eckart theorem (Edmonds, 1957) to get

$$\int Y_\lambda^{*\mu}(\theta,\varphi)\, Y_{l_2}^{m_2}(\theta,\varphi)\, Y_{l_1}^{m_1}(\theta,\varphi)\, \mathrm{d}\Omega = \left[\frac{(2l_1+1)(2l_2+1)}{4\pi(2\lambda+1)}\right]^{1/2} C^{l_1\, l_2\, \lambda}_{m_1\, m_2\, \mu}\, C^{l_1\, l_2\, l_3}_{0\;\; 0\;\; 0}. \tag{5.11}$$

Equation (5.11) tells us that the states $Y_{l_1}^{m_1}$ and $Y_{l_2}^{m_2}$ on the unit sphere may be connected only by a virtual photon of multipolarity λ which satisfies the conditions $m_1 + m_2 = \mu$ and $\Delta(l_1 l_2 \lambda)$, the well-known triangle rule.

Using this result we get the spin-angular matrix element

$$\langle J_f M_f j_f j_f' | P_\lambda(\cos\theta_{12}) | J_i M_i j_i j_i' \rangle_{\text{direct}}$$

$$= \delta_{m_i - m_{l_i},\, m_f - m_{l_f}}\, \delta_{M_i - m_i - m_{l_i'},\, M_f - m_f - m_{l_f'}} \left[\frac{(2l_f+1)(2l_f'+1)}{(2l_i+1)(2l_i'+1)}\right]^{1/2} C^{l_f\, \lambda\, l_i}_{0\;\; 0\;\; 0}\, C^{l_f'\, \lambda\, l_i'}_{0\;\; 0\;\; 0}.$$

$$\cdot \sum_{m_i} \sum_{m_f} \sum_{m_{l_f}} \sum_{m_{l_i}} \sum_{m_{l_f'}} \sum_{m_{l_i'}} \sum_{\mu} (-)^\mu\, C^{l_f\;\; \lambda\;\; l_i}_{m_{l_f}\, \mu\, m_{l_i}}\, C^{l_f'\;\; \lambda\;\; l_i'}_{m_{l_f'}\, \mu\, m_{l_i'}}\, C^{j_i\;\; j_i'\;\; J_i}_{m_i\; M_i - m_i\; M_i}\, C^{j_f\;\; j_f'\;\; J_f}_{m_f\; M_f - m_f\; M_f}$$

$$\cdot\, C^{l_i\;\; 1/2\;\; j_i}_{m_{l_i}\; m_i - m_{l_i}\; m_i}\, C^{l_f\;\; 1/2\;\; j_f}_{m_{l_f}\; m_f - m_{l_f}\; m_f}\, C^{l_i'\;\; 1/2\;\; j_i'}_{m_{l_i'}\; M_f - m_f - m_{l_f'}\; m_i'}\, C^{l_f'\;\; 1/2\;\; j_f'}_{m_{l_f}'\; M_f - m_f - m_{l_f}'\; m_f'}\,. \tag{5.12}$$

The two Clebsch–Gordan coefficients $C^{l_f\; \lambda\; l_i}_{m_{l_f}\, \mu\, m_{l_i}}$ and $C^{l_f'\; \lambda\; l_i'}_{m_{l_f'}\, -\mu\, m_{l_i'}}$ appearing just after the summation signs in (5.12) imply that

$$\mu = m_{l_i} - m_{l_f} = m_{l_f'} - m_{l_i'}, \tag{5.13}$$

so that

$$m_{l_f} + m_{l_f'} = m_{l_i} + m_{l_i'}. \tag{5.14}$$

This just says that the sum of the orbital projection quantum numbers is conserved. But the two Kronecker deltas in the right-hand member of eqn (5.12) imply

$$M_f - m_f - m_{l_f'} = M_i - m_i - m_{l_i'}, \tag{5.15}$$

$$m_f - m_{l_f} = m_i - m_{l_i}. \tag{5.16}$$

Adding eqns (5.14) and (5.16), we get

$$m_f + m_{l_f'} = m_i + m_{l_i'}. \tag{5.17}$$

Substituting eqn (5.17) in eqn (5.15),

$$M_f = M_i,$$

i.e. the total angular momentum projection quantum number is also conserved, a result we have already inferred from the nature of the interaction. Hence eqn (5.12) may be written as

$$\begin{aligned}
&\langle J_f M_f j_f j'_f | P_\lambda(\cos\theta_{12}) | J_i M_i j_i j'_i \rangle_{\text{direct}} \\
&= \delta_{M_f M_i} [(2l_i+1)(2l'_i+1)/(2l_f+1)(2l'_f+1)]^{1/2} C^{l_i\,\lambda\,l_f}_{0\,0\,0} C^{l'_i\,\lambda\,l'_f}_{0\,0\,0} \cdot \\
&\sum_{m_f}\sum_{\mu} (-)^{\mu} C^{j_i\;\;j'_i\;\;J_i}_{\mu+m_f\;M_f-\mu-m_f\;M_f} C^{j_f\;\;j'_f\;\;J_f}_{m_f\;M_f-m_f\;M_f} \sum_{m_{l_f}} (-)^{\lambda+l_f-l_i} C^{\lambda\;l_f\;\;l_i}_{\mu\;m_{l_f}\;\mu+m_{l_f}} \\
&\cdot C^{l_i\;\;1/2\;\;j_i}_{\mu+m_{l_f}\;m_f-m_{l_f}\;\mu+m_f} C^{l_f\;1/2\;\;j_f}_{m_{l_f}\;m_f-m_{l_f}\;m_f} \sum_{m_{l'_f}} (-)^{\lambda+l'_f-l'_i} C^{\lambda\;\;l'_f\;\;l'_i}_{-\mu\;m_{l'_f}\;-\mu+m_{l'_f}} \\
&\cdot C^{l_i\;\;1/2\;\;j'_i}_{m'_{l_f}\,-\mu\;M_f-m_f-m_{l'_f}\;M_f-m_f-\mu} \cdot C^{l'_f\;1/2\;\;j'_f}_{m_{l'_f}\;M_f-m_f-m_{l'_f}\;M_f-m_f}.
\end{aligned} \tag{5.18}$$

In eqn (5.18) we have made use of the symmetry relation

$$C^{j_1\,j_2\,j_3}_{m_1\,m_2\,m_3} = (-)^{j_1+j_2-j_3} C^{j_2\,j_1\,j_3}_{m_2\,m_1\,m_3}. \tag{5.19}$$

Now the summations over m_{l_f} and $m_{l'_f}$ involve products of three Clebsch–Gordan coefficients each, implying Racah coefficients. Using the relation (Rose, 1957; Edmonds, 1957)

$$\begin{aligned}
\sum_{\beta} C^{a\;b\;\;e}_{\alpha\;\beta\;\alpha+\beta} C^{e\;\;\;d\;\;\;c}_{\alpha+\beta\;\delta\;\alpha+\beta+\delta} C^{b\;d\;\;f}_{\beta\;\delta\;\beta+\delta} &= [(2e+1)(2f+1)]^{1/2} W(abcd;ef) \cdot \\
&\cdot C^{a\;\;f\;\;\;c}_{\alpha\;\beta+\delta\;\alpha+\beta+\delta},
\end{aligned} \tag{5.20}$$

these two summations yield

$$\begin{aligned}
&(-)^{(l_f+l'_f)-(l_i+l'_i)} [(2l_i+1)(2l'_i+1)(2j_f+1)(2j'_f+1)]^{1/2} W(\lambda l_f j_i \tfrac{1}{2}; l_i j_f) \\
&W(\lambda l'_f j'_i \tfrac{1}{2}; l'_i j'_f) \cdot \sum_{m_f} C^{j_f\;\;j'_f\;\;J_f}_{m_f\;M_f-m_f\;M_f} (-)^{\lambda+J_i-(j_f+j'_f)} \left[\frac{2j_i+1}{2j_f+1}\right]^{1/2} \\
&\sum_{\mu} C^{j'_f\;\;\lambda\;\;j'_i}_{M_f-m_f\;-\mu\;M_f-m_f-\mu} C^{j'_i\;\;\;j_i\;\;\;J_i}_{M_f-\mu-m_f\;\mu+m_f\;M_f} C^{\lambda\;\;j_i\;\;j_f}_{-\mu\;\mu+m_f\;m_f}.
\end{aligned} \tag{5.21}$$

In this expression we have utilized the symmetry relation (5.19) and also the relations (Rose, 1957),

$$C^{j_1\,j_2\,j_3}_{m_1\,m_2\,m_3} = (-)^{j_1-m_1} \left[\frac{2j_3+1}{2j_2+1}\right]^{1/2} C^{j_1\;\;j_3\;\;j_2}_{m_1\,-m_3\,-m_2} \tag{5.22}$$

and

$$C^{j_1\,j_2\,j_3}_{m_1\,m_2\,m_3} = (-)^{j_1+j_2-j_3} C^{j_1\;\;\;j_3\;\;\;j_2}_{-m_1\,-m_3\,-m_2}. \tag{5.23}$$

Using again the addition theorem for Clebsch–Gordan coefficients, i.e. eqn (5.21), and the symmetry relation (5.19), the summation over μ in (5.21) gives

$$(-)^{j_i+j_f'-J_i}[(2j_i'+1)(2j_f+1)]^{1/2}\, W(j_f'\lambda J_i j_i; j_i' j_f)\, C^{j_f\ j_f'\ \ J_i}_{m_f\, M_f-m_f\, M_f}. \tag{5.24}$$

Also, the orthogonality relation of the Clebsch–Gordan coefficients gives

$$\sum_{m_f} C^{j_f\ j_f'\ \ J_f}_{m_f\, M_f-m_f\, M_f}\, C^{j_f\ j_f'\ \ J_i}_{m_f\, M_f-m_f\, M_f} = \delta_{J_f\, J_i}, \tag{5.25}$$

i.e. the total angular momentum is conserved.

Using eqns (5.18), (5.21), (5.24) and (5.25), we finally get for the spin-angular matrix element

$$\begin{aligned}
&\langle J_f\, M_f j_f j_f' | P_\lambda(\cos\theta_{12}) | J_i\, M_i j_i j_i'\rangle_{\text{direct}} \\
&= \delta_{J_f\, J_i}\delta_{M_f\, M_i}(-)^{\lambda+l_i+l_i'+l_f+l_f'}[(2l_f+1)(2l_f'+1)]^{1/2}\cdot \\
&\quad\cdot\ [(2j_i+1)(2j_i'+1)(2j_f+1)(2j_f'+1)]^{1/2}\cdot \\
&\quad\cdot\ C^{l_i\,\lambda\, l_f}_{0\ 0\ 0}\, C^{l_i'\,\lambda\, l_f'}_{0\ 0\ 0}\, W(\lambda l_f j_i \tfrac{1}{2}; l_i j_f)\, W(\lambda l_f' j_i' \tfrac{1}{2}; l_i' j_f')\, W(j_f'\lambda J_f j_i; J_i' j_f).
\end{aligned} \tag{5.26}$$

In eqn (5.26) the Racah Wigner algebra has taken care of all the conserved quantities. All geometrical information, i.e. information about individual magnetic substates which do not manifest themselves in an experiment, has been summed over. The dynamical part of the information is contained in the Clebsch–Gordan and W-coefficients. The Racah coefficients can be put in a more symmetrical form by using the relation

$$W(abcd;\, ef) = (-)^{a+d-e-f}\, W(bfec;\, da). \tag{5.27}$$

Then the direct matrix element becomes

$$\begin{aligned}
&\langle J_f\, M_f j_f j_f' | e^2/r_{12} | J_i\, M_i j_i j_i'\rangle_{\text{direct}} \\
&= \delta_{J_f\, J_i}\delta_{M_f\, M_i}(-)^{j_f+j_f'+l_f+l_f'+1}[(2l_f+1)(2l_f'+1)(2j_i+1)\cdot \\
&\qquad\cdot\ (2j_i'+1)(2j_f+1)(2j_f'+1)]^{1/2}\cdot \\
&\qquad\cdot\ \sum_\lambda (-)^\lambda\, C^{l_f\,\lambda\, l_i}_{0\ 0\ 0}\, C^{l_f'\,\lambda\, l_i'}_{0\ 0\ 0}\, W(l_f j_f l_i j_i; \tfrac{1}{2}\lambda)\, W(l_f' j_f' l_i' j_i'; \tfrac{1}{2}\lambda)\cdot \\
&\qquad\cdot\ W(j_f'\lambda J_f j_i; j_i' j_f)(2\lambda+1)\, D(\lambda),
\end{aligned} \tag{5.28}$$

where $D(\lambda)$ is the direct radial integral corresponding to the virtual photon of multipolarity λ:

$$\begin{aligned}
D(\lambda) &= \frac{1}{2\lambda+1}\, e^2 \int_{r_2=0}^{\infty}\int_{r_1=0}^{\infty} \gamma_\lambda(r_1, r_2)\, R_{N_i\, l_i}(r_1)\, R_{N_f l_f}(r_1)\, R_{N_i'\, l_i'}(r_2)\, R_{N_f' l_f'}(r_2) \\
& r_1^2 r_2^2\, dr_1\, dr_2 = \frac{1}{(2\lambda+1)} R^\lambda(N_i\, l_i,\, N_i'\, l_i';\, N_f\, l_f,\, N_f'\, l_f').
\end{aligned} \tag{5.29}$$

The R^λ has been defined in eqn (4.18).

Similarly, the exchange matrix element becomes

$$\begin{aligned}
&\langle J_f M_f j_f j_f' | e^2/r_{12} | J_i M_i j_i j_i' \rangle_{\text{exchange}} \\
&= \delta_{J_f J_i} \delta_{M_f M_i} (-)^{J_i + l_f + l_f' - 1} [(2l_f + 1)(2l_f' + 1)(2j_i + 1) \cdot \\
&\quad \cdot (2j_i' + 1)(2j_f + 1)(2j_f' + 1)]^{1/2} \cdot \\
&\quad \cdot \sum_\lambda (-)^\lambda C^{l_f\ \lambda\ l_i'}_{0\ 0\ 0} C^{l_f'\ \lambda\ l_i}_{0\ 0\ 0} W(l_f' j_f' l_i j_i; \tfrac{1}{2}\lambda)\, W(j_f \lambda J_f j_i; j_i' j_f') \cdot \\
&\quad \cdot W(l_f j_f\, l_i' j_i'; \tfrac{1}{2}\lambda)(2\lambda + 1)\, E(\lambda),
\end{aligned} \tag{5.30}$$

where $E(\lambda)$ is the exchange radial integral

$$\begin{aligned}
E(\lambda) &= \frac{1}{2\lambda + 1} e^2 \int_{r_2=0}^{\infty} \int_{r_1=0}^{\infty} \gamma_\lambda(r_1, r_2)\, R_{N_f' l_f'}(r_1)\, R_{N_i l_i}(r_1)\, R_{N_f l_f}(r_2) \cdot \\
&\quad \cdot R_{N_i' l_i'}(r_2)\, r_1^2 r_2^2\, \mathrm{d}r_1\, \mathrm{d}r_2 = \frac{1}{(2\lambda + 1)} R^\lambda(N_i l_i, N_i' l_i'; N_f' l_f', N_f l_f)
\end{aligned} \tag{5.31}$$

Note that the exchange matrix element may be obtained from the direct matrix element via the exchange $N_f, l_f, j_f \rightleftarrows N_f', l_f', j_f'$, i.e. by interchanging the outgoing lines in the Feynman graph.

Summing over all total projection quantum numbers and averaging over all possible states of the initial vacancy, we get the Auger transition probability

$$\begin{aligned}
&\frac{1}{2j_i + 1} \frac{2\pi}{\hbar} \sum_M |\langle JMj_f j_f' | e^2/r_{12} | JMj_i j_i' \rangle|^2 \\
&= \tau^2 (2J_f + 1)(2l_f + 1)(2l_f' + 1)(2j_f + 1)(2j_f' + 1)(2j_i' + 1) \cdot \\
&\quad \cdot \Bigg[\sum_\lambda (-)^\lambda (2\lambda + 1)\, C^{l_f\ \lambda\ l_i}_{0\ 0\ 0} C^{l_f'\ \lambda\ l_i'}_{0\ 0\ 0} W(l_f j_f l_i j_i; \tfrac{1}{2}\lambda)\, W(l_f' j_f' l_i' j_i'; \tfrac{1}{2}\lambda) \cdot \\
&\quad \cdot W(j_f' \lambda J_f j_i; j_i' j_f)\, D(\lambda) - (-)^{J_f - (j_f + j_f')} \sum_{\lambda'} (-)^{\lambda'} (2\lambda' + 1)\, C^{l_f\ \lambda'\ l_i'}_{0\ 0\ 0} C^{l_f'\ \lambda'\ l_i}_{0\ 0\ 0} \cdot \\
&\quad \cdot W(l_f' j_f' l_i j_i; \tfrac{1}{2}\lambda')\, W(l_f j_f l_i' j_i'; \tfrac{1}{2}\lambda')\, W(j_f \lambda' J_f j_i; j_i' j_f')\, E(\lambda') \Bigg]^2.
\end{aligned} \tag{5.32}$$

In eqn (5.32), we have used $J_i = J_f = J$ and $M_i = M_f = M$. The normalizing factor τ^2 is given by

$$\tau = \begin{cases} 1 & \text{for non-equivalent electrons,} \\ \dfrac{1}{\sqrt{2}} & \text{for equivalent electrons in the initial state.} \end{cases}$$

Certain restrictions on the allowed values of the orbital angular momentum of the Auger electron are embedded in this expression for the transition probability.
The Clebsch–Gordan coefficient $C_{0\ 0\ 0}^{l_f\ \lambda\ l_i} = 0$ unless $l_f + l_i + \lambda$ is even and $l_f l_i \lambda$ satisfy the triangle rule. Thus for a given initial vacancy l_i, and a given vacancy configuration $l_f l_f'$ of the doubly ionized atom, the summation over λ must be taken such that

$$\text{(a) } l_f + l_i + \lambda \quad \text{is even and} \quad \text{(b) } \Delta(l_f\, l_i\, \lambda) \quad \text{is satisfied.} \tag{5.33}$$

This fixes the allowed values of λ. For these values of λ, the allowed values of l_i' are given by

$$\text{(i) } l_f' + \lambda + l_i' \quad \text{even and} \quad \text{(ii) } \Delta(l_f'\, l_i'\, \lambda), \tag{5.34}$$

so that the direct matrix element may not vanish. Similarly, one obtains restrictions on l_i from the second sum.

Alternatively, we can use the Z-coefficients defined by

$$Z(abcd;\, ef) = (-)^{(1/2)(f-a+c)}[(2a+1)(2b+1)(2c+1)(2d+1)]^{1/2}$$
$$C_{0\ 0\ 0}^{a\ c\ f}\, W(abcd;\, ef) \tag{5.35}$$

to put eqn (5.32) in a more concise form:

$$\frac{1}{2j_i+1}\frac{2\pi}{\hbar}\sum_M |\langle JMj_f j_f'|e^2/r_{12}|JMj_i j_i'\rangle|^2$$
$$= \tau^2 \frac{2\pi(2J+1)}{\hbar(\ldots\ldots)2j_i+1(\ldots\ldots)}\Bigg[\sum_\lambda Z(l_f j_f l_i j_i;\, \tfrac{1}{2}\lambda)\, Z(l_f' j_f' l_i' j_i';\, \tfrac{1}{2}\lambda) W(j_f' \lambda J j_i;\, j_i' j_f) D(\lambda) -$$
$$- (-)^{J-(j_f+j_f')}\sum_{\lambda'} Z(l_f' j_f' l_i j_i;\, \tfrac{1}{2}\lambda')\, Z(l_f j_f l_i' j_i';\, \tfrac{1}{2}\lambda')\, W(j_f \lambda' J j_i;\, j_i' j_f')\, E(\lambda')\Bigg]^2. \tag{5.36}$$

Equation (5.32) or (5.36) may be used, along with the tables for the Clebsch–Gordan, Racah and Z-coefficients, to obtain the allowed Auger transition probabilities in *jj* coupling in terms of the radial integrals $D(\lambda)$ and $E(\lambda)$. We should also like to give here an alternative notation for these integrals introduced by Asaad and Burhop (1958), namely,

$$D(\lambda) = \frac{1}{2\lambda+1}\{(N_f\, l_f),\, (N_f'\, l_f'),\, \lambda,\, l_i'\} \tag{5.37}$$

and

$$E(\lambda) = \frac{1}{2\lambda+1}\{(N_f'\, l_f'),\, (N_f\, l_f),\, \lambda,\, l_i'\}. \tag{5.38}$$

The calculation of transition probabilities is straightforward though somewhat laborious. For any given transition the allowed values of λ' and l_f' must be determined from the relevant Clebsch–Gordan coefficients.

There is one class of transitions for which the calculation is very simple. This is the case of the initial $s_{1/2}$ vacancy, corresponding to the K, L_1, M_1, etc. series. Putting $l_i = 0$ and $j_i = \frac{1}{2}$ in eqn (5.32), we see that the only allowed values of λ and λ' are given by $\lambda = l_f$ and $\lambda' = l_f'$.

Under the jj coupling scheme, to each l-value there corresponds two possible values of j, namely $l+\frac{1}{2}$, and $l-\frac{1}{2}$. Asaad devised a notation to cover this situation. For a given shell l, he used the symbol l for the state $j = l+\frac{1}{2}$ and the symbol $\bar{l}$ for the state $j = l-\frac{1}{2}$. Thus a $p_{3/2}$ state is denoted by p, whereas a $p_{1/2}$ state is written as $\bar{p}$. Using eqn (5.32), he tabulated expressions for Auger transition probabilities in terms of the direct and exchange radial integrals $D(\lambda)$ and $E(\lambda)$ for initial s and p vacancies and for final vacancies ranging from s to d. These expressions may be used to compute jj coupling transition rates with any potential.

Asaad (1963a) used the screened hydrogenic radial integrals computed by Callan (1961) to obtain KLL transition rates in jj coupling for atoms in the range $12 \leqslant Z \leqslant 80$. The general trend was for the intensity of each line to increase with Z and gradually reach saturation. For KL_1L_2 and KL_1L_3 lines this happened at $Z \approx 45$. For the L_2L_3 and L_3L_3 lines this happened only at $Z > 60$. An important finding was that extreme jj coupling is not actually reached even at $Z = 80$, the experimental situation being better described by intermediate coupling. In other words, the spin–orbit interaction does not dominate to an extent that the electrostatic interaction between electrons may be neglected.

Expressions for transitions involving an initial s or p hole and final f holes were obtained by McGuire (1970) and by Crasemann *et al.* (1971). McGuire used an approximation to the self-consistent field approach which we shall describe in the next chapter. Crasemann *et al.* used non-relativistic screened hydrogenic functions which lead to radial integrals of the type described in Chapter 3. Their expressions for Auger transition probabilities in jj coupling to $s_{1/2}$ and $p_{1/2}$ initial vacancies have been listed by Bambynek *et al.* (1972).

The total XYZ Auger transition rate for a given initial vacancy in the shell X is of course independent of the coupling scheme. For example, one can calculate individual transition rates for each final-state term ${}^{2S+1}L$ in LS coupling and use the unitary transformation (4.31) to obtain the transition rate for a given final-state configuration (N_f, l_f, j_f) and $(N_f' l_f' j_f')$ in jj coupling. Using the above notation we get

$$|\langle N_f l_f j_f, N_f' l_f' j_f'; JM | e^2/r_{12} | N_i l_i j_i, N_i' l_i' j_i'; JM \rangle|^2$$
$$= \sum_{SLS'L'} [(2S+1)(2L+1)(2S'+1)(2L'+1)\,(2j_i+1)(2j_i'+1)(2j_f+1)(2j_f'+1)].$$

$$\cdot\begin{Bmatrix}\frac{1}{2} & l_i & j_i\\ \frac{1}{2} & l_i' & j_i'\\ S' & L' & J\end{Bmatrix}\begin{Bmatrix}\frac{1}{2} & l_f & j_f\\ \frac{1}{2} & l_f' & j_f'\\ S' & L' & J\end{Bmatrix}\begin{Bmatrix}\frac{1}{2} & l_i & j_i\\ \frac{1}{2} & l_i' & j_i'\\ S & L & J\end{Bmatrix}\begin{Bmatrix}\frac{1}{2} & l_f & j_f\\ \frac{1}{2} & l_f' & j_f'\\ S & L & J\end{Bmatrix}\cdot$$

$$\cdot\langle N_f l_f, N_f' l_f'; SLJM|e^2/r_{12}|N_i l_i, N_i' l_i'; SLJM\rangle\cdot$$

$$\cdot\langle N_f l_f, N_f' l_f'; S'L'JM|\,e^2/r_{12}|N_i l_i, N_i' l_i'; S'L'JM\rangle. \qquad (5.39)$$

To get the total transition rate XYZ for a given initial vacancy $(N_i l_i j_i)$ in the shell X, one sums both sides over l_i', j_i', j_f, j_f', J and M to get (Kostroun *et al.*, 1971)

$$P_{XYZ} = \frac{1}{2j_i+1}\frac{2\pi}{\hbar}\sum_J (2J+1)\sum_{l_i', j_i'}\sum_{j_f j_f'} |\langle N_f l_f j_f, N_f' l_f' j_f'; J|\,e^2/r_{12}|\cdot$$

$$\cdot N_i l_i j_i, N_i' l_i' j_i'; J\rangle|^2$$

$$= \frac{2\pi}{\hbar}\frac{1}{2(2l_i+1)}\sum_{SL}(2S+1)(2L+1)\sum_{l_i}|\langle N_f l_f, N_f' l_f'; SL|\,e^2/r_{12}|\cdot$$

$$\cdot N_i l_i, N_i' l_i'; SL\rangle|^2. \qquad (5.40)$$

Equation (5.40) just states that the total transition rate is independent of the coupling scheme.

However, there is one class of transitions where the calculation does not go through like this, namely transitions of the type XXY, where the initial vacancy $N_i l_i j_i$ is included in the final vacancy configuration. Coster–Kronig transitions of the type L_2L_3X belong to this class. It was pointed out by Kostroun *et al.* (1971) that here the left-hand side of eqn (5.39) cannot be summed over all values of j_f, and the right-hand side of eqn (5.39) does not yield the simple expression on the right-hand side of eqn (5.40). For example, in a calculation of L_2L_3X transitions it would be wrong to include on the left-hand side of eqn (5.39) transitions of the type $\langle 2p_{1/2}, N_f' l_f' j_f', J|e^2/r_{12}|2p_{1/2}, N_i' l_i' j_i', J\rangle$, which do not occur. Obviously, *jj* coupling is the natural scheme for calculations involving such transitions. Crasemann and his co-workers used *jj* coupling to compute L_1 subshell Auger yields as well as L_1L_2X and L_1L_3X Coster–Kronig yields for 14 elements in the range $33 \leqslant Z \leqslant 85$ (Crasemann *et al.*, 1971).

Similarly, they computed L_2 and L_3 subshell Auger yields and L_2L_3X Coster–Kronig transition probabilities for 21 elements with $26 \leqslant Z \leqslant 93$ (Chen *et al.*, 1971). Their values of L_1L_2X, L_1L_3X and L_2L_3X transition probabilities are shown in Figs 5.1, 5.2 and 5.3 respectively. Also shown are theoretical values obtained by McGuire (1971) on the basis of an approximate self-consistent-field SCF potential and experimental results obtained by various workers (for detailed references, see the papers of Crasemann and his co-

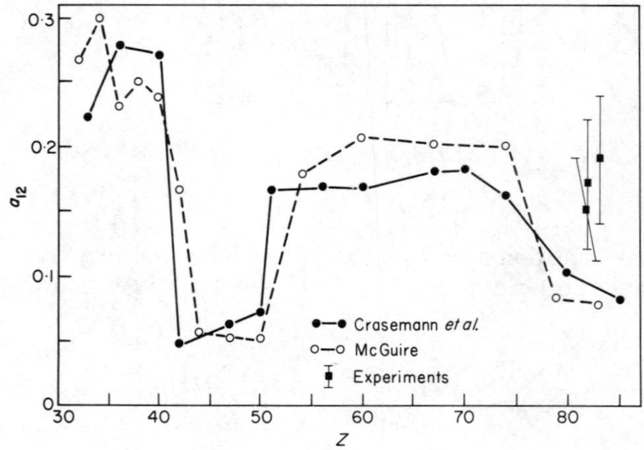

Fig. 5.1. L_1L_2X Coster–Kronig transition probabilities a_{12} calculated by Crasemann *et al.* (1971) and by McGuire (1971) as functions of atomic number Z. For references to the experimental points, see the former (from Crasemann *et al.*, 1971, courtesy of American Institute of Physics).

workers). It may be noted here that the corresponding radiative transitions L_1L_2, L_1L_3 and L_2L_3 have low probability. The last of these has a magnetic dipole (spin-flip) character.

One difficulty is that experimental data are not very extensive. Figure 5.3 is particularly instructive. A sharp discontinuity in the L_2L_3X transition probability occurs at $Z \approx 30$. Above this atomic number, $L_2L_3M_{4,5}$ radiationless transitions become energetically prohibited, because the $M_{4,5}$ binding energy

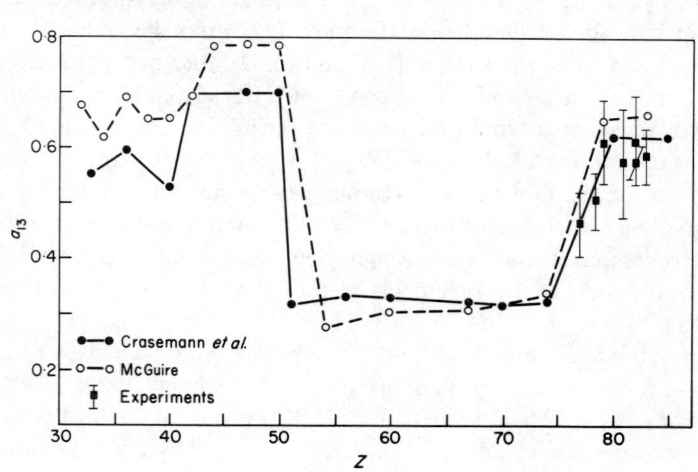

Fig. 5.2. L_1L_3X Coster–Kronig transition probabilities a_{13} from Crasemann *et al.* (1971) and by McGuire (1971). For references to experimental points, see the former (from Crasemann *et al.*, 1971, courtesy of American Institute of Physics).

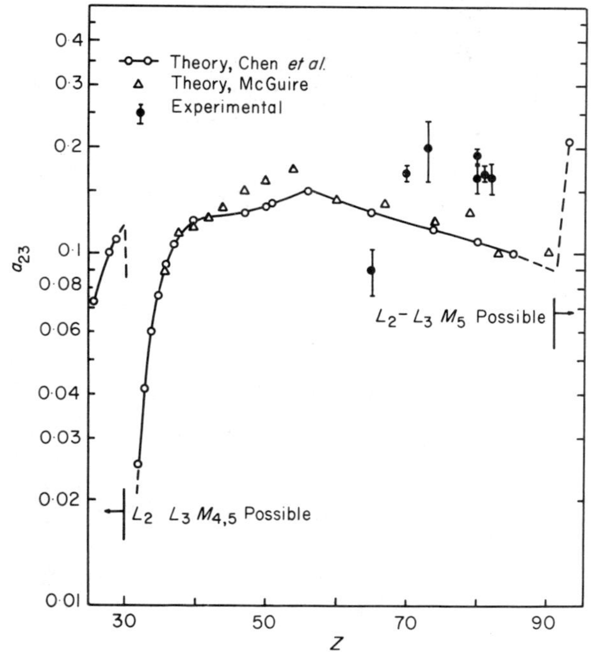

Fig. 5.3. L_2L_3X Coster–Kronig transition probabilities a_{23} calculated by Chen *et al.* (1971) and McGuire (1971). For references to experimental points see former (from Chen *et al.*, 1971, courtesy of American Institute of Physics).

in an atom ionized in the L_2 subshell exceeds the spin–orbit coupling energy-splitting between the L_2 and L_3 subshells (i.e. the $2p_{1/2} - 2p_{3/2}$ energy difference). A sharp increase in the transition rate takes place at $Z \approx 91$. From this point onward, $L_2L_3M_5$ transitions are again energetically possible. Another point about these results is the disagreement by more than 35% with the experimental data available in this region. On the other hand, the fairly close agreement between the SCF results of McGuire and the screened hydrogenic calculations of these authors illustrates the fact that the largest contribution to Auger matrix elements comes from a region where the two types of wave functions do not differ very much, namely, the region around $\bar{r}$, the average radius of the orbital. This is particularly true of shells with low principal quantum number, as shown in Fig. 5.4, where $2p$, $3p$ and $4p$ screened hydrogenic radial functions are compared with the corresponding SCF functions of Mann (1968).

Since spin–orbit interaction is intrinsically a relativistic effect, the *jj* coupling scheme provides a natural framework for relativistic calculations. Here the splitting in energy between different subshells with the same l value cannot be ignored. In fact, for a given j value there is a mixing of l-values, namely,

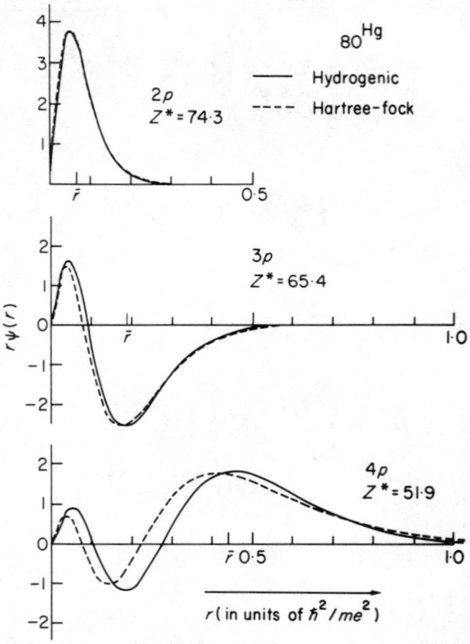

Fig. 5.4. Comparison of hydrogenic and Hartree–Fock $l = 1$ radial wave functions for mercury. Screening for the hydrogenic wave functions was computed according to Hartree (1957). The mean self-consistent-field radii $\bar{r}$ and the Hartree–Fock wave functions are from Mann (1968) (from Chen *et al.*, 1971, courtesy of American Institute of Physics).

$l = j - \frac{1}{2}$ and $l = j + \frac{1}{2}$. As a result, the ejected electron can now have l-values which would not be allowed in the non-relativistic case. The first relativistic calculation using *jj* coupling (Massey and Burhop, 1936) has already been discussed in Chapter 3. Later calculations include those by Asaad (1959) for the *KLL* spectrum of mercury ($Z = 80$), by Listengarten (1961, 1962) for the *KLL* spectrum with $Z = 65$, 81 and 92, and the more recent work by Bhalla *et al.* (1970a, b, c, d).

Listengarten's work was based on the Thomas–Fermi–Dirac model of screening, while those of Asaad as well as Bhalla *et al.* utilized self-consistent-field techniques. We shall discuss these calculations in the next chapter. Calculations based on relativistic screened hydrogenic functions by Chattarji and Talukdar (1968, 1970) for KL_1L_1 transitions covering $72 \leqslant Z \leqslant 80$ and $L_1L_2M_{4,5}$ transitions covering $32 \leqslant Z \leqslant 41$ also used *jj* coupling.

4. Transition rates with *LS* coupling

The *LS* coupling scheme is strictly valid only for low atomic numbers. However, as we have seen, it forms a natural basis for the description of the more general situation, namely, the intermediate coupling. Because in the non-relativistic limit the Auger transition is caused by the electrostatic interaction, this scheme is particularly suitable for its description.

In the *SLJM* representation, the antisymmetrized and normalized two-particle wave functions are of the form given in eqn (4.10):

$$\Psi_a(Nl, N'l'; SLJM) = \frac{1}{\sqrt{2}} \sum_{M_S M_L} C^{S\ L\ J}_{M_S M_L M} [\varphi(NlN'l'LM_L) + \\ + (-1)^{l+l'-L+S} \varphi(N'l'NlLM_L)]\, \chi(ss'SM_S), \quad (5.41)$$

where

$$\varphi(NlN'l'LM_L) = \sum_{m_l m_l'} C^{l\ l'\ L}_{m_l m_{l'} M_L} Y_l^{m_l}(\theta_1, \varphi_1)\, Y_{l'}^{m_{l'}}(\theta_2, \varphi_2)\, R_{Nl}(r_1)\, R_{N'l'}(r_2), \quad (5.42)$$

and

$$\chi(ss'SM_S) = \sum_{m_s, m_{s'}} C^{1/2\ 1/2\ S}_{m_s\ m_{s'}\ M_S}\, \chi_s^{m_s}(1)\, \chi_{s'}^{m_{s'}}(2). \quad (5.43)$$

As before $R_{Nl}(r)$, $Y_l^{m_l}(\theta, \varphi)$ and $\chi_m^{m_s}$ are the single-particle angular, radial and spin wave functions of an electron with quantum numbers Nlm_l and m_s. We have modified the notation slightly from that in Chapter 4 to bring it in line with the previous section of this chapter. Separation of the matrix element into radial and spin-angular parts proceeds exactly as before. One gets direct and exchange radial integrals as in the last section. In the expression for the total matrix element, these radial integrals are multiplied by the corresponding angular factors. The latter, of course, are obtained by using the Wigner–Eckart theorem. After going through the usual Wigner algebra, one gets an expression for the total transition probability into all possible states of L and S for a given final configuration of the atom

$$P = \frac{2\pi}{\hbar} \sum_{L,S} \frac{(2S+1)(2L+1)}{2(2l_i+1)} \sum_{l_l} \Big| \sum_{\lambda} [d_\lambda D(\lambda) \pm (-)^{l_f+l_f'} e_\lambda E(\lambda)] \Big|^2, \quad (5.44)$$

where the plus sign goes with even $L+S$ and the minus sign with odd $L+S$. The angular factors d_λ and e_λ are given by

$$d_\lambda = (-)^{l_f+l_f'+L} \left[\frac{(2l_f+1)(2l_f'+1)}{(2l_i+1)(2l_i'+1)}\right]^{1/2} C^{l_f\ \lambda\ l_i}_{0\ 0\ 0}\, C^{l_f'\ \lambda\ l_i'}_{0\ 0\ 0} \begin{Bmatrix} l_i & l_i' & L \\ l_f' & l_f & \lambda \end{Bmatrix} \quad (5.45)$$

and

$$e_\lambda = (-)^{l_i'+l_f'+L} \left[\frac{(2l_f+1)(2l_f'+1)}{(2l_i+1)(2l_i'+1)}\right]^{1/2} C^{l_f'\ \lambda\ l_i}_{0\ 0\ 0}\, C^{l_f\ \lambda\ l_i'}_{0\ 0\ 0} \begin{Bmatrix} l_i & l_i' & L \\ l_f & l_f' & \lambda \end{Bmatrix}, \quad (5.46)$$

the notation being the same as in the last section. *KLL* transition probabilities in *LS* coupling were given by Asaad and Burhop (1958) in terms of the radial integrals $D(\lambda)$ and $E(\lambda)$. These results are shown in Table 4. It will be seen that there are only five lines.

Table 4. *KLL*-transition probabilities in Russel–Saunders coupling

Transition	Final vacancies	l_i' (orbital ang. mon. of Auger electron)	Transition probability
$KL_1L_1(^1S)$	$(2s)^{-2}, {}^1S_0$	0	$\|D(0)\|^2 \equiv K\{(2s)^{-2}\,{}^1S_0\}$
$KL_1L_{2,3}(^1P)$	$(2s2p)^{-1}, {}^1P_1$	1	$\frac{3}{2}\|D(0)+E(1)\|^2 \equiv K\{(2s)^{-1}(2p)^{-1}\,{}^1P_1\}$
$KL_1L_{2,3}(^3P_J)$	$(2s2p)^{-1}, {}^3P_J$	1	$(J+\frac{1}{2})\|D(0)-E(1)\|^2 \equiv K\{(2s)^{-1}(2p)^{-1}\,{}^3P_J\}$
$KL_{2,3}L_{2,3}(^1S_0)$	$(2p)^{-2}, {}^1S_0$	0	$\frac{1}{3}\|D(1)\|^2 \equiv K\{(2p)^{-2}\,{}^1S_0\}$
$KL_{2,3}L_{2,3}(^1D_2)$	$(2p)^{-2}, {}^1D_2$	2	$\frac{3}{2}\|D(1)\|^2 \equiv K\{(2p)^{-2}\,{}^1D_2\}$
$KL_{2,3}L_{2,3}(^3P_J)$	$(2p)^{-2}, {}^3P_J$		0

The ${}^3P_J$ terms corresponding to the $(2p)^{-2}$ vacancy configuration in the final state cannot be connected to the initial state because parity and total orbital angular momentum cannot be conserved simultaneously in such a transition. Körber and Mehlhorn (1966) indeed found only a five-line spectrum in neon ($Z = 10$). However, Siegbahn *et al.* and Carlson and Krause observed a rather small intensity for the sixth line:

$$I(KL_{2,3}L_{2,3}(^3P_J)) = 2 \times 10^{-3}\, I(KL_{2,3}L_{2,3}(^1D_2)).$$

This arises from the fact that the spin–orbit interaction is not actually zero, so that intermediate coupling wave functions need to be used for the final state. This would couple the following Russell–Saunders states:

$$\pi = +1, J = 0: (2p)^{-2}, {}^1S_0 \quad \text{and} \quad {}^3P_0,$$
$$J = 2: (2p)^{-2}, {}^1D_2 \quad \text{and} \quad {}^3P_2.$$

The coupling coefficients are of course small.

More general expressions for Auger transition probabilities to an initial *K*-shell vacancy with final-state configurations extending to $(nd)(n'f)$ were given by Kostroun *et al.* (1971). These are shown in Table 5. In this table the radial integrals have been expressed in the forms of eqns (5.37) and (5.38). Using these expressions, these authors obtained theoretical *K*-shell transition probabilities with hydrogenic functions in the manner already described in the last section. For their binding energies, Kostroun *et al.* used values taken from the compilation by Bearden and Burr (1967). Following Callan (1961), they used neutral atom binding energies for $E_{N_il_i}$ and $E_{N_fl_f}$. To take account of

Table 5. Auger transition probabilities to an initial $1s$ vacancy, in LS coupling, in terms of radial matrix elements $\{(nl)(n'l'), \lambda, l_f\}$. (From Kostroun *et al.*, 1971)

Final-state configuration	Term	Transition probability[a]
$ns \quad n's$	${}^1S_0; {}^3S_1$	$(1/2)(2J+1)\,\tau^2\lvert\{(ns)(n's),0,0\} \pm \{(n's)(ns),0,0\}\rvert^2$
$ns \quad n'p$	${}^1P_1; {}^3P_{012}$	$(1/2)(2J+1)\lvert\{(ns)(n'p),0,1\} \pm \frac{1}{3}\{(n'p)(ns),1,1\}\rvert^2$
$np \quad n'p$	${}^1S_0; {}^3S_1$	$(1/6)(2J+1)\,\tau^2\lvert\{(np)(n'p),1,0\} \pm \{(n'p)(np),1,0\}\rvert^2$
	${}^1D_2; {}^3D_{123}$	$(1/15)(2J+1)\,\tau^2\lvert\{(np)(n'p),1,2\} \pm \{(n'p)(np),1,2\}\rvert^2$
$ns \quad n'd$	${}^1D_2; {}^3D_{123}$	$(1/2)(2J+1)\lvert\{(ns)(n'd),0,2\} \pm \frac{1}{5}\{(n'd)(ns),2,2\}\rvert^2$
$np \quad n'd$	${}^1P_1; {}^3P_{012}$	$(1/9)(2J+1)\lvert\{(np)(n'd),1,1\} \pm \frac{3}{5}\{(n'd)(np),2,1\}\rvert^2$
	${}^1F_3; {}^3F_{234}$	$(1/14)(2J+1)\lvert\{(np)(n'd),1,3\} \pm \frac{3}{5}\{(n'd)(np),2,3\}\rvert^2$
$ns \quad n'f$	${}^1F_3; {}^3F_{234}$	$(1/2)(2J+1)\lvert\{(ns)(n'f),0,3\} \pm \frac{1}{7}\{(n'f)(ns),1,3\}\rvert^2$
$np \quad n'f$	${}^1D_2; {}^3D_{123}$	$(1/10)(2J+1)\lvert\{(np)(n'f),1,2\} \pm \frac{3}{7}\{(n'f)(np),3,2\}\rvert^2$
	${}^1G_4; {}^3G_{345}$	$(1/27)(4J+2)\lvert\{(np)(n'f),1,4\} \pm \frac{3}{7}\{(n'f)(np),3,4\}\rvert^2$
$nd \quad n'd$	${}^1S_0; {}^3S_1$	$(1/10)(2J+1)\,\tau^2\lvert\{(nd)(n'd),2,0\} \pm \{(n'd)(nd),2,0\}\rvert^2$
	${}^1D_2; {}^3D_{123}$	$(1/35)(2J+1)\,\tau^2\lvert\{(nd)(n'd),2,2\} \pm \{(n'd)(nd),2,2\}\rvert^2$
	${}^1G_4; {}^3G_{345}$	$(1/35)(2J+1)\,\tau^2\lvert\{(nd)(n'd),2,4\} \pm \{(n'd)(nd),2,4\}\rvert^2$
$nd \quad n'f$	${}^1P_1; {}^3P_{012}$	$(1/50)(6J+3)\lvert\{(nd)(n'f),2,1\} \pm \frac{5}{7}\{(n'f)(nd),3,1\}\rvert^2$
	${}^1F_3; {}^3F_{234}$	$(1/75)(4J+2)\lvert\{nd)(n'f),2,3\} \pm \frac{5}{7}\{(n'f)(nd),2,3\}\rvert^2$
	${}^1H_5; {}^3H_{456}$	$(1/33)(2J+1)\lvert\{(nd)(n'f),2,5\} \pm \frac{5}{7}\{(n'f)(nd),3,5\}\rvert^2$

[a] As in the last section, $\tau^2 = \frac{1}{2}$ if $n = n'$ and $\tau^2 = 1$ if $n \neq n'$.
$\pm$ means $+$ for singlet and $-$ for triplet states.

decreased screening due to inner-shell ionization, for the binding energy $E_{N_f' l_f'}$ of the ejected electron, they used the value for the next-higher atomic number $Z + 1$.

These authors compared their results with experiment in regard to the following.

1. Total K-shell fluorescence yields computed by combining their Auger yields with Scofield's (1969) X-ray emission rates. These are shown as a function of Z in Fig. 5.5. The experimental points correspond to most reliable data based on the work of various experimentalists (see Bambynek *et al.*, 1972, for references). There is good agreement down to $Z = 10$. Above $Z = 50$, the theoretical values are a little too high. The radiative rates in this region were calculated relativistically.
2. Ratios of transition probabilities of the type $(KLX)/(KLL)$ and $(KXY)/(KLL)$ as a function of Z (Fig. 5.6). Also shown in this figure are the ratios obtained by Geffrion and Nadeau (1959) on the basis of unscreened hydro-

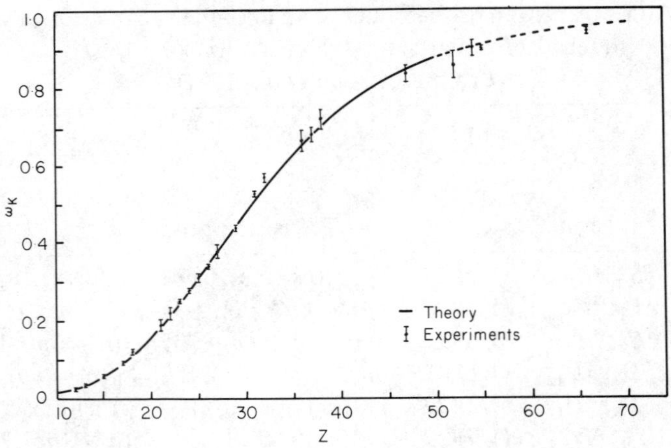

Fig. 5.5. Theoretical K-shell fluorescence yield ω_K computed by Kostroun *et al.* (1971) as a function of atomic number Z. The data points represent "most reliable" critically evaluated experimental results as discussed by Bambynek *et al.* (1972) (from Kostroun *et al.*, 1971, courtesy of American Institute of Physics).

genic functions and those obtained by McGuire (1970) using the approximate SCF method already mentioned. The results of Geffrion and Nadeau are far too high. McGuire's results agree quite well with experiment in the

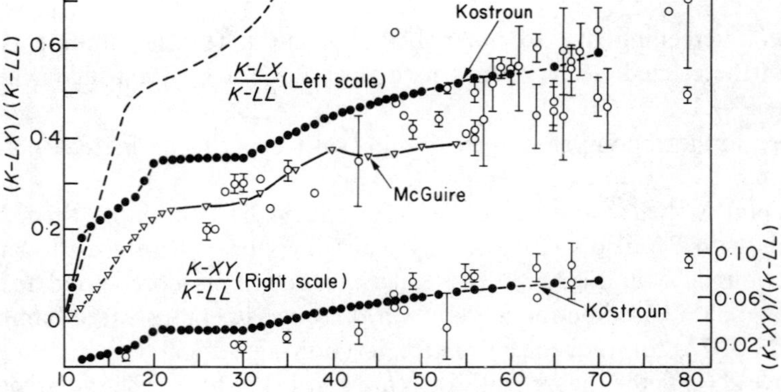

Fig. 5.6. Theoretical and experimental Auger transition-probability ratios $(KLX)/(KLL)$ and $(KXY)/(KLL)$ as a function of Z. Calculated ratios are from Kostroun *et al.* (1971), McGuire (1970) and Geffrion and Nadeau (1959). For references to measured values see Kostroun *et al.* (1971; from Kostroun *et al.* courtesy of American Institute of Physics).

relevant region. Since these group intensities are independent of the coupling scheme, their ratio provides a sensitive test for wave functions.

In spite of these large differences in the Auger transition probabilities, *K*-shell fluorescence yields computed by different authors seem to agree quite well. This can be seen from Fig. 5.7, which presents the work of Burhop (1935), Rubenstein (1955), Callan (1963), McGuire (1970) and Kostroun *et al.* (1971). The last-named authors attribute this to the fact that those calculations which lead to exceptionally large Auger transition probabilities also appear to result

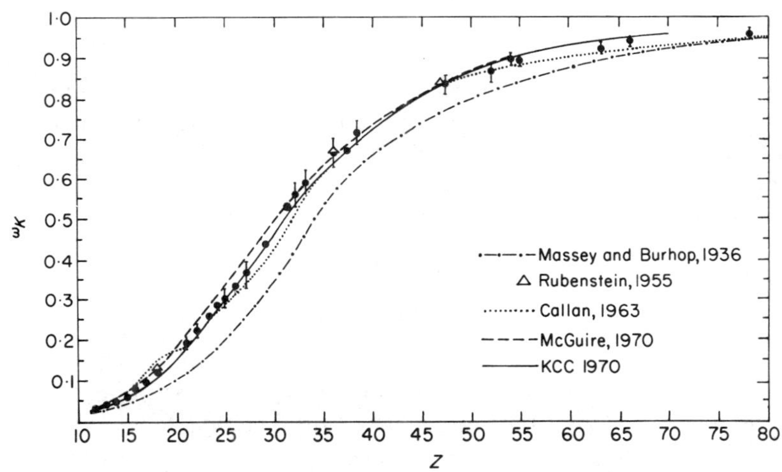

Fig. 5.7. Theoretical *K*-shell fluorescence yield as a function of *Z* according to various authors and experimental data as listed by Bambynek *et al.* 1972.

in large radiative transition probabilities. This is what makes the ratio of group intensities a particularly important criterion. Also, the agreement of non-relativistic results for fluorescence yields at high *Z* may be due to the fact that here radiative transitions dominate, so that this is not a sensitive test for Auger calculations.

5. Transition probabilities with intermediate coupling

We have already seen how the wave function for a given term is modified by intermediate coupling. Let us now see how the calculation of Auger transition probabilities proceeds in this case (Asaad and Burhop, 1958).

Take, for example, the *KLL* transitions. The final state in each transition can be expressed in terms of the two-vacancy configurations

$$(2s)^{-2}, (2s)^{-1}(2p)^{-1} \quad \text{and} \quad (2p)^{-2}.$$

We now need the matrices of the total Hamiltonian for these two-electron configurations in order to find the intermediate-coupling wave functions and, finally, the transition probabilities. The quantities of interest are

1. the electrostatic energies, i.e. matrix elements of the electrostatic perturbation $H_I' = \sum_{i<j} e^2/r_{ij}$, and
2. the energy of the spin–orbit interaction, i.e. matrix elements of the perturbation $H_{II}' = \sum_i \xi(r_i)\mathbf{l}_i \cdot \mathbf{s}_i$ for different values of the total angular momentum J.

We write the total Hamiltonian as

$$H = H_0 + H_I' + H_{II}'. \tag{5.47}$$

As already outlined in Section 4.4, H_{II}' is regarded as a perturbation of the Russell–Saunders system and we write out the perturbed eigenfunction for each state ${}^{2S+1}L_J$ as a linear combination of Russell–Saunders functions belonging to the particular J value. For example, for the ${}^3P_1$ and ${}^1P_1$ states with $J = 1$, belonging to the $(2s)(2p)^5$ configuration, we write

$$\begin{aligned} \Psi_1({}^1P_1) &= C_{11}\,\varphi_1({}^1P_1) + C_{12}\,\varphi_2({}^3P_1), \\ \Psi_2({}^3P_1) &= C_{21}\,\varphi_1({}^1P_1) + C_{22}\,\varphi_2({}^3P_1). \end{aligned} \tag{5.48}$$

Here Ψ_1 and Ψ_2 are the eigenfunctions with intermediate coupling for the $J = 1$, $L = 1$ subspace of the $(2s)(2p)^5$ configuration. We shall henceforth use the notation $(2s)^{-1}(2p)^{-1}$ for this configuration. The functions φ are Russell–Saunders functions of the type defined in eqn (4.9).

We can at once see that this problem is formally analogous to the problem of finding mixing coefficients as described in Section 4.5. Following the same procedure, we get the eigenvalue equation for the two-dimensional subspace:

$$\begin{pmatrix} H_{11} & H_{12} \\ H_{21} & H_{22} \end{pmatrix} \begin{pmatrix} C_{i1} \\ C_{i2} \end{pmatrix} = E_i(J=1) \begin{pmatrix} C_{i1} \\ C_{i2} \end{pmatrix} (i = 1, 2). \tag{5.49}$$

The secular equation is

$$\begin{Vmatrix} H_{11} - E_i & H_{12} \\ H_{21} & H_{22} - E_i \end{Vmatrix} = 0, \tag{5.50}$$

and the eigenvalues are

$$E_i = \frac{H_{11} + H_{22}}{2} \pm \sqrt{\left(\frac{(H_{11} - H_{22})^2}{4} + H_{12}\,H_{21}\right)}. \tag{5.51}$$

The two signs correspond to the two eigenvalues. These would be degenerate if

$$(H_{11} + H_{22})^2 = 4(H_{11}\,H_{22} - H_{12}\,H_{21}), \tag{5.52}$$

in which case two linearly independent solutions may be found by using the Schmidt orthogonalization procedure. Assuming that they are non-degenerate, we write

$$E_1 = \frac{H_{11}+H_{22}}{2} + \sqrt{\left(\frac{(H_{11}-H_{22})^2}{4} + H_{12}H_{21}\right)},$$

$$E_2 = \frac{H_{11}+H_{22}}{2} - \sqrt{\left(\frac{(H_{11}-H_{22})^2}{4} + H_{12}H_{21}\right)}.$$

Using the orthonormalization conditions on Ψ_1 and Ψ_2, we get

$$C_{11} = \frac{H_{12}(E_2 - H_{11})\,e^{i\delta}}{\sqrt{((H_{12}H_{21})^2 + H_{12}H_{21}(E_2 - H_{11})^2)}},$$

$$C_{12} = -\frac{H_{12}H_{21}\,e^{i\delta}}{\sqrt{((H_{12}H_{21})^2 + H_{12}H_{21}(E_2 - H_{11})^2)}},$$

$$C_{21} = \frac{H_{12}\,e^{i\delta}}{\sqrt{(H_{12}H_{21} + (E_2 - H_{11})^2)}},$$

$$C_{22} = \frac{(E_2 - H_{11})\,e^{i\delta}}{\sqrt{(H_{12}H_{21} + (E_2 - H_{11})^2)}}. \tag{5.53}$$

Here δ is an arbitrary phase factor. Equation (5.53) assumes a simpler form in the special case where the off-diagonal matrix elements are real, i.e. $H_{12} = H_{21}$, and the phase δ is chosen to be zero. Evidently, the first condition holds here and the second does not detract anything from the physics of the situation. Then

$$C_{11} = \frac{1}{\sqrt{(1+\alpha^2)}}, \qquad C_{12} = -\frac{\alpha}{\sqrt{(1+\alpha^2)}},$$

$$C_{21} = \frac{\alpha}{\sqrt{(1+\alpha^2)}} \quad \text{and} \quad C_{22} = \frac{1}{\sqrt{(1+\alpha^2)}}, \tag{5.54}$$

where

$$\alpha = \frac{H_{12}}{E_2 - H_{11}} = \frac{2H_{12}/(H_{22}-H_{11})}{1 - \sqrt{\left(1 + \frac{4H_{12}^2}{(H_{22}-H_{11})^2}\right)}}. \tag{5.55}$$

Since coefficients C_{12} and C_{21} provide a measure of the mixing, we can see from eqn (5.54) and (5.55) that the mixing depends on the quantity

$$R = \frac{H_{12}}{H_{22} - H_{11}}, \tag{5.56}$$

i.e. the ratio of the off-diagonal element H_{12} to the difference between the diagonal elements. This is an important result for any situation involving mixing.

To obtain the mixing coefficients, we must now determine the matrix elements of H between the Russell–Saunders states in question.

Consider first the matrix elements of the electrostatic interaction H'_{I}, namely,

$$H'^{\mathrm{I}}_{11} = \left\langle \varphi_1(^1P_1) \left| \sum_{i<j} \frac{e^2}{r_{ij}} \right| \varphi_1(^1P_1) \right\rangle, \qquad H'^{\mathrm{I}}_{22} = \left\langle \varphi_2(^3P_1) \left| \sum_{i<j} \frac{e^2}{r_{ij}} \right| \varphi_2(^3P_1) \right\rangle,$$

and

$$H'^{\mathrm{I}}_{21} = H'^{\mathrm{I}}_{12} = \left\langle \varphi_1(^1P_1) \left| \sum_{i<j} \frac{e^2}{r_{ij}} \right| \varphi_3(^1P_1) \right\rangle. \tag{5.57}$$

To compute the matrix elements in eqn (5.57), we expand the Russell–Saunders wave functions in terms of the Clebsch–Gordan series given in eqn (4.9):

$$\left\langle SLJM \left| \sum_{i<j} \frac{e^2}{r_{ij}} \right| S'LJM \right\rangle = \sum_{M_L M_S} \sum_{M'_L M_{S'}} C^{S\ L\ J}_{M_S M_L M} C^{S'\ L\ J}_{M'_S M'_L M} \cdot \left\langle LM_L, SM_S \left| \sum_{i<j} \frac{e^2}{r_{ij}} \right| LM'_L, S'M_{S'} \right\rangle, \tag{5.58}$$

where $|LM_L, SM_S\rangle$ is a product function obtained from eqns (4.10) and (4.11). Each term in the antisymmetrized expression (4.9) makes an equal contribution, so the factor $1/\sqrt{2}$ is removed. The matrix element on the right-hand side of eqn (5.58) may be written in terms of single-electron functions. The indices i and j range over the values 1 and 2 only, corresponding to the two vacancies. We get two-body matrix elements like

$$\left\langle ab \left| \frac{e^2}{r_{12}} \right| cd \right\rangle = \iint \Psi^\dagger_a(1)\, \Psi^\dagger_b(2) \frac{e^2}{r_{12}} \Psi_c(1)\, \Psi_d(2)\, \mathrm{d}^3 r_1\, \mathrm{d}^3 r_2. \tag{5.59}$$

The right-hand side of this equation may be written in the form

$$\iint R_{N_a l_a}(r_1) \overset{*}{Y}{}^{m_{l_a}}_{l_a}(\theta_1, \varphi_1) R_{N_b l_b}(r_2) \overset{*}{Y}{}^{m_{l_b}}_{l_b}(\theta_2, \varphi_2) \overset{\dagger}{\chi}{}^{\tau_a}_{1/2}(1) \overset{\dagger}{\chi}{}^{\tau_b}_{1/2}(2) \frac{e^2}{r_{12}} R_{N_c l_c}(r_1) \cdot$$

$$\cdot\, Y^{m_{l_c}}_{l_c}(\theta_1, \varphi_1) R_{N_d l_d}(r_2) Y^{m_{l_d}}_{l_d}(\theta_2, \varphi_2) \chi^{\tau_c}_{1/2}(1) \chi^{\tau_d}_{1/2}(2) \mathrm{d}^3 r_1 \, \mathrm{d}^3 r_2.$$

Making the expansion

$$\frac{1}{r_{12}} = \sum_{k=0}^{\infty} \frac{r_<^k}{r_>^{k+1}} \overset{*}{Y}{}^{\nu}_{k}(\theta_1, \varphi_1) Y^{\nu}_{k}(\theta_2, \varphi_2) \frac{4\pi}{2k+1},$$

and performing the angular part of the integration, only values of k satisfying the triangle rule embodied in the Clebsch–Gordan coefficients $C^{l_c k l_a}_{0\,0\,0}$ and $C^{l_a k l_b}_{0\,0\,0}$ contribute. The result may be written in terms of the Slater integrals F^k and G^k introduced in the previous chapter. In the present case, we get

$$H'^{\mathrm{I}}_{11} = F^0 + \tfrac{1}{3} G^1(2s, 2p), \quad H'^{\mathrm{I}}_{22} = F^0 - \tfrac{1}{3} G^1(2s, 2p),$$
$$H'^{\mathrm{I}}_{12} = H'^{\mathrm{I}}_{21} = 0. \tag{5.60}$$

Matrix elements of the spin–orbit interaction can be evaluated by means of the Wigner–Eckart theorem, since $\mathbf{l} \cdot \mathbf{s}$ is an irreducible tensor. Consider, for example, a diagonal element

$$\langle SLJM \Big| \sum_i \xi(r_i)\, \mathbf{l}_i \cdot \mathbf{s}_i \Big| SLJM \rangle.$$

Proceeding as above, we write

$$\langle SLJM \Big| \sum_i \xi(r_i)\, \mathbf{l}_i \cdot \mathbf{s}_i \Big| SLJM \rangle = \sum_{M_L M_S} \sum_{M_L' M_S'} C^{L\;S\;J}_{M_L M_S M} C^{L\;S\;J}_{M_L' M_S' M} \cdot$$
$$\cdot \langle LM_L', SM_S' \Big| \sum_i \xi(r_i)\, \mathbf{l}_i \cdot \mathbf{s}_i \Big| LM_L, SM_S \rangle. \tag{5.61}$$

Using the Wigner–Eckart theorem, we get (Bethe and Jackiw, 1965)

$$\langle LM_L', SM_S' | \sum_i \xi(r_i)\, \mathbf{l}_i \cdot \mathbf{s}_i | LM_L, SM_S \rangle$$
$$= \frac{1}{2m^2 c^2} \xi_{Nl} \sum_i \langle L \| l_i \| L \rangle \langle S \| s_i \| S \rangle \langle LM_L', SM_S' | \mathbf{L} \cdot \mathbf{S} | LM_L, SM_S \rangle$$
$$= \frac{1}{2m^2 c^2} \xi_{Nl} \gamma_{LS} [J(J+1) - L(L+1) - S(S+1)], \tag{5.62}$$

where

$$\gamma_{LS} = \sum_i \langle L \| l_i \| L \rangle \langle S \| s_i \| S \rangle, \tag{5.63}$$

and the expressions $\langle L\|l_i\|L\rangle$ and $\langle S\|s_i\|S\rangle$ are reduced matrix elements. They are given by Condon and Shortley (1959)

$$\langle L\|l_1\|L\rangle = \frac{L(L+1) - l_2(l_2+1) + l_1(l_1+1)}{2L(L+1)}\hbar,$$

$$\langle L\|l_2\|L\rangle = \frac{L(L+1) + l_2(l_2+1) - l_1(l_1+1)}{2L(L+1)}\hbar,$$

$$\langle S\|s_1\|S\rangle = \langle S\|s_2\|S\rangle = \tfrac{1}{2}\hbar. \tag{5.64}$$

For off-diagonal elements, we have

$$\left.\begin{array}{l}\langle L\|l_1\|L-1\rangle\\ \langle L\|l_2\|L-1\rangle\end{array}\right\} =$$

$$= \pm\hbar\left[\frac{(L-l_1+l_2)(L+l_1-l_2)(l_1+l_2+L+1)(l_1+l_2-L+1)}{4L^2(2L+1)(2L-1)}\right]^{1/2},$$

$$\langle S\|s_1\|S-1\rangle = -\langle S\|s_2\|S-1\rangle = \tfrac{1}{2}\hbar.$$

The matrix elements of the spin–orbit interaction in the above example turn out to be

$$H'^{\text{II}}_{11} = 0,\quad H'^{\text{II}}_{22} = \tfrac{1}{2}\xi_{2p},\quad H'^{\text{II}}_{12} = -\frac{1}{\sqrt{2}}\xi_{2p},\quad H'^{\text{II}}_{21} = -\frac{1}{\sqrt{2}}\xi_{2p}. \tag{5.65}$$

Since we are within a single configuration, we can use an energy scale where the expectation value of H_0 is zero. Then the complete matrix elements are

$$H_{11} = F^0 + \tfrac{1}{3}G^1(2s,2p),\quad H_{22} = F^0 - \tfrac{1}{3}G^1(2s,2p) + \tfrac{1}{2}\xi_{2p}$$

and

$$H_{21} = H_{12} = -\frac{1}{\sqrt{2}}\xi_{2p}. \tag{5.66}$$

The mixing coefficients are

$$C_{11} = \frac{\gamma}{(\gamma^2+2\beta^2)^{1/2}},\qquad C_{12} = \frac{-\sqrt{2}\beta}{(\gamma^2+2\beta^2)^{1/2}},$$

$$C_{21} = -C_{12},\qquad C_{22} = C_{11}, \tag{5.67}$$

where

$$\gamma = \tfrac{1}{3}\{G^1(2s,2p) - \beta/2\} + \sqrt{(\tfrac{1}{3}\{G^1(2s,2p) - \beta/2\}^2 + 2\beta^2)} \tag{5.68}$$

and

$$\beta = \tfrac{1}{2}\xi_{2p}.$$

In general, following a similar procedure, the transition probabilities to a given J value in the final-state configuration may be written out as a linear combination of transition probabilities for individual Russell–Saunders terms (Asaad and Burhop, 1958)

$$P\{(N_f l_f), (N_f' l_f'), J\} = \frac{\sum_i \{C_i(^{2S+1}L_J)\}^2 K\{(N_f l_f)(N_f' l_f')(^{2S+1}L_J)\}}{\sum_i \{C_i(^{2S+1}L_J)\}^2}, \quad (5.69)$$

where $K\{(N_f l_f)(N_f' l_f')(^{2S+1}L_J)\}$ is the pure Russell–Saunders transition probability to the term $^{2S+1}L_J$. We have already introduced these objects for KLL transitions in Table 4. We list below the transition probabilities in intermediate coupling given by Asaad and Burhop (1958) for KLL spectra.

$$(2s)^{-2} L_1 L_1(^1S_0) : K\{(2s)^{-2}\,{}^1S_0\}.$$

$$\begin{aligned}
(2s)^{-1}(2p)^{-1} L_1 L_2(^1P_1) &: K\{(2s)^{-1}(2p)^{-1}\,{}^1P_1\} - \\
&\quad - \frac{\alpha^2}{1+\alpha^2}[K\{(2s)^{-1}(2p)^{-1}\,{}^1P_1\} - K\{(2s)^{-1}(2p)^{-1}\,{}^3P_1\}], \\
L_1 L_2(^3P_0) &: K\{(2s)^{-1}(2p)^{-1}\,{}^3P_0\}, \\
L_1 L_3(^3P_1) &: K\{(2s)^{-1}(2p)^{-1}\,{}^3P_1\} + \\
&\quad + \frac{\alpha^2}{1+\alpha^2}[K\{(2s)^{-1}(2p)^{-1}\,{}^1P_1\} - K\{(2s)^{-1}(2p)^{-1}\,{}^3P_1\}], \\
L_1 L_3(^3P_2) &: K\{(2s)^{-1}(2p)^{-1}\,{}^3P_2\}.
\end{aligned}$$

$$\begin{aligned}
(2p)^{-2} L_2 L_2(^1S_0) &: \frac{1}{1+\alpha_1^2} K\{(2p)^{-2}\,{}^1S_0\}, \\
L_2 L_3(^1D_2) &: \frac{1}{1+\alpha_2^2} K\{(2p)^{-2}\,{}^1D_2\}, \\
L_3 L_3(^3P_0) &: \frac{\alpha_1^2}{1+\alpha_1^2} K\{(2p)^{-2}\,{}^1S_0\}, \\
L_3 L_3(^3P_2) &: \frac{\alpha_2^2}{1+\alpha_2^2} K\{(2p)^{-2}\,{}^1D_2\}.
\end{aligned} \quad (5.70)$$

For the quantities K in eqn (5.70), the reader is referred to Table 4. The mixing coefficients α, α_1 and α_2 are given by

$$\alpha = \frac{1}{\sqrt{2}}\,\xi_{2p}/[\tfrac{1}{3}G^1 - \tfrac{1}{4}\xi_{2p} + \sqrt{((\tfrac{1}{3}G^1 - \tfrac{1}{4}\xi_{2p})^2 + \tfrac{1}{2}\xi_{2p}^2)}],$$

$$\alpha_1 = \sqrt{2}\xi_{2p}/[\tfrac{3}{10}F^2 - \tfrac{1}{2}\xi_{2p} + \sqrt{((\tfrac{3}{10}F^2 - \tfrac{1}{2}\xi_{2p})^2 + 2\xi_{2p}^2)}],$$

$$\alpha_2 = \frac{1}{\sqrt{2}}\,\xi_{2p}/[\tfrac{3}{25}F^2 + \tfrac{1}{4}\xi_{2p} + \sqrt{((\tfrac{3}{25}F^2 + \tfrac{1}{4}\xi_{2p})^2 + \tfrac{1}{2}\xi_{2p}^2)}]. \qquad (5.71)$$

Here

$$F^2 \equiv F^2(2p, 2p), \quad G^1 \equiv G^1(2s, 2p).$$

In their early work with intermediate coupling, Asaad and Burhop (1958) used the transition amplitudes obtained by Burhop (1935). These were based on non-relativistic screened hydrogenic wave functions, with the screening parameters given by Slater's rule. In a later calculation, Asaad (1963a) used the LS coupling transition amplitudes computed by Callan (1961) for the KLL Auger spectrum. Callan's calculations covered all atomic numbers from $Z = 12$ to $Z = 41$ and also ten atomic numbers in the range from $Z = 44$ to $Z = 80$. He used screened hydrogenic wave functions with screening parameters derived from Hartree–Fock self-consistent field calculations. For the $2s$ and $2p$ electrons of the same atom, different parameters were used. In calculating the mixing coefficient α for the $(2s)^{-1}(2p)^{-1}$ configuration, Asaad (1963a) took screened hydrogenic functions in which the effective nuclear charge $\bar{Z}$ was the geometric mean of the effective nuclear charges Z_{L_1} and $Z_{L_{2,3}}$ given by Callan. He thus got expressions for G^1 and ξ_{2p} in terms of $\bar{Z}$ for this configuration,

$$G^1 = \tfrac{45}{512}\bar{Z}, \quad \xi_{2p} = \bar{Z}^4/(48 \times 137^2), \qquad (5.72)$$

using atomic units. For the $(2p)^{-2}$ configuration, the effective charge $Z_{L_{2,3}}$ was used directly to calculate α_1 and α_2. For F^2 and ξ_{2p} he got, in terms of $Z_{L_{2,3}}$,

$$F^2 = \tfrac{45}{512} Z_{L_{2,3}} \quad \text{and} \quad \xi_{2p} = Z_{L_{2,3}}^4/(48 \times 137^2). \qquad (5.73)$$

Several interesting features emerged from these intermediate coupling calculations.

1. It was found that deviations from pure LS coupling begin practically from $Z = 25$, while the extreme jj coupling is not fully approached, even at $Z = 80$. Not only is the nine-line KLL spectrum obtained over this range of Z-values, but the absolute line strengths of this spectrum begin to differ from extreme LS coupling line strengths for Z-values as low as 25 or less.

2. An interesting aspect is the variation of line intensities with Z (see Fig. 4.2). The $L_1L_1(^1S_0)$ line has the same strength on the basis of any kind of

coupling—it remains unsplit. Also, its variation with Z is slow, with a flat maximum at $Z = 23$. It is, therefore, appropriate that relative line strengths are given by experimentalists in terms of this line.

For the configuration $(2s)^{-1}(2p)^{-1}$, the ^3P line expected in LS coupling is split, as intermediate coupling is approached, into three components $^3P_0$, $^3P_1$ and $^3P_2$. The fourth line for this configuration is the ^1P line. In the limit of jj coupling, $^3P_1$ and $^3P_2$ combine to form the L_1L_3 line, and $^3P_0$ combines with $^1P_1$ to form the L_1L_2 line. In intermediate coupling, the sum of the intensities of the $^3P_1$ and $^1P_1$ lines remains the same as in LS coupling, although their individual intensities vary. In fact, with increasing Z, the strength of the $^3P_1$ increases while that of $^1P_1$ decreases.

For the configuration $(2p)^{-2}$, LS coupling gives only two lines $^1S_0$ and $^1D_2$. The strength of the ^3P line is zero in this limit because of the selection rules. With intermediate coupling, the components $^3P_0$ and $^3P_2$ begin to build up at the expense of $^1S_0$ and $^1D_2$ respectively. The third ^3P component, namely $^3P_1$, still has zero intensity, giving four lines for this configuration. With increasing Z, extreme jj coupling is approached, when the $^1S_0$ line becomes the L_2L_2 line, the $^1D_2$ line becomes the L_2L_3 line and the $^3P_0$ and $^3P_2$ combine to form the L_3L_3 line.

The most intense line of the KLL Auger spectrum, in this theory, is the $L_2L_3(^1D_2)$ line, for all $Z > 17$. For heavy elements, the decreasing order of intensity is as follows: $L_2L_3(^1D_2)$, $L_1L_3(^3P_1)$, $L_3L_3(^3P_2)$, $L_1L_2(^1P_1)$, $L_1L_1(^1S_0)$, $L_1L_3(^3P_2)$, $L_3L_3(^3P_0)$, $L_2L_2(^1S_0)$ and $L_1L_2(^3P_0)$.

3. For comparison with experiment, there are three possible candidates so far as intensities or transition probabilities are concerned (Melhorn, 1969).

(a) Relative line intensities, e.g. $I(KL_xL_y, {}^{2S+1}L_J)/I(KL_1L_1)$.

(b) Relative group intensities, e.g. $I(KL_1L_{2,3})/I(KL_1L_1)$ and $I(KL_{2,3}L_{2,3})/I(KL_1L_1)$.

(c) Absolute transition probabilities.

The intensity of the KL_1L_1 line remains unaffected by the coupling scheme and shows only small variation with Z, so that it serves as a convenient basis for (a) and (b) throughout the periodic table. Also, (a) and (b) are easier to measure than (c). Hence the usefulness of relative intensities.

As for group intensities, the group $KL_1L_{2,3}$ belongs to the configuration $(2s)^{-1}(2p)^{-1}$ and the group $KL_{2,3}L_{2,3}$ to the configuration $(2p)^{-2}$. Apart from configuration interaction, these group intensities are independent of the coupling scheme. This is one advantage of using them as a criterion for comparison with experiment.

Intermediate coupling implies the mixing of states within a group with the same J value. The mixing coefficients, as we have already seen, depend on the Slater integrals F and G, and on the spin–orbit interaction parameter ξ_{2p}. The integrals F and G are usually obtained by fitting to observed Auger energies and inner-shell binding energies (see next section). Inaccuracies in the latter would also be reflected as inaccuracies in the mixing coefficients. Hence, relative line intensities would be sensitive to wave functions as well as these other parameters. This makes the group intensities a better candidate.

With the availability of more accurate numerical wave functions, some of these difficulties of course become less formidable. Also, for a given element, the accuracy of transition amplitudes and hence of wave functions can be tested fairly well by comparing relative line intensities with experiment. In fact, we have already used relative line intensities as a criterion in a different context, namely to see the effectiveness of configuration interaction in neon (see Section 4.5).

With the inclusion of configuration interaction, the intensity of the $^1S_0$ line in the group $KL_{2,3}L_{2,3}$ increases at the expense of the KL_1L_1, $^1S_0$ line. Hence the relative group intensity as a whole will increase. This increase depends on the mixing coefficients and the transition amplitudes. In eqn (5.56) we saw that the mixing depends, in general, on the ratio R of the off-diagonal matrix element of the Hamiltonian to the difference between the diagonal elements. In this case, from eqns (4.45) and (4.46a), we get

$$R = \frac{\dfrac{1}{\sqrt{3}} G^1(2s, 2p)}{2[E(L_1) - E(L_{2,3})] + F^0(2s, 2s) - F^0(2p, 2p) - \frac{10}{25} F^2(2p, 2p)}. \quad (5.74)$$

Here, $E(L_{2,3})$ is the average binding energy of the L_2 and L_3 subshells. It may be shown that the ratio R decreases with increasing Z (Mehlhorn, 1969). Using the amplitudes given in Table 4 and the resulting mixing coefficients, it has been estimated that the increase in relative intensity of the $KL_{2,3}L_{2,3}$ group ranges from about 18% in neon ($Z = 10$) to about 25% in krypton ($Z = 36$).

Figures 5.8(a) and (b) compare experimental relative group intensities for the $(2p)^{-2}$ and $(2s)^{-1}(2p)^{-1}$ configurations with theoretical values (Cleff and Mehlhorn, 1969). References to the experimental literature may be found in the original paper by these authors. The theoretical results are taken from Krause *et al.* (1964), Archard (1963), Rubenstein (1955), Callan (1961, 1968), Burhop (1935), Asaad (1965b) and Mehlhorn and Asaad (1966). The last work is more important because it includes configuration interaction with improved values of the binding energy $E(L_1)$ and of the Coulomb and spin–orbit interaction energies. There is good agreement for large and medium Z, but large deviations occur for low Z, between theory and experiment, as well as between different theoretical results. It has been estimated by Mehlhorn and Asaad (1966) that not more than 10% of this discrepancy may be caused by inaccuracies of the parameters $G^1(2s, 2p)$, ξ_{2p} and the diagonal elements of the matrix in eqn (4.45). The rest of the deviation must be attributed to (a) inaccuracies in the wave functions used to calculate transition amplitudes and (b) neglect of correlation effects.

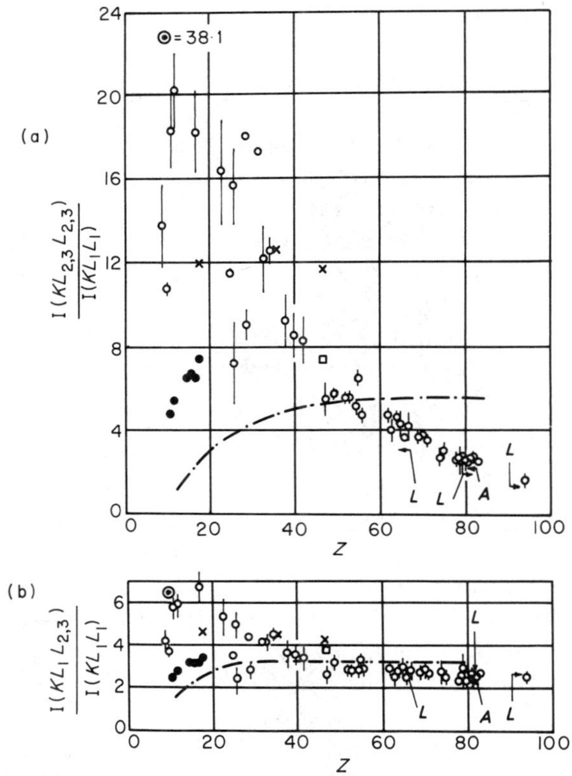

Fig. 5.8. Theoretical and experimental relative group intensities (a) $I(KL_{2,3}L_{2,3})/I(KL_1L_1)$ and (b) $I(KL_1L_{2,3})/I(KL_1L_1)$. For references see text (from Cleff and Mehlhorn, 1969, courtesy of Springer Verlag).

6. Electron energies of the *K*-Auger spectrum

For given initial and final state configurations, with parity and angular momentum conserved, only those transitions which are energetically possible actually take place. The condition for energy conservation is

$$E_{N_i l_i} - E_{N_f l_f} - E_{N'_f l'_f} = E_{\eta l'_i} > 0, \tag{5.75}$$

where we have used our earlier notation to designate the quantum numbers of the vacancies in the initial and final states. $E_{N_i l_i}$ is the binding energy of the electron corresponding to the initial vacancy. $E_{N_f l_f}$ and $E_{N'_f l'_f}$ are the binding energies of the two participating electrons, i.e. corresponding to the two final vacancies in bound states. These two binding energies relate to the state of the atom which is singly ionized in the $N_i l_i$ shell. Assuming the $N_f l_f$ electron to jump in and fill the vacancy, the usual practice is to correct for the binding energy of the other electron $N'_f l'_f$, which is ejected into the continuum and, therefore, supposed to "feel" the reduced screening due to the inner-shell ionization. The correction is made by taking $E_{N'_f l'_f}$ corresponding to an atom of the next higher atomic number, $Z+1$. Needless to say, this is a rough approximation. From the viewpoint of quantum electrodynamics, no statement can be made about which electron jumps in "first" and which one is ejected "later" and so on. All we have is the direct Feynman graph and its exchange counterpart. Since the lifetime of an inner-shell vacancy $N_i l_i$ is actually comparable to the time taken for an Auger transition to take place, there is no reason why both the $N_f l_f$ and $N'_f l'_f$ electrons should not see the reduced screening before they go their own ways via the Auger transition.

A more reasonable and symmetric way of writing the equation of energy conservation for the Auger electron is

$$E_{nl'_i} = E_{N_i l_i} - E_{N_f l_f} - E_{N'_f l'_f} - \Delta E_{(N_f l_f)(N'_f l'_f)}. \tag{5.76}$$

Here, assuming the electron from the $N_f l_f$ shell to fill the vacancy $N_i l_i$ and that from the $N'_f l'_f$ shell to be emitted, $E_{N_i l_i} - E_{N_f l_f}$ is the energy liberated in the transition, and $E_{N'_f l'_f} + \Delta E_{(N_f l_f)(N'_f l'_f)}$ is the binding energy of an electron in the $N'_f l'_f$ shell of an atom ionized in the $N_f l_f$ shell. Due to the indistinguishability of the electrons, we could equally well consider $E_{N_i l_i} - E_{N'_f l'_f}$ to be the energy liberated and $E_{N_f l_f} + \Delta E_{(N_f l_f)(N'_f l'_f)}$ the binding energy of the electron in the $N_f l_f$ shell of the atom ionized in the $N'_f l'_f$ shell. Here one does not have to worry about which transition takes place first.

The difficult part of this ansatz is to find $\Delta E_{(N_f l_f)(N'_f l'_f)}$. Bergstrom and Hill (1954) suggested the following empirical method. The idea is that when the atom is ionized in the $N_f l_f$ shell, $E_{N'_f l'_f}$ increases as though it were due to some increase $\Delta Z_{N_f l_f}$ in the atomic number Z of the atom. We thus write

$$\Delta E_{(N_f l_f)(N'_f l'_f)}(Z) = \Delta Z_{N_f l_f}[E_{N'_f l'_f}(Z+1) - E_{N'_f l'_f}(Z)]. \tag{5.77}$$

Choosing a reasonable value for $\Delta Z_{N_f l_f}$, we can find $\Delta E_{(N_f l_f)(N'_f l'_f)}(Z)$ according to eqn (5.77) and then use eqn (5.76) to find the Auger electron energy. For all its simplicity, this empirical method gives reasonable results for most groups and series (except the *KLL* series). In eqn (5.77), the parameter ΔZ corresponds to ionization in the $N_f l_f$ shell. The same energy increment ΔE should be obtained by considering the change $\Delta Z'_{N_f l_f}$ in the effective nuclear charge seen by the $N_f l_f$ electron for ionization in the $N'_f l'_f$ shell. This is done by making the interchange $N_f l_f \rightleftarrows N'_f l'_f$ on the right-hand side of eqn (5.77), a more symmetric expression for the energy results from using the average of these two expressions for ΔE in eqn (5.76) (Albridge and Hollander, 1961).

The values of ΔZ given by Bergstrom and Hill for the *KLL* spectrum in the region $80 \leqslant Z \leqslant 85$ are as follows. $\Delta Z = 0.54$ for the L_1 subshell, $=0.55$ for the L_2 subshell and $=0.76$ for L_3 electrons. These parameters were taken to be the same for all lines in the *KLL* group.

Generally speaking, first principle calculations of Auger-electron energies $E_{nl'_i}$ have not been very successful. The result depends on the coupling scheme, on spin–orbit interaction, relativistic effects and even on configuration interaction. Experimental measurements of Auger electron energies being quite accurate, this remains a rather sensitive test for theory. Where electron binding energies are required for calculation of Auger transition rates, compilations giving experimental values are the best sources. Important among these are those by Bearden and Burr (1967), Siegbahn *et al.* (1967a) and Sandstrom (1959). A recent compilation by Packer and Wilson (1971), based mainly on Bearden and Burr and on Siegbahn *et al.*, gives both level energies and Auger electron energies for all observed transitions between 0 and 1500 eV for elements up to $Z = 88$.

We shall briefly discuss here the energies in the *K*-Auger spectrum in intermediate coupling without configuration interaction. That is to say, we shall outline the determination of the eigenvalues of H'_{I} and H_{II}. Once this is done, the energy of the *K*-Auger line is found by subtracting the eigenvalues of the total Hamiltonian H from $E(K)$.

Consider first the matrix elements of the Hamiltonian $H_0 + H'_{\text{I}}$, i.e. without spin–orbit interaction (Condon and Shortley, 1959). The terms occurring in any given configuration are obtained as in Chapter 4, where we constructed tables giving M_L and M_S values for the different states occurring in LS coupling. To be specific, we shall consider the p^2 configuration as given in Table 1. This is relevant to the vacancy configuration $(2p)^{-2}$ we have already discussed. There are no matrix elements connecting states of different M_L and M_S. Therefore, the secular equation for the whole configuration breaks up into a number of secular equations, one for each value of $M_S M_L$. In other words, each value of $M_S M_L$ defines an invariant subspace for the configuration. Thus for each cell in Table 1 there is one secular equation with as many roots as there are states in the cell. For example, for $M_L = 1$, $M_S = 1$, there is one state (1^+0^+) which must belong to $L = 1$, $S = 1$, i.e. the triplet ^3P. For this particular cell, this single state forms the complete set and the secular equation has only one root. On the other hand, for $M_L = 0$, $M_S = 0$, there are three states, $(1^+ - 1^-)$, $(1^- - 1^+)$ and (0^+0^-), belonging to ^1D, ^3P and ^1S respectively. Hence for this cell, the complete set consists of these three states and the secular equation has, correspondingly, three roots.

To obtain the eigenvalues, i.e. the roots of the secular equation, we make use of the so-called diagonal-sum rule. This states that the sum of the roots of a secular equation is equal to the sum of the diagonal matrix elements occurring in it. This follows from the well-known fact that the trace of a matrix is invariant under a similarity transformation. In the absence of spin–orbit interaction, all the members of a multiplet have the same energy. So we can use this rule to write a set of linear equations, one for each $M_S M_L$ cell, equating the sum of the energies of the multiplets having states in that cell to the sum of the diagonal matrix elements of the Hamiltonian for the complete set in that cell. For example, we can denote the energy of the ${}^1D$ term by $E({}^1D)$, this energy being degenerate in M_S and M_L. Correspondingly, the cell $M_L=2$, $M_S=0$ contains the complete set (1^+1^-). This implies only one diagonal matrix element which we write, for brevity, in the form $H(1^+1^-)$. In general, this would stand for the diagonal element H_{ii} where the ith state in the cell is (1^+1^-).

Using this notation, we write the linear equations for the p^2 configuration

$$
\begin{aligned}
&E({}^1D)=H(1^+1^-),\\
&E({}^3P)=H(1^+0^+),\\
&E({}^3P)+E({}^1D)=H(1^+0^-)+H(1^-0^+),\\
&E({}^3P)=H(1^-0^-),\\
&E({}^3P)=H(1^+-1^-),\\
&E({}^3P)+E({}^1D)+E({}^1S)=H(1^+-1^-)+H(1^--1^+)+H(0^+0^-),\\
&E({}^3P)=H(1^--1^-),\\
&E({}^3P)=H(0^+-1^+),\\
&E({}^3P)+E({}^1D)=H(0^+-1^-)+H(0^--1^+),\\
&E({}^3P)=H(0^--1^-),\\
&E({}^1D)=H(-1^+-1^-).
\end{aligned}
\tag{5.78}
$$

Solving these equations for the term energies, we get

$$
\begin{aligned}
&E({}^1D)=H(1^+1^-),\ E({}^3P)=H(1^+0^+) \quad \text{and}\\
&E({}^1S)=H(1^+-1^-)+H(1^--1^+)+H(0^+0^-)-H(1^+0^+)-H(1^+1^-).
\end{aligned}
\tag{5.79}
$$

The problem now reduces to evaluating the diagonal matrix elements occurring on the right-hand side. The procedure for the electrostatic interaction has been indicated in the previous section.

The diagonal matrix element for H_0+H_1' for the (1^+1^-) state may be written

$$H(1^+1^-)=E_0+F^0(Np,Np)+\tfrac{1}{25}F^2(Np,Np). \tag{5.80}$$

Here E_0 is the contribution from H_0, i.e. due to the nucleus and the other electrons. It has a fixed value for each configuration which comes out in terms of the integrals $I(Nl)$ defined in eqn (4.16). In the present case we have

$$E_0=2I(Np). \tag{5.81}$$

Similarly,

$$H(1^+0^+)=E_0+F^0(Np,Np)-\tfrac{1}{25}F^2(Np,Np). \tag{5.82}$$

Proceeding in this manner and solving for $E(^1D)$, $E(^3P)$ and $E(^1S)$, we get, finally, for the p^2 configuration,

$$\begin{aligned} E(^1S) &= E_0 + F^0 + \tfrac{2}{5}F^2, \\ E(^1D) &= E_0 + F^0 + \tfrac{1}{25}F^2, \\ E(^3P) &= E_0 + F^0 - \tfrac{1}{5}F^2. \end{aligned} \tag{5.83}$$

This is immediately applicable to the $(2p)^{-2}$ configuration for KLL Auger spectra if we write $N = 2$. Calculation of the spin–orbit interaction energy proceeds along the lines already indicated.

The results obtained by Asaad and Burhop (1958) for the energies of the nine lines of the KLL spectrum using intermediate coupling are as follows:

1. $KL_1L_1(^1S_0): E(K) - 2E(L_1) - F^0(2s, 2s)$;

2. $KL_1L_2(^1P_1): E(K) - E(L_1) - E(L_2) - F^0(2s, 2p) -$
   $- \{[(\tfrac{1}{3}G^1(2s, 2p) - \tfrac{1}{4}\xi_{2p})^2 + \tfrac{1}{2}\xi_{2p}^2]^{1/2} - \tfrac{3}{4}\xi_{2p}\}$;

3. $KL_1L_3(^3P_1): E(K) - E(L_1) - E(L_3) - F^0(2s, 2p) +$
   $+ \{[(\tfrac{1}{3}G^1(2s, 2p) - \tfrac{1}{4}\xi_{2p})^2 + \tfrac{1}{2}\xi_{2p}^2]^{1/2} - \tfrac{3}{4}\xi_{2p}\}$;

4. $KL_1L_2(^3P_0): E(K) - E(L_1) - E(L_2) - F^0(2s, 2p) + \tfrac{1}{3}G^1(2s, 2p)$;

5. $KL_1L_3(^3P_2): E(K) - E(L_1) - E(L_3) - F^0(2s, 2p) + \tfrac{1}{3}G^1(2s, 2p)$

6. $KL_2L_3(^1D_2): E(K) - E(L_2) - E(L_3) - F^0(2p, 2p) + \tfrac{2}{25}F^2(2p, 2p) -$
   $- \{[(\tfrac{3}{25}F^2(2p, 2p) + \tfrac{1}{4}\xi_{2p})^2 + \tfrac{1}{2}\xi_{2p}^2]^{1/2} - \tfrac{3}{4}\xi_{2p}\}$;

7. $KL_3L_3(^3P_2): E(K) - E(L_3) - E(L_3) - F^0(2p, 2p) + \tfrac{2}{25}F^2(2p, 2p) +$
   $+ \{[(\tfrac{3}{25}F^2(2p, 2p) + \tfrac{1}{4}\xi_{2p})^2 + \tfrac{1}{2}\xi_{2p}]^{1/2} - \tfrac{3}{4}\xi_{2p}\}$.

8. $KL_3L_3(^3P_0): E(K) - E(L_3) - E(L_3) - F^0(2p, 2p) - \tfrac{1}{10}F^2(2p, 2p) +$
   $+ \{[(\tfrac{3}{10}F^2(2p, 2p) - \tfrac{1}{2}\xi_{2p})^2 + 2\xi_{2p}^2]^{1/2} - \tfrac{3}{2}\xi_{2p}\}$;

9. $KL_2L_2(^1S_0): E(K) - E(L_2) - E(L_2) - F^0(2p, 2p) - \tfrac{1}{10}F^2(2p, 2p) -$
   $- \{[\tfrac{3}{10}F^2(2p, 2p) - \tfrac{1}{2}\xi_{2p})^2 + 2\xi_{2p}^2]^{1/2} - \tfrac{3}{2}\xi_{2p}\}$. (5.84)

By fitting the expressions (5.84) to observed K-Auger energies of $Z = 83$ (Mladjenovic and Slätis, 1955), Asaad and Burhop (1958) sought to determine the Slater integrals $F^0(2s, 2s)$, $F^0(2s, 2p)$, $F^0(2p, 2p)$, $F^2(2p, 2p)$ and $G^1(2s, 2p)$ as functions of Z. Taking relativity and screening into account, they suggested a Z-dependence of the type $A(Z - \sigma)(1 - \alpha Z^2)$, where σ is a screening parameter and $(1-\alpha Z^2)$ the relativistic correction. Expressing all energies in keV, this gives

$$\begin{aligned} F^0(20,20) &= 0.00408(Z-3.8)(1-1.76\times 10^{-5}Z^2); \\ F^0(21,20) &= 0.00441(Z-4.0)(1-3.25\times 10^{-5}Z^2); \\ G^1(21,20) &= 0.00239(Z-2.6); \\ F^0(21,21) &= 0.00494(Z-4.5)(1-2.5\times 10^{-5}Z^2); \\ F^2(21,21) &= 0.00239(Z-5.1). \end{aligned} \tag{5.85}$$

It was shown by Hörnfeldt *et al.* (1963) and by Fahlman and Nordling (1964) that eqn (5.85) does not account with sufficient accuracy for the variation of *KLL*-Auger energies with Z. Hörnfeldt (1963) showed that better agreement could be obtained with experimental data by using a factor $(1+\beta Z^3)$ instead of $(1-\alpha Z^2)$ in these expressions. Using a least-squares fit of the constants σ, A and β shown in Table 6 good agreement was obtained between theory and available experimental data for the range $20 \leqslant Z \leqslant 100$. Agreement for the $KL_2L_3(^1D_2)$ line was better than 0.05% in most cases.

Table 6. Parameters for Slater integrals used in semi-empirical calculations of *KLL* Auger energies (Hörnfeldt, 1963)

Integral	σ	A (in keV)	β
$F^0(2s,2s)$	3.8	2.10×10^{-3}	1.40×10^{-6}
$F^0(2s,2p)$	4.0	2.30×10^{-3}	1.10×10^{-6}
$G^1(2s,2p)$	2.6	2.66×10^{-3}	—
$F^0(2p,2p)$	4.5	2.70×10^{-3}	0.80×10^{-6}
$F^2(2p,2p)$, $G^2(2p,2p)$	5.1	2.75×10^{-3}	—

Calculations have also been made using the relativistic self-consistent-field method for $Z=12$, 19 and 29 (Siegbahn *et al.*, 1967b). The difference with experiment decreases with increasing Z. For $Z=12$, the difference is about 10 eV. Part of this discrepancy may be due to solid state effects, since measurements were carried out for atoms in a solid chemical compound. This is borne out by the rather close agreement for neon (to within about 4 eV) shown in Table 7. In contrast, the difference with semi-empirical values is about 13 eV.

Asaad (1967) calculated the Slater integrals relevant to *K*- and *L*-shell Auger transitions using unscreened relativistic hydrogenic wave functions. He showed that these integrals may be expressed in the form $AZ(1+\alpha Z^2)$. The factor $(1+\alpha Z^2)$ appears to be more plausible than $(1+\beta z^3)$, particularly in view of the fact that most expansions of relativistic wave functions in terms of the parameter αZ have $\alpha^2 Z^2$ as the first-order term (see Chapter 3).

An interesting point was made by Nestor *et al.* (1966), who compared the importance of relativity with that of exchange on the calculation of *Ns*-shell binding energies for a number of elements in the range $10 \leqslant Z \leqslant 79$. For binding energies above 2 keV, they found that the effect of relativity was the dominant one. Even for outer shells, this effect was not negligible. This seems to indicate that it is preferable to compute Auger energies and transition probabilities with relativistic wave functions even for light atoms.

Table 7. Comparison of experimental values of *KLL* Auger energies of neon. (From K. D. Sevier, *Low Energy Electron Spectrometry*)

Transition	Experiment		Theory		Semi-empirical
	Körber and Mehlhorn (1966)	Siegbahn *et al.* (1967b)	Nestor *et al.* (1966)	Siegbahn *et al.* (1967b)	Sevier (1972)
$KL_1L_1(^1S)$	748.05 (42)	748.1 (1)	746.91	748.13	761
$KL_1L_{2,3}(^1P)$	771.45 (42)	771.6 (1)	770.81	775.81	781
$KL_1L_{2,3}$	781.95 (42)	781.9 (1)	782.95	786.36	794
$KL_{2,3}L_{2,3}(^1S)$	800.45 (42)	800.5 (1)	800.90	804.47	808
$KL_{2,3}L_{2,3}(^1D)$	804.15 (40)	804.15	805.98	808.52	813

Chapter Six
The Many-Electron Atom

1. Introduction

Throughout our account of Auger transitions so far, we have been dealing with many-electron systems without explicitly bringing in some of the features peculiar to such systems. We could do this because we emphasized an approximate symmetry of the many-electron atom, namely rotational symmetry. This in its turn helped us to deal with at least one many-body aspect of the problem—the addition of angular momenta. We must now look into the features which distinguish this system from the hydrogen atom and make the radial part of its solution much more complicated. At the same time, we must consider a symmetry which belongs specifically to a many-particle system, namely, symmetry under permutations of the particles. In atomic physics, this symmetry is embodied in the Pauli exclusion principle and the resulting Fermi–Dirac statistics.

An important question about the many-electron atom is: what happens to the assumption of a central potential? So far as an atom with a complete-shell configuration is concerned, i.e. for an atom with no incomplete electronic shells, the central-field character continues to hold. Delbrück (1930) showed that, if we adopt the approximation of writing the many-electron wave function as a product of one-electron wave functions, then for a complete-shell configuration the use of central-field wave functions is required by the rotational symmetry of the atom as a whole. The proof is simple. A configuration of complete shells gives a ^1S state, and such a state is non-degenerate. Let us anticipate matters a little and write the total wave function Ψ as a Slater determinant,

$$\Psi = \begin{Vmatrix} \varphi_1(x_1) & \cdots\cdots & \varphi_1(x_n) \\ \cdots & \cdots & \cdots \\ \cdots & \cdots & \cdots \\ \cdots & \cdots & \cdots \\ \cdots & \cdots & \cdots \\ \cdots & \cdots & \cdots \\ \varphi_n(x_1) & \cdots\cdots & \varphi_n(x_n) \end{Vmatrix}, \tag{6.1}$$

where the $\varphi_i(x)$ are one-electron wave functions. If $\Psi(\mathbf{r}_1,\mathbf{r}_2,\ldots,\mathbf{r}_n)$ represents a ${}^1S$ state, then it must be spherically symmetrical. Otherwise the states $O\Psi = \Psi'(\mathbf{r}_1',\mathbf{r}_2',\ldots\mathbf{r}_n') \neq \Psi(\mathbf{r}_1',\mathbf{r}_2',\ldots\mathbf{r}_n')$ (where O is any rotation operator and $\mathbf{r}_1', \mathbf{r}_2', \ldots \mathbf{r}_n'$ are the rotated coordinates of the electrons) would belong to the same eigenvalue of energy and therefore would have to be space-degenerate, which contradicts the definition of a ${}^1S$ state. In fact, Delbrück showed that the transformation properties of the spherical harmonics under a rotation are just such as to give the determinant (6.1) this isotropic property, provided of course that the one-electron functions $\varphi_1(x), \ldots, \varphi_n(x)$ are of the central-field type (so that one may separate them in the form $R(r)\, Y_l^m(\theta\varphi)\,\chi^{1/2}$).

So much for a closed-shell configuration. For an n-electron atom with an incomplete shell, both L and S can be different from zero. Neglecting spin–orbit interaction, we get a number W of linearly independent states, given by

$$W = \sum_{LS} (2L+1)(2S+1), \tag{6.2}$$

where the sum extends over allowed values of L and S. If we write one of these states as a single determinant Ψ, the operator $\mathcal{O}$ will change it into a function Ψ' which is a linear combination of determinants:

$$\mathcal{O}\Psi = \Psi' = \sum_{i=1}^{W} c_i \Psi_i. \tag{6.3}$$

To obtain a complete set of basis functions for the system, we must include W such linearly independent combinations of determinants. Obviously, such a system does not have rotational symmetry. For example, the carbon atom with the ground state configuration $(1s)^2(2s)^2(2p)^2$ has an incomplete $2p$ shell. Of the six different single-particle orbitals for the $2p$ shell, only two can be included in a single determinant of the type Ψ_i. We have, therefore, $6!/2!4! = 15$ different ways of choosing these $2p$ orbitals. Correspondingly, we get 15 different determinantal wave functions Ψ_i. This is exactly what we get by summing over the allowed terms ${}^1D$, ${}^3P$ and ${}^1S$ in eqn (6.2).

For open-shell configurations, one must, therefore, either try to work with a non-central potential, or perform a spherical averaging over the incomplete shell to obtain one-electron orbitals of the central-field type. This is done by writing

$$V(r) = \frac{1}{4\pi}\int U(\mathbf{r})\,\mathrm{d}\Omega, \tag{6.4}$$

where $U(\mathbf{r})$ is a non-central potential which incorporates the effect of the presence of other electrons. Because of the advantages of the central-field approximation, this is the more popular procedure; for atomic systems, this is a much milder approximation than in nuclei (Bethe and Jackiw, 1968).

2. Symmetries of the many-electron system

Consider a many-electron atom described by the Hamiltonian

$$H = -\frac{\hbar^2}{2\mu}\sum_j \nabla_j^2 - \sum_j \frac{Ze^2}{r_j} + \sum_{i<j} \frac{e^2}{r_{ij}}. \tag{6.5}$$

The Schrödinger equation is

$$H\Psi = E\Psi, \tag{6.6}$$

Ψ being a many-electron wave function. To be able to write Ψ in terms of single-electron wave functions, we make the independent-particle approximation, namely that each electron may be regarded as being subject to an equivalent potential due to the other electrons and the nucleus. In other words, we approximate, for a given j, the terms $-Ze^2/r_j + \sum_{i\neq j} e^2/r_{ij}$ by an equivalent central potential $V(r_j)$. We now write a trial function for the many-electron system:

$$\Phi = \varphi_a(1)\,\varphi_b(2)\ldots\varphi_p(n), \tag{6.7}$$

where $\varphi_a, \varphi_b, \ldots, \varphi_p$ are single-particle orbitals including spin and the numbers 1 to n label the electrons. At this stage we can only say that Ψ may be expressed in terms of product functions of the type Φ.

Looking at the Hamiltonian H, we can see that it is invariant under permutations of the particle labels, i.e. under the operation $P_{ij}: i \rightleftarrows j$. Hence the many-electron function Ψ should depend on the single-electron orbitals $\varphi_a(j)$ in such a way that permuting the space and spin coordinates of two electrons i and j would give another many-electron function $P_{ij}\Psi$ belonging to the same energy eigenvalue E. Apparently, this would imply an equivalence degeneracy of the system. Corresponding to the $n!$ different permutations, there would seem to be an $n!$-fold degeneracy of each eigenvalue E. However, this is not the case. To see how this goes through, we consider first the non-degenerate case

$$P_{ij}\Psi = c_{ij}\Psi, \tag{6.8}$$

where c_{ij} is a constant. Iterating eqn (6.8) with P_{ij},

$$P_{ij}(P_{ij}\Psi) = c_{ij}P_{ij}\Psi = c_{ij}^2\Psi. \tag{6.9}$$

But the effect of this iteration should be to reproduce the original function Ψ. Hence

$$c_{ij}^2 = 1, \quad \text{i.e.} \quad c_{ij} = \pm 1. \tag{6.10}$$

Thus we get either symmetric or antisymmetric states depending on whether $c_{ij} = 1$ or -1. For degenerate eigenvalues, $P_{ij}\Psi \neq \pm\Psi$, and we can form the non-vanishing linear combinations $\Psi + P_{im}\Psi$ and $\Psi - P_{im}\Psi$. One of them is symmetric and the other antisymmetric, so that again $c_{ij} = \pm 1$.

If we choose $c_{ij} = -1$, and write

$$\Phi_{12} = \varphi_a(1)\,\varphi_b(2)\ldots\varphi_p(n),$$
$$\Phi_{21} = \varphi_a(2)\,\varphi_b(1)\ldots\varphi_p(n),$$

then

$$\Phi_{21} = -\Phi_{12}.$$

Now, if we take the states a and b to be identical, then

$$\Phi_{21} = \Phi_{12},$$

and

$$\Phi_{12} = 0.$$

This is the Pauli exclusion principle. From statistical considerations, we know that electrons obey this principle and have antisymmetric wave functions. The unique many-electron wave function which is totally antisymmetric under permutations and contains products of single-particle orbitals as in eqn (6.7) is the Slater determinant

$$\Psi_A = \left\| \begin{matrix} \varphi_a(1)\,\varphi_a(2)\ldots\varphi_a(n) \\ \varphi_b(1)\,\varphi_b(2)\ldots\varphi_b(n) \\ \vdots \quad \cdot \quad \cdots \quad \vdots \\ \varphi_p(1)\,\varphi_p(2)\ldots\varphi_p(n) \end{matrix} \right\| . \tag{6.11}$$

The Pauli principle removes the apparent equivalence degeneracy of the system by admitting only one linear combination of products of the type (6.7) as a valid wave function, namely, the antisymmetric combination (6.11). Obviously, Ψ_A is an approximation to the wave function Ψ. It vanishes if $\varphi_a = \varphi_b$, for any two single-particle states a and b. The latter may be taken to form an orthonormal set without vitiating any of the above arguments.

Apart from permutation symmetry, all many-electron functions of a definite symmetry type, for all values of L and its projection M_L, transform irreducibly under the linear group. Consider, for example, the single-electron space orbital Ψ_{lm}. Under a three-dimensional rotation $\mathcal{O}$ this transforms as

$$\mathcal{O}\Psi_{lm} = \Psi'_{lm} = \sum_{m_1} \Psi_{lm_1}\, D^{(l)}_{m_1 m}(\alpha, \beta, \gamma), \tag{6.12}$$

where $D^{(l)}_{m_1 m}(\alpha, \beta, \gamma)$ are elements of the rotation matrix corresponding to a rotation described by the Euler angles (α, β, γ). Since the transformation (6.12) is linear, it induces a linear transformation of the many-electron function Ψ_A. Also, because $\mathcal{O}$ commutes with the permutation P_{ij}, the resulting linear transformation preserves the symmetry type.

3. Techniques for the many-electron atoms

Having accepted the separable wave function approximation with the central field model and the Slater determinant involving central-field orbitals, we must start worrying about the shape of the single-particle orbital. The angular part is fixed by the central-field approximation, so we have to consider the radial function $R_{Nl}(r)$. In a many-electron atom this means we have to realistically determine the effective screening potential seen by the individual electron. To approximate the latter by means of a screened Coulomb potential is no more adequate.

The search for better radial functions, therefore, constitutes the heart of the many-electron approach. This is a more open-ended game than that of Wigner algebra and coupling schemes, because here symmetry-principles by themselves do not provide adequate guidance. Nevertheless, we do have certain basic principles to fall back on. Essentially, they are:

(a) self-consistency, leading to iterative procedures, and

(b) the variation principle.

A. Self-consistency

In the central-field approximation, the radial part of the single-particle orbital satisfies the Schrödinger equation

$$\frac{\hbar^2}{2\mu}\frac{\mathrm{d}^2 R_{Nl}(r)}{\mathrm{d}r^2}+\left[\varepsilon_{Nl}-V_{Nl}(r)-\frac{l(l+1)}{2\mu r^2}\hbar^2\right]R_{Nl}(r)=0. \tag{6.13}$$

Here ε_{Nl} is the single particle energy eigenvalue. The interesting point about this equation is that the potential $V_{Nl}(r)$ seen by an electron in the orbital (Nl) depends on this orbital. To solve this equation, Hartree (1928) suggested the following procedure. By associating a charge density $\rho_{Nl}(r)=-4\pi e R_{Nl}^2(r)$ with each electron, we could construct a tentative potential from the total charge distribution due to the other electrons and the nucleus. Solving eqn (6.13), we could obtain solutions $R_{Nl}(r)$ which now determine a new potential. This procedure is continued until the final wave function is self-consistent to a high order of accuracy.

B. Variation principle

It was subsequently shown by Hartree (1928, 1957) that the criterion of self-consistency is equivalent to a variation calculation which minimizes the energy of the system. The trial function for this minimization is taken to be a product of single-particle orbitals, so that correlations between electrons are neglected.

The minimization implies that the resulting many-particle function represents the ground state of the system. Calculation of excited states is more complicated. For the ground state, there are two alternative but equivalent ways of effecting the minimization.

1. One varies the parameter

$$E' = \frac{\int \Phi^* H \Phi \, d\tau}{\int \Phi^* \Phi \, d\tau}, \tag{6.14a}$$

setting

$$\delta E' = 0,$$

where Φ is a trial function for the many-particle system. If Φ be the correct choice, E' represents the ground-state energy of the system.

2. Alternatively, one demands that

$$\delta\left[\int \Phi^* H \Phi \, d\tau - \lambda \int \Phi^* \Phi \, d\tau\right] = 0, \tag{6.14b}$$

λ being a Lagrange parameter which may be identified with the energy for the correct choice Φ.

In Hartree's original treatment, the requirement of antisymmetry was not included, the Pauli principle being satisfied by choosing the single-electron orbitals in an appropriate manner. However, antisymmetry can be explicitly incorporated by imposing the variation condition

$$\delta \int \Phi^* H \Phi \, d\tau = 0, \tag{6.14c}$$

where the trial function Φ is the Slater determinant

$$\Phi = \frac{1}{(n!)^{1/2}} \begin{Vmatrix} \varphi_i(1) & \varphi_i(2) \dots \varphi_i(n) \\ \varphi_j(1) & \varphi_j(2) \dots \varphi_j(n) \\ \vdots & \vdots \\ \varphi_n(1) & \varphi_n(2) \dots \varphi_n(n) \end{Vmatrix}. \tag{6.15}$$

The indices i, j, etc. stand for the complete sets of quantum numbers labelling the single-particle orbitals φ. The factor $1/(n!)^{1/2}$ ensures that Φ is automatically normalized, provided that the orbitals φ_i are orthonormal.

$$\int \varphi_i^*(1)\, \varphi_j(1)\, d\mathbf{r}_1 = \delta_{ij}. \tag{6.16}$$

Actually, since the spin-functions which form part of φ_i are orthogonal for different spins, eqn (6.16) reduces to the condition that the space functions corresponding to the same spin be orthonormal. Equation (6.14) then yields the non-linear integro-differential equation

$$-\frac{\hbar^2}{2\mu}\nabla^2\varphi_i(\mathbf{r})-\frac{Ze^2}{r}\varphi_i(\mathbf{r})+\left[\sum_j\int d\mathbf{r}'\frac{e^2}{|\mathbf{r}-\mathbf{r}'|}|\varphi_j(\mathbf{r}')|^2\right]\varphi_i(\mathbf{r})-$$

$$-\sum_j\delta_{m_{s_i},m_{s_j}}\left[\int d\mathbf{r}'\frac{e^2}{|\mathbf{r}-\mathbf{r}'|}\varphi_j^*(\mathbf{r}')\varphi_i(\mathbf{r}')\right]\varphi_j(\mathbf{r})=\varepsilon_i\varphi_i(\mathbf{r}), \tag{6.17}$$

where $\varphi_i(\mathbf{r})$ is the space-part of the orbital corresponding to the single-particle state i. This is the Hartree–Fock equation for the single-particle orbitals (Slater, 1960). The first term within square brackets on the left-hand side represents the potential energy of the ith electron due to the other electrons. The second term within square brackets represents the "exchange" potential energy which arises specifically from antisymmetry, i.e. from use of the Slater determinant (6.15). The Kronecker symbol, $\delta_{m_{s_i},m_{s_j}}$, tells us that exchange effects do not exist for identical particles with different spin components. It can be seen that the exchange term represents a non-local potential. In fact, eqn (6.17) may be written in the form of a Schrödinger equation:

$$\frac{\hbar^2}{2\mu}\nabla^2\varphi_i(\mathbf{r})+[\varepsilon_i-V(\mathbf{r})]\varphi_i(\mathbf{r})-\int U(\mathbf{r},\mathbf{r}')\varphi_i(\mathbf{r}')\,d\mathbf{r}'=0, \tag{6.18}$$

where $V(\mathbf{r})$ is a local potential energy

$$V(\mathbf{r})=-\frac{Ze^2}{r}+\sum_{j=1}^{n}\int|\varphi_j(\mathbf{r}')|^2\frac{e^2}{|\mathbf{r}-\mathbf{r}'|}\,d\mathbf{r}', \tag{6.19}$$

and $U(\mathbf{r},\mathbf{r}')$ the non-local potential energy

$$U(\mathbf{r},\mathbf{r}')=-\frac{e^2}{|\mathbf{r}-\mathbf{r}'|}\sum_{j=1}^{n}\varphi_j^*(\mathbf{r}')\varphi_j(\mathbf{r}). \tag{6.20}$$

The Kronecker delta is suppressed with the understanding that $U\neq0$ only if $m_{s_i}=m_{s_j}$. Obviously,

$$U(\mathbf{r},\mathbf{r}')=U^*(\mathbf{r}',\mathbf{r}). \tag{6.21}$$

i.e. U is Hermitian. It can be easily shown that the solutions belonging to different eigenvalues are orthogonal. We can give a physical interpretation of this exchange potential energy which ultimately helps to simplify the problem (Bethe and Jackiw, 1968).

We define the density matrix

$$\rho(\mathbf{r},\mathbf{r}')=-\sum_{j=1}^{n}\varphi_j^*(\mathbf{r}')\varphi_j(\mathbf{r}). \tag{6.22}$$

If all the orbitals are occupied, we have the closure relation

$$\rho(\mathbf{r},\mathbf{r}') = -\sum_{j=1}^{\infty} \varphi_j^*(\mathbf{r}')\,\varphi_j(\mathbf{r}) = -\delta(\mathbf{r}'-\mathbf{r}), \tag{6.23}$$

so that

$$\int \rho(\mathbf{r},\mathbf{r}')\,\mathrm{d}\mathbf{r}' = -1. \tag{6.24}$$

Even when all the single-particle states are not occupied, this turns out to be a good approximation, $\rho(\mathbf{r},\mathbf{r}')$ being a function which peaks sharply at $\mathbf{r}=\mathbf{r}'$.

We can, therefore, write

$$U(\mathbf{r},\mathbf{r}') = -\frac{(-e)}{|\mathbf{r}-\mathbf{r}'|}\,e\rho(\mathbf{r},\mathbf{r}'), \tag{6.25}$$

i.e. the exchange potential energy may be attributed to a charge density $-e\rho(\mathbf{r},\mathbf{r}')$ which integrates to e, i.e. to the opposite of an electronic charge. In other words, the total potential energy of the electron in the orbital φ_i is as follows:

1. The potential energy $-Ze^2/r$ due to the Coulomb field of the nucleus,
2. the local potential energy $V(\mathbf{r})$ due to all the electrons with spin parallel or antiparallel to it, and
3. the non-local "exchange" potential energy $U(\mathbf{r},\mathbf{r}')$ due to a "hole" e with spin parallel to it.

Thus the exchange potential is a direct result of the Pauli principle which excludes two electrons of the same spin from being in the immediate neighbourhood of each other. It has the nature of a Fermi hole.

It was suggested by Slater (1951) that this fact may be used to simplify the solution of eqn (6.17). To do this, one defines an effective potential energy $V_{\text{eff}}(\mathbf{r})$ by

$$V_{\text{eff}}(\mathbf{r})\,\varphi_i(\mathbf{r}) = \int U(\mathbf{r},\mathbf{r}')\,\varphi_i(\mathbf{r}')\,\mathrm{d}\mathbf{r}'. \tag{6.26}$$

This implies that $V_{\text{eff}}(\mathbf{r})$ has the form

$$V_{\text{eff}}(\mathbf{r}) = e\int \frac{\rho(\mathbf{r},\mathbf{r}')}{|\mathbf{r}-\mathbf{r}'|}\frac{\varphi_i(\mathbf{r}')}{\varphi_i(\mathbf{r})}\,\mathrm{d}\mathbf{r}'. \tag{6.27}$$

Obviously, this is a local operator.

The effective potential energy $V_{\text{eff}}(\mathbf{r})$ still depends on the orbital i, so that electrons in different orbitals see different potentials. A universal exchange potential for all the orbitals in an atom may be introduced by using the Thomas–Fermi–Dirac model which we shall discuss in Section 6. The exchange potential seen by electrons in a large atom may be approximated by that in a free-electron gas. Consider electrons moving in a constant potential, the wave functions φ_1 become plane waves,

$$\varphi_i(\mathbf{r}) = \frac{1}{\sqrt{v}} \, \mathrm{e}^{i\mathbf{k}_i \cdot \mathbf{r}}, \tag{6.28}$$

v being the normalization volume. Then

$$\rho(\mathbf{r}, \mathbf{r}') = -\frac{1}{v} \sum_i \mathrm{e}^{i\mathbf{k}_i \cdot (\mathbf{r}-\mathbf{r}')} \xrightarrow[\text{large } v]{} -\frac{1}{(2\pi)^3} \int \mathrm{e}^{i\mathbf{k}_i \cdot \mathbf{R}} \, d\mathbf{k} =$$

$$= -\frac{4\pi}{(2\pi)^3} \int_0^{k_F} \frac{\sin kR}{kR} k^2 \, dk = -\frac{1}{2\pi^2 R^3} (\sin k_F R - k_F R \cos k_F R), \tag{6.29}$$

where $\mathbf{R} = \mathbf{r} - \mathbf{r}'$, $R = |\mathbf{R}|$ and k_F is the Fermi wave number.

It may be seen that this quantity behaves like a Dirac delta-function, peaking sharply at $R = 0$. Thus

$$\rho(\mathbf{r}, \mathbf{r}') = -\frac{1}{2\pi^2} \lim_{R \to 0} \frac{1}{R^3} (\sin k_F R - k_F R \cos k_F R) =$$

$$= -\frac{1}{2\pi^2} \lim_{R \to 0} \frac{k_F^2 R \sin k_F R}{3R^2} = -\frac{k_F^3}{6\pi^2} = -\frac{\rho}{2}, \tag{6.30}$$

where ρ is the Fermi gas density at $\mathbf{r} = \mathbf{r}'$. We now define the averaged effective potential

$$\bar{V}_{\text{eff}} = \frac{\int_0^{k_F} V_{\text{eff}} \, d\mathbf{k}}{\int_0^{k_F} d\mathbf{k}} = -\frac{2e^2}{\rho} \int \frac{[\rho(\mathbf{r}, \mathbf{r}')]^2}{R} \, d\mathbf{r}',$$

where we have used eqns (6.27) and (6.28). Using eqns (6.29) and (6.30), this gives

$$\bar{V}_{\text{eff}} = -\frac{3e^2}{2\pi} k_F = -\frac{3}{2} e^2 \left(\frac{3}{\pi} \rho \right)^{1/3}. \tag{6.31}$$

Using this averaged quantity in eqn (6.18) we get

$$\left[-\frac{\hbar^2}{2\mu} \nabla^2 - \frac{Ze^2}{r} + e^2 \int d\mathbf{r}' \frac{\rho(\mathbf{r}')}{|\mathbf{r} - \mathbf{r}'|} - \frac{3}{2} e^2 \left(\frac{3}{\pi} \rho(\mathbf{r}) \right)^{1/3} - \varepsilon_i \right] \varphi_i(\mathbf{r}) = 0. \tag{6.32}$$

This is the Hartree–Fock–Slater equation with a common local potential for all orbitals.

So far we have been writing the self-consistent-field equations in their three-dimensional form, $\varphi_i(\mathbf{r})$ being the space part of the single-particle orbital. To be able to separate the radial and angular parts, we recall our discussion of Section 1, namely, that a closed-shell atom is spherically symmetric (this may be explicitly shown for the Hartree–Fock equations) and that for incomplete shells we have to perform the spherical averaging defined in eqn (6.4) to get a central potential.

Once a central potential is obtained, we can write

$$\begin{aligned}\varphi_i(\mathbf{r}) &= R_{N_i l_i}(r)\, Y_{l_i}^{m_{li}}(\theta,\varphi)\\ &= \frac{P_{N_i l_i}(r)}{r}\, Y_{l_i}^{m_{li}}(\theta,\varphi).\end{aligned} \tag{6.33}$$

The complete single-particle orbital is obtained by adjoining the spinor $\chi_{s_i}^{m_{si}}$ to this function. The function $P_{N_i l_i}(r)$ is normalized to

$$\int_0^\infty P_{N_i l_i}^2(r)\,\mathrm{d}r = 1. \tag{6.34}$$

With the orbitals given by eqn (6.33), we can formally separate the radial and angular parts of the Hartree–Fock eqn (6.17) to get the radial eqn (Slater, 1960),

$$\begin{aligned}&\left[-\frac{\hbar^2}{2\mu}\frac{\mathrm{d}^2}{\mathrm{d}r^2}-\frac{Ze^2}{r}+\frac{l_i(l_i+1)\hbar^2}{2\mu r^2}+\sum_{N_j l_j} 2(2l_j+1)\frac{1}{r}\,Y_0(N_j l_j, N_j l_j; r)\right]P_{N_i l_i}(r)-\\ &-\sum_{N_l l_l}\left[\frac{2l_j+1}{2l_i+1}\right]^{1/2}\sum_k c^k(l_i 0, l_j 0)\frac{1}{r}\,Y_k(N_i l_i, N_j l_j; r)P_{N_j l_j}(r)=\varepsilon_i P_{N_i l_i}(r).\end{aligned} \tag{6.35}$$

In eqn (6.35), the quantity $Y_k(N_i l_i, N_j l_j, r)$ is defined by

$$Y_k(N_i l_i, N_j l_j; r) = e^2 r\int_0^\infty \gamma_k(r,r')P_{N_i l_i}(r')P_{N_j l_j}(r')\,\mathrm{d}r'. \tag{6.36}$$

γ_k has already been defined in eqn (5.9). The coefficient $c^k(l_i 0, l_j 0)$ arises from the angular integration and is given by

$$\begin{aligned}c^k(lm, l'm') &= \left(\frac{4\pi}{2k+1}\right)^{1/2}\int \overset{*}{Y_l^m}(\theta,\varphi)\, Y_{l'}^{m'}(\theta,\varphi)\, Y_k^{m-m'}(\theta,\varphi)\,\mathrm{d}\Omega\\ &= (-)^{m'}\frac{[(2l+1)(2l'+1)]^{1/2}}{2k+1}\,C^{l\ \ l'\ \ k}_{-m\, m'\, m'-m}\,C^{l\, l'\, k}_{0\, 0\, 0}.\end{aligned} \tag{6.37}$$

Equation (6.35) is the more commonly used form of the Hartree–Fock equations, because one is really interested in the radial part of the wave function. The last term on the left-hand side represents exchange. This is the term which makes eqn (6.35) inhomogeneous. It can also be clearly seen that for different occupied orbitals there is a different exchange potential.

Analogously, for the Hartree–Fock–Slater eqn (6.32) one can write, either for the inherently central potential in a closed-shell atom or for the spherically averaged potential in an atom with incomplete shells,

$$V(r) = \frac{Ze}{r} U(r), \tag{6.38}$$

where

$$U(r) = 1 - \frac{1}{Z} \sum_{N_l l_l} 2(2l_j + 1)\, Y_0(N_j l_j, N_j l_j; r) + \frac{3}{2Z}\left(\frac{3}{\pi}\rho(r)\right)^{1/3}. \tag{6.39}$$

Equation (6.38) gives the total potential $V(r)$ experienced‡ by an electron in any one of the orbitals. It is both local and central in character. Orbitals $P_{Nl}(r)$ and $P_{N'l}(r)$ with the same l and different N are orthogonal. This simplified Hartree–Fock–Slater potential was used by Herman and Skillman (1963) in an extensive calculation of single-particle orbitals for the entire periodic table up to $Z = 103$.

The first self-consistent-field calculation of Auger transitions was carried out by Rubenstein and Snyder (1955). They used Hartree wave functions with LS coupling to compute K, L and M shell transition rates for $Z = 18$, 36 and 47. We included their results for KLL group intensities in our comparison of theory and experiment given in Chapter 5. Mehlhorn (1968) used radial integrals given by Rubenstein (1955) to obtain $L_1L_{2,3}M_1$, $L_1L_{2,3}M_{2,3}$ and L_1MM transition rates in argon with intermediate coupling and configuration interaction. Coefficients for the latter were computed with Herman–Skillman wave functions. The line intensities within a group agreed reasonably well with experiment. But intensities of the groups relative to one another differed considerably from experiment. More recently, Walters and Bhalla (1971a,b,c) used the Hartree–Fock–Slater technique to obtain non-relativistic K and L shell transition rates in LS coupling for a large number of atoms. Their K-shell calculation covers all Z-values from 4 to 54. They use an approximate exchange scheme given by Kohn and Sham (1965) and Gaspar (1954) (to be called the KSG approximation in the following) which gives a free-electron exchange density quite close to the true Hartree–Fock exchange density. This approximation consists in replacing the factor $\frac{3}{2}$ in eqn (6.31) by unity.§ Equation (6.32) is then solved numerically using the Numerov method.

‡ A word of caution is necessary at this point. In eqn (6.38), we have the potential field seen by an electron and not its potential energy.

§ This really amounts to using the expression in eqn (6.31) in the Hamiltonian and varying it to obtain a minimum rather than using $\bar{V}_{\text{eff}}$ directly in the Hartree–Fock equation (Bethe and Jackiw, 1968).

McGuire (1967, 1968, 1969a,b, 1970, 1971a,b, 1972a,b) developed an approach to the Herman–Skillman potential of eqn (6.38) which he utilized to obtain analytic solutions for Auger calculations. His idea was to approximate this potential by a piecewise continuous potential given by

$$ZU(r) \equiv Z(r) = Z_i - \Delta_i r; \; r_{i-1} \leqslant r \leqslant r_i; \quad (i = 1, 2, \ldots, N). \tag{6.40}$$

Here $r_0 = 0$, $r_N = \infty$, $\Delta_N = 0$ and $Z_N = Z$, the nuclear charge. In other words, $U(r)$ does not vary smoothly, but is approximated by a series of straight lines with varying slope Δ_i over a number of N steps from $r = 0$ to $r = \infty$. To ensure that the potential be continuous, these straight lines are required to join up at the points r_i, i.e.

$$Z_i - \Delta_i r_i = Z_{i+1} - \Delta_{i+1} r_i, \tag{6.41}$$

so that

$$Z_i - Z_{i+1} = (\Delta_i - \Delta_{i+1}) r_i. \tag{6.41a}$$

The derivative of $U(r)$ is not continuous. The step-length $r_i - r_{i-1}$ is a variable parameter, and so also is the slope Δ_i. The parameters Z_i are fitted to get the correct screening at $r = 0$ and $r = \infty$. The boundary conditions on screening are obtained from the fact that $U(r)$ represents the partial shielding of the nucleus by the other electrons. Thus $U(r)$ goes from unity at $r = 0$ (no shielding) to a much smaller value for large values of r. Finally, the parameter r_{N-1} is fitted to reproduce the energy eigenvalue of the outermost filled orbital given by Herman and Skillman (1963).

Once the parameters are fixed, the radial equation for the ith step assumes the form

$$\left(\frac{\hbar^2}{2\mu} \frac{d^2}{dr^2} + \frac{Z_i e^2}{r} - e^2 \Delta_i - \frac{l(l+1)}{r^2} + E \right) P^i_{Nl}(r) = 0, \; r_{i-1} \leqslant r \leqslant r_i. \tag{6.42}$$

This is the radial Schrödinger eqn for the Coulomb field with a constant term $e^2 \Delta_i$ which goes into the energy. The solution is expressed in terms of confluent hypergeometric functions and the methods of Chapter 3 are applicable.

McGuire's model was originally devised to calculate photoionization cross-sections (McGuire, 1967, 1968). Although his results were close to those obtained by Manson and Cooper (1968) with numerical wave functions, both calculations disagreed with experiment for elements heavier than argon near the optical threshold. The cross-sections also dropped off faster than experiment with decreasing wavelength. At higher energies, i.e. energies corresponding to K-shell Auger transitions, they agreed with experiment to within 20%.

McGuire used his approximation to compute KLL and KLM transition probabilities in LS coupling for $4 \leqslant Z \leqslant 54$ (McGuire, 1969, 1970) and L-shell transitions up to LNN and LNO in LS and jj coupling covering $11 \leqslant Z \leqslant 90$

(McGuire, 1971a). He also computed M-shell Auger rates for various Z-values in the range $20 \leqslant Z \leqslant 90$ (McGuire, 1972a), and N-shell rates for $38 \leqslant Z \leqslant 103$ (McGuire, 1974d).

4. Non-relativistic self-consistent-field calculations—Theory and experiment

Figure 6.1, taken from Walters and Bhalla (1971b), compares their theoretical values of KL_1L_1, $KL_1L_{2,3}$ and $KL_{2,3}L_{2,3}$ group intensities relative to the total KLL Auger rates with experiment. Also shown are the results of Rubenstein and Snyder (1955). These results correspond to LS coupling. For $Z \leqslant 10$, the $2p$ shell is being filled and the group intensities show a strong dependence on Z. This is a situation where the usual stipulation of a two-vacancy final state configuration in an otherwise complete shell does not hold. If the final vacancies occur in different shells, and one of these shells has n electrons rather than the number m required to complete it, the full-shell rate is reduced by a factor n/m. If the final vacancies are in the same shell, the full-shell rate is reduced by $n(n-1)/m(m-1)$ (McGuire, 1969). A more complete treatment of partially filled shells will be given in the next chapter. The agreement between

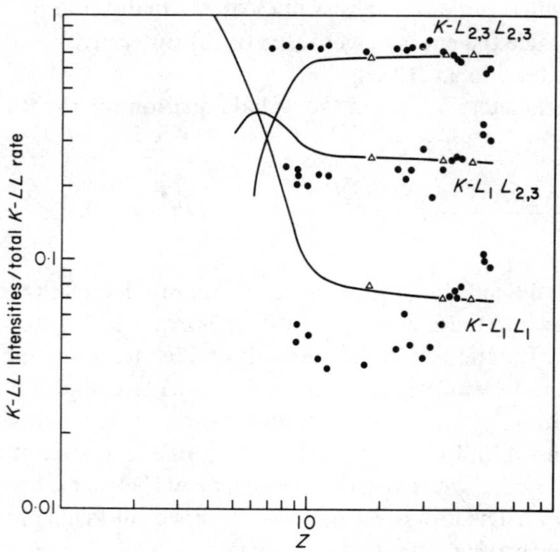

Fig. 6.1. Theoretical and experimental intensities of the groups KL_1L_1, $KL_1L_{2,3}$ and $KL_{2,3}L_{2,3}$ relative to the total KLL intensities. Solid lines represent nonrelativistic results obtained by Walters and Bhalla (1971b) using LS coupling. Triangles indicate calculations of Rubenstein (1955) and solid points experimental values (for references see Walters and Bhalla, 1971a). (From Walters and Bhalla, 1971b, courtesy of American Institute of Physics.)

theory and experiment is not very good in this region. This may be partly due to the fact that theoretical rates relate to free atoms, whereas actually the electrons involved are under the influence of chemical bonding, particularly in solids.

With the L shells filled, the theoretical relative intensities vary smoothly with increasing Z. This is in general agreement with the hydrogenic results of Callan (1961) discussed in Chapter 3. There is very good agreement between the two sets of theoretical values in the region $10 \leqslant Z \leqslant 54$. The disagreement with experiment is not unexpected, since the theoretical results do not include configuration interaction. Actually, there is a strong mixing between the $L_1L_1(^1S_0)$ and $L_2L_2(^1S_0)$ configurations so that the latter is enhanced at the expense of the former. This effect is more prominent in the region $10 \leqslant Z \leqslant 25$. The agreement with experiment is better in the region $30 \leqslant Z \leqslant 40$. At higher Z values relativistic effects begin to manifest themselves.

Figure 6.2 shows the total theoretical K-shell X-ray and Auger transition rates as functions of Z. The non-relativistic results are from the Hartree–Fock–Slater (HFS) calculations of Walters and Bhalla (1971b) and the approximate Herman–Skillman potential approach of McGuire (1970). Also included

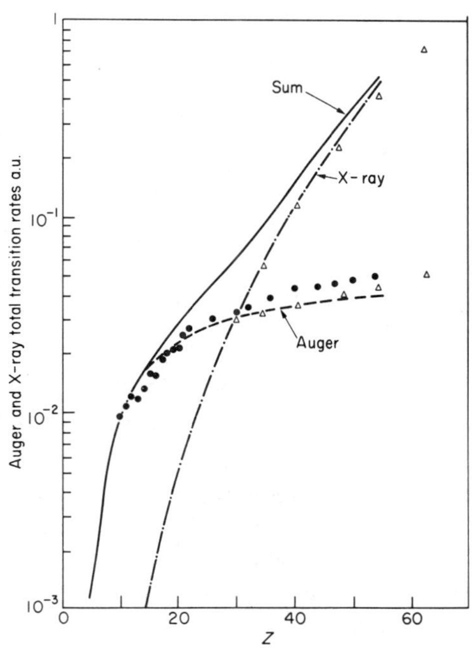

Fig. 6.2. K-shell X-ray and Auger transition rates and their sum as functions of Z. Curves represent nonrelativistic results of Walters and Bhalla (1971b), triangles the relativistic results of Bhalla and Ramsdale (1970b) and Rosner and Bhalla (1970) and solid points refer to calculations by McGuire (1970) (from Walters and Bhalla, 1971b, courtesy of American Institute of Physics).

for comparison are relativistic HFS results of Bhalla and Ramsdale (1970) and of Rosner and Bhalla (1970). The strong Z-dependence of radiative transitions as compared to Auger rates is clearly brought out. In the low Z region, Auger transitions are particularly dominant. The HFS results of Walters and Bhalla show a very smooth variation with Z, whereas those of McGuire show fluctuations. For high Z-values the difference is about 22%. The total radiative rates of McGuire are also larger by about 20% at $Z = 10$. The difference slowly decreases as we go to higher Z values.

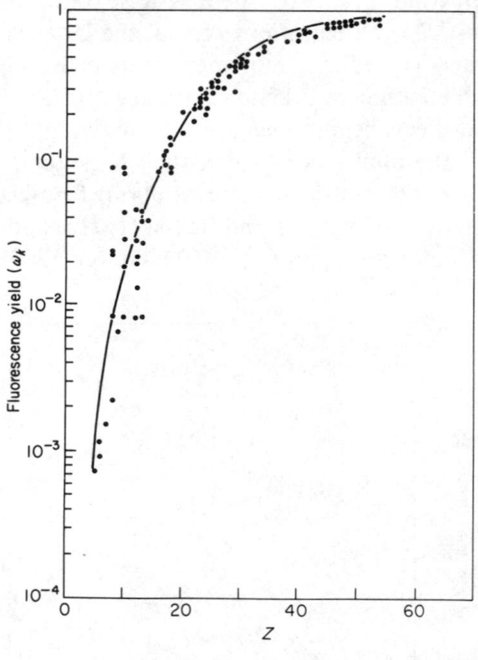

Fig. 6.3. Theoretical K-shell fluorescence yields ω_K (solid line) calculated by Walters and Bhalla (1971b) and experimental results from Fink *et al.*, (1966) and Dick and Lucas (1970) (from Walters and Bhalla, 1971b, courtesy of American Institute of Physics).

A word may be in order here about the possible implications of McGuire's model. Obviously, it amounts to a local restoration of O_4 symmetry for individual orbitals. This is brought about at the cost of discontinuities in the derivative of the potential at the points r_i. Within each potential shell $r_i < r < r_{i+1}$, the screening increases linearly. This implies a kind of uniform charge distribution which is more likely to be realized in heavier atoms. On the other hand, discontinuities at the boundaries of these shells may affect transition probabilities through their implications for charge conservation. Interestingly, McGuire's results for K-shell fluorescence yields agree quite closely with experiment, i.e. to within 8%. Agreement with regard to both Auger rates and fluorescence yields is better for heavier atoms ($18 \leqslant Z \leqslant 54$) than for lighter ones ($5 \leqslant Z \leqslant 18$).

Figure 6.3 presents *K*-shell fluorescence yields computed by Walters and Bhalla (1971b) along with experimental values (Fink *et al.* 1966; Dick and Lucas, 1970). For low atomic numbers, theoretical yields are generally larger than experimental values. The latter relate to solids and it appears that radiative rates are more strongly influenced by the environment than Auger rates. For elements up to $Z = 40$, theoretical fluorescence yields agree quite well with experiment. For heavier elements, relativistic effects become prominent. Interestingly, McGuire's results are in better agreement with the HFS results in this region (about 1% for $Z = 54$).

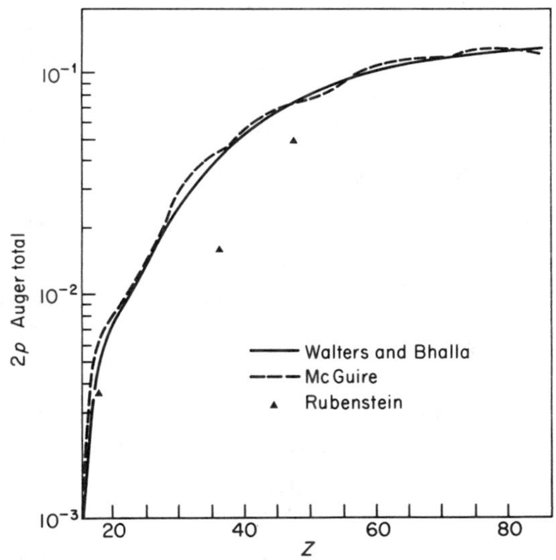

Fig. 6.4. Theoretical total $2p$ Auger transition rates in atomic units as function of Z. Solid curve represents calculations of Walters and Bhalla (1971c) and dashed curve those of McGuire (1971). Triangles represent results of Rubenstein (1955) (from Walters and Bhalla, 1971c, courtesy of American Institute of Physics).

Figure 6.4 shows total Auger transition rates for an initial vacancy in the $2p$ shell as computed by Walters and Bhalla (1971c), McGuire (1971a) and Rubenstein (1955). There is closer agreement between the former two in this case and their results differ substantially from those of Rubenstein. This may be due to errors in the central potential used by Rubenstein. Figure 6.5 shows theoretical $2p$ fluorescence yields obtained by Walters and Bhalla (1971c) and by McGuire (1971a) along with experimental data. Here too there is broad agreement between the two on the one hand and with experiment on

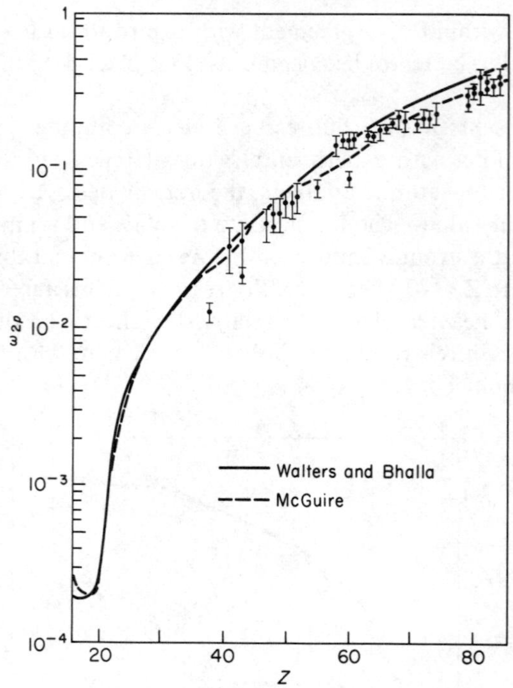

Fig. 6.5. Theoretical $2p$ fluorescence yields of Walters and Bhalla (1971c) (solid curve) and McGuire (1971) compared with experiment (for references, see Walters and Bhalla) (from Walters and Bhalla, 1971c, courtesy of American Institute of Physics).

the other. Above $Z = 50$ there is an overestimation of X-ray yields which may be corrected by a relativistic calculation. At $Z = 18$ there is a dip caused by the use of a free atom valence electron configuration. Experimental results relate to solids and are expected to be affected by the environment.

The over-all improvement caused by the use of self-consistent-field techniques can be seen by comparing these results with the earlier hydrogenic calculations.

5. The relativistic Hartree–Fock method

One of the interesting features of recent theoretical work on Auger transitions is the good agreement obtained with experiment for heavier atoms using relativistic self-consistent-field techniques. Since the relativistic analogue of the Hartree–Fock equations are not very popularly known, we shall briefly discuss them in this section.

The derivation of relativistic Hartree–Fock equations from a variation principle was first given by Swirles (1935). The first step is, as before, to write down a Hamiltonian for the many-electron system. To do this, Swirles neglected the magnetic interaction for closed shells as well as the effect of retardation, leaving only the electrostatic interaction between electrons. The many-electron Hamiltonian is then written

$$H = H_0 + H_{\text{int}}, \tag{6.43}$$

where (with $n = Z$ for a normal atom)

$$H_0 = \sum_{i=1}^{n} \left(-\frac{Ze^2}{r_i} U_i + c\boldsymbol{\alpha}\cdot\mathbf{p}_i + mc^2 \beta \right) \prod_{j \neq i}^{n} U_j \tag{6.44a}$$

and

$$H_{\text{int}} = \sum_{i \neq j}^{n} \sum_{j} \frac{e^2}{r_{ij}} \prod_{k=1}^{n} U_k. \tag{6.44b}$$

Here U_i is a 4×4 unit matrix corresponding to the fact that all operators belong to the 4×4 Dirac space. The indices i, j, k, etc. are particle labels, and $\boldsymbol{\alpha}$ and β are the usual Dirac operators. We write the single-electron orbitals in the form

$$\Psi_{\rho_i}(\alpha_p|i) \quad (\rho_i = 1, 2, 3, 4), \tag{6.45}$$

which stands for the ith electron being in the state $\alpha_p \equiv (N_p l_p j_p m_p)$. The index ρ_i refers to the ρth spinor component of the ith electron. Then we have to minimize the quantity

$$E' = \int d\tau_1\, d\tau_2 \ldots d\tau_n \left[\sum_{P} (-1)^P \sum_{\rho_1 \cdots \rho_n = 1}^{4} \Psi^*_{\rho_1}(\alpha_1|1) \ldots \Psi^*_{\rho_n}(\alpha_n|n) \right] H$$
$$\left[\sum_{P'} (-1)^{P'} \sum_{\sigma_1 \cdots \sigma_n = 1}^{4} \Psi_{\sigma_1}(\beta_1|1) \ldots \Psi_{\sigma_n}(\beta_n|n) \right] \tag{6.46}$$

$$= \int d\tau_1\, d\tau_2 \ldots d\tau_n \sum_{\rho_1, \ldots, \rho_n = 1}^{4} \sum_{\sigma_1, \ldots, \sigma_n = 1}^{4} \sum_{P} (-1)^P \Psi^*_{\rho_1}(\alpha_1|1) \ldots \Psi^*_{\rho_n}(\alpha_n|n) \cdot$$
$$\cdot H_{\rho_1 \cdots \rho_n, \sigma_1 \cdots \sigma_n} \sum_{P'} (-1)^{P'} \Psi_{\sigma_1}(\beta_1|1) \ldots \Psi_{\sigma_n}(\beta_n|n). \tag{6.46a}$$

The expressions within square brackets in eqn (6.46) represent Slater determinants. P and P' represent permutation of the particle indices. The symbol $H_{\rho_1 \ldots \rho_n, \sigma_1 \ldots \sigma_n}$ stands for the general matrix element of H, each row or column being labelled by the set of single-particle indices $(\rho_1 \ldots \rho_n)$ or $(\sigma_1 \ldots \sigma_n)$. It may be shown that E' is real, provided that all the spinor components Ψ_p are $\mathcal{O}(1/r)$ for large r, and that it represents the energy of the system when the variation condition $\delta E' = 0$ is satisfied.

We write the orbital

$$\Psi \equiv \Psi_{\kappa}^{\mu} = \begin{pmatrix} i f_{\kappa}(r)\, \chi_{-\kappa}^{\mu} \\ g_{\kappa}(r)\, \chi_{\kappa}^{\mu} \end{pmatrix}, \tag{6.47}$$

where

$$f_{\kappa}(r) = \frac{P_{N,\kappa}(r)}{r}, \quad g_{\kappa}(r) = \frac{Q_{N,\kappa}(r)}{r}, \tag{6.48}$$

κ being the Dirac quantum number introduced in Chapter 3 and N the principal quantum number relevant to the orbital. The normalization condition on the radial functions P and Q is

$$\int_0^{\infty} (P_{N,\kappa}^2 + Q_{N,\kappa}^2)\, \mathrm{d}r = 1. \tag{6.49}$$

As in the non-relativistic case, we can define radial integrals

$$F_k(N\kappa; N'\kappa') = e^2 \int_0^{\infty}\int_0^{\infty} [P_{N,\kappa}^2(r_1) P_{N',\kappa'}^2(r_2) + P_{N,\kappa}^2(r_1)\, Q_{N',\kappa'}^2(r_2) + \\ + Q_{N,\kappa}^2(r_1) P_{N',\kappa'}^2(r_2) + Q_{N,\kappa}^2(r_1)\, Q_{N',\kappa'}^2(r_2)]\, \gamma_k(r_1, r_2)\, \mathrm{d}r_1\, \mathrm{d}r_2 \tag{6.50}$$

and

$$G_k(N\kappa; N'\kappa') = e^2 \int_0^{\infty}\int_0^{\infty} [P_{N,\kappa}(r_1) P_{N',\kappa'}(r_1)\, P_{N,\kappa}(r_2) P_{N',\kappa'}(r_2) + \\ + P_{N,\kappa}(r_1) P_{N',\kappa'}(r_1)\, Q_{N,\kappa}(r_2)\, Q_{N',\kappa'}(r_2) + \\ + Q_{N,\kappa}(r_1)\, Q_{N',\kappa'}(r_1) P_{N,\kappa}(r_2) P_{N',\kappa'}(r_2) + \\ + Q_{N,\kappa}(r_1)\, Q_{N',\kappa'}(r_1) Q_{N,\kappa}(r_2)\, Q_{N',\kappa'}(r_2)]\, \gamma_k(r_1, r_2)\, \mathrm{d}r_1\, \mathrm{d}r_2. \tag{6.51}$$

γ_k has the usual meaning given in eqn (5.9).

On minimizing E' by using the variation principle, we get first-order integro-differential equations for the radial orbitals which are the relativistic analogues of the Hartree–Fock equations. These equations contain direct and exchange potentials of the form

$$V_k(N\kappa; r_1) = e^2 \int_0^{\infty} [P_{N,\kappa}^2(r_2) + Q_{N,\kappa}^2(r_2)]\, \gamma_k(r_1, r_2)\, \mathrm{d}r_2, \tag{6.52}$$

and

$$V_k(N\kappa, N'\kappa'; r_1) = e^2 \int_0^{\infty} [P_{N,\kappa}(r_2) P_{N',\kappa'}(r_2) + Q_{N,\kappa}(r_2)\, Q_{N',\kappa'}(r_2)]\, \gamma_k(r_1, r_2)\, \mathrm{d}r_2. \tag{6.53}$$

For $2p$ electrons, Swirles obtained the following equations.‡

$$\frac{dP_{2,1}}{dr}-\frac{2}{r}P_{2,1}+\left(mc+\frac{\varepsilon_{2,1}}{c}+\frac{Ze^2}{cr}\right)Q_{2,1}+\frac{1}{c}\Bigg[2\sum_{N=1}^{3}V_0(N,0|r)+$$

$$+V_0(4,0|r)+3V_0(2,1|r)+2V_0(2,-2|r)-\tfrac{1}{25}V_2(2,1|r)+4V_0(3,1|r)+$$

$$+2V_0(3,-2|r)+6V_0(3,2|r)+4V_0(3,-3|r)\Bigg]Q_{2,1}-$$

$$-\frac{1}{c}\Bigg[\tfrac{1}{3}\sum_{N=1}^{3}V_1(N,0;2,1|r)\,Q_{N,0}+\tfrac{1}{6}V_1(4,0;2,1|r)\,Q_{4,0}+$$

$$+\tfrac{2}{25}V_2(2,1;2,1|r)Q_{2,1}+\tfrac{1}{5}V_2(2,1;2,-2|r)\,Q_{2,-2}+\{V_0(3,1;2,1|r)+$$

$$+\tfrac{1}{5}V_2(3,1;2,1|r)\}\,Q_{3,1}+\tfrac{1}{5}V_2(3,-2;2,1|r)\,Q_{3,-2}+\{\tfrac{3}{5}V_1(3,2;2,1|r)+$$

$$+\tfrac{6}{35}V_3(3,2;2,1|r)\}\,Q_{3,2}+\{\tfrac{1}{15}V_1(3,-3;2,1|r)+$$

$$+\tfrac{9}{35}V_3(3,-3;2,1|r)\}\,Q_{3,-3}\Bigg]=0, \tag{6.54}$$

and

$$\frac{dP_{2,-2}}{dr}+\frac{1}{r}P_{2,-2}+\left(mc+\frac{\varepsilon_{2,-2}}{c}+\frac{Ze^2}{cr}\right)Q_{2,-2}+\frac{1}{c}\Bigg[2\sum_{N=1}^{3}V_0(N,0|r)+$$

$$+V_0(4,0|r)+4V_0(2,1|r)+V_0(2,-2|r)+4V_0(3,1|r)+$$

$$+2V_0(3,-2|r)+6V_0(3,2|r)+4V_0(3,-3|r)\Bigg]Q_{2,-2}-$$

$$-\frac{1}{c}\Bigg[\tfrac{1}{3}\sum_{N=1}^{3}V_1(N,0;2,1|r)\,Q_{N,0}+\tfrac{1}{6}V_1(4,0;2,1|r)\,Q_{4,0}+$$

$$+\tfrac{2}{5}V_2(2,1;2,-2|r)\,Q_{2,1}+V_0(3,-2;2,-2|r)\,Q_{3,-2}+$$

$$+\tfrac{2}{5}V_2(3,1;2,-2|r)\,Q_{3,1}+\tfrac{2}{3}V_1(3,-3;2,-2|r)\,Q_{3,-3}+$$

$$+\tfrac{3}{7}V_3(3,2;2,-2|r)\,Q_{3,2}\Bigg]=0. \tag{6.55}$$

For $2p$ electrons, the κ values are 1 and -2. Thus, the radial functions $P_{2,1}$, $P_{2,-2}$, $Q_{2,1}$ and $Q_{2,-2}$ relate to the $2p$ electrons themselves. Radial functions such as $Q_{3,1}$, $Q_{4,0}$, $Q_{3,-2}$, $Q_{3,-3}$, etc. relate to other electrons belonging to higher major shells and represent exchange effects. Omitting these terms together with the exchange potentials gives us the SCF equations without exchange.

‡ Actually, each equation obtained by Swirles contains an additional inhomogeneous term of the form $-(1/c)\varepsilon_{N,\kappa;N',\kappa'}\,Q_{N',\kappa'}$. However, this may be removed by a unitary transformation which leaves the physics unaffected.

These equations are derived in the limit of a static interaction between electrons which is described by an unretarded potential. Hence they are valid for low energies only. The only difference with non-relativistic theory is that the electrons are described by the Dirac equation. In Auger calculations, this deficiency is partly removed by including the current–current interaction described in Chapter 2, although the wave functions used in computing this interaction are themselves based on the static approximation.

The first relativistic self-consistent-field calculation of KLL Auger transitions was made by Asaad (1959) for mercury ($Z = 80$). He used Hartree wave functions (i.e. without exchange) and chose a potential originally calculated non-relativistically by Hartree and Hartree (1935). Comparing his results with the earlier screened hydrogenic results of Massey and Burhop (1936), he found that the intensities of the KL_1L_2 and KL_1L_3 lines relative to the KL_1L_1 line obtained by the latter were about four times too large. This was felt by Asaad as due to a numerical error in the KL_1L_1 intensity and not due to the hydrogenic approximation. His calculation successfully accounted for the fact that the observed intensity of the KL_1L_2 line is greater than that of the KL_1L_3 line. He also showed that the intensities of the KL_2L_3 and KL_3L_3 lines relative to the KL_1L_1 line came out in close agreement with the experimental values of 1.4 and 0.6 observed by Bergstrom and Hill (1954). More recently, Bhalla and his colleagues (Bhalla and Ramsdale, 1970a,b; Bhalla, 1970a,b; Bhalla *et al.*, 1970) have made extensive calculations using the relativistic Hartree–Fock–Slater method to compute Auger transition rates. We shall briefly describe their work in this section.

The single-particle Dirac equation in the presence of a spherically symmetric potential $V(r)$ may be written in the form

$$[c\boldsymbol{\alpha}\cdot\mathbf{p} + \beta mc^2 + V(r) - \varepsilon]\,\Psi^{\mu}_{\kappa}(r) = 0. \tag{6.56}$$

Substituting from eqns (6.47) and (6.48), this gives the first order coupled differential equations

$$\begin{aligned}\frac{\mathrm{d}}{\mathrm{d}r}P_{N,\kappa}(r) &= \frac{\kappa}{r}P_{N,\kappa}(r) - [\varepsilon - mc^2 - V(r)]\,Q_{N,\kappa}(r),\\ \frac{\mathrm{d}}{\mathrm{d}r}Q_{N,\kappa}(r) &= [\varepsilon + mc^2 - V(r)]P_{N,\kappa}(r) - \frac{\kappa}{r}Q_{N,\kappa}(r).\end{aligned} \tag{6.57}$$

Comparing these equations with eqns (6.54) and (6.55), we can see that the latter are of this general form, provided that $V(r)$ is taken to include both the direct and exchange potentials due to other electrons along with the nuclear Coulomb potential.

Bhalla *et al.* used the Slater approximation for the exchange term so as to simplify the calculation. They also included finite nuclear size effects, which become appreciable particularly for inner atomic shells, by using

$$\left.\begin{aligned} V(r) &= -\frac{Ze^2}{2R}\left\{3-\left(\frac{r}{R}\right)^2\right\} \qquad &\text{for } r \leqslant R, \\ &= V_1(r) &\text{for } r > R. \end{aligned}\right\} \tag{6.58}$$

Here $V_1(r)$ is the self-consistent-field (SCF) potential with exchange discussed above. The choice of eqn (6.58) implies a uniformly charged spherical model for the nucleus, R being its radius. To initiate the iterative procedure for the SCF calculation, they used the non-relativistic Hartree–Fock–Slater potential given by Herman and Skillman (1963). The nuclear radius R was taken equal to 1.2 $A^{1/3}$ fm. The above choice of potential also removes the singularity of the Coulomb potential at $r = 0$.

The coupled differential eqns (6.57) have to be integrated for all orbitals in the atom with the appropriate boundary conditions. For bound states, the functions P and Q behave like r^{l+1} near the origin and like $\exp(-\sqrt{-2E}\,r)$ for large r. To start the integration one generates the ratio P/Q by power series methods near the origin and integrates outwards by the Runge–Kutta method. One of the functions, either P or Q is chosen arbitrarily and the other fixed to yield the correct behaviour. Outward integration proceeds up to a suitable matching radius r_0. At a large distance from r_0, inward integration is started, again using a power series for the ratio P/Q. The output potential $V'(r)$ at the end of such a cycle is computed from P and Q by summing eqn (6.52) over all orbitals. Rather than using $V'(r)$ as the input potential for the next stage, Bhalla used the average of $V'(r)$ and $V_1(r)$. The procedure was repeated till self-consistency was achieved.

In his relativistic Hartree–Fock–Slater calculations, Bhalla included the effect of retardation in both the charge–charge and current–current interactions. We have already seen in Chapter 3 how this affects the charge–charge term. In the current–current term one uses the expansion

$$\frac{\boldsymbol{\alpha}_1\cdot\boldsymbol{\alpha}_2}{|\mathbf{r}_1-\mathbf{r}_2|}\,e^{iq|\mathbf{r}_1-\mathbf{r}_2|} = 4\pi iq \sum_{LM\lambda} \Phi_{L+\lambda}(r_1, r_2)\,(\boldsymbol{\alpha}_1\cdot\mathbf{Y}^{(\lambda)}_{LM}(\theta_1,\varphi_1))\times$$

$$\times\,(\boldsymbol{\alpha}_2\cdot\Gamma^{(\lambda)}_{LM}(\theta_2,\varphi_2)), \tag{6.59}$$

where $\mathbf{Y}^{(\lambda)}_{LM}(\theta,\varphi)$ is the vector spherical harmonic with spherical components given by

$$[\mathbf{Y}^{(\lambda)}_{LM}(\theta,\varphi)]_\nu = -\,C^{L+\lambda\ \ 1\ \ L}_{M-\nu\ \ \nu\ \ M}\,Y^{M-\nu}_{L+\lambda}(\theta,\varphi), \quad (\nu = 1, 0, -1). \tag{6.60}$$

The relativistic calculations described in this section covered KLL transition rates in jj coupling for $Z = 21$, 30, 35, 41, 48, 55, 63, 70, 80, 81 and 93, as well as KLM and KMM rates for all these atoms except $Z = 81$. Bhalla also computed theoretical values of K-shell fluorescence yields by using these results along with radiative widths calculated by the same method (Bhalla, 1970c).

6. The Thomas–Fermi–Dirac method

This is an application of the Fermi–Dirac statistics which provides a simple approximation to the potential and charge distribution in an atom. The results are less accurate than those obtained from the Hartree–Fock equations, but are useful in obtaining trial potentials for the latter. The original Thomas–Fermi model, which does not include exchange, was used by Ramberg and Richtmeyer (1937) to calculate the radiative and Auger widths of different atomic levels in gold ($Z = 79$). The relation between the level width and mean life of a state was just being recognized at the time and they wanted to ascertain if the observed width of X-ray lines could be obtained by considering the two types of transitions. Subsequently, Listengarten (1961, 1962) used the Thomas–Fermi–Dirac model with exchange to compute relativistic Auger rates for $Z = 65$, 81 and 92.

We shall first outline the Thomas–Fermi model and then indicate how exchange is incorporated. Consider a system of weakly interacting electrons moving freely in a box of finite volume v. The number of electrons with momenta lying between k and $k + \mathrm{d}k$ is obtained by using the Pauli principle and is given by

$$\mathrm{d}n = 2\,\frac{v}{(2\pi)^3}\,4\pi k^2\,\mathrm{d}k. \tag{6.61}$$

The maximum allowed momentum, i.e. the Fermi momentum k_F, is obtained from the condition that $\mathrm{d}n$ should integrate to give n, the total number of electrons in the box. This gives the density

$$\rho = \frac{n}{v} = \frac{k_F^3}{3\pi^2} = \frac{8\pi}{3}\,\frac{(2mE_F)^{3/2}}{h^3}, \tag{6.62}$$

E_F being the Fermi energy. The system so far has a uniform density.

Let us now switch on a spherically symmetric electrostatic potential $V(r)$, with the stipulation that $V(r) \to 0$ only as $r \to \infty$. Such a system will have Fermi energy $E_F = 0$, so that at any point $\mathbf{r}$, electrons having the maximum kinetic energy are just prevented from escaping out of the system. Obviously this maximum kinetic energy depends on r, being given by

$$\frac{\hbar^2 k_F^2(r)}{2m} = eV(r). \tag{6.63}$$

The crucial assumptions of the Thomas–Fermi model are:

1. The density $\rho(r)$ of the electron-distribution at a point $\mathbf{r}$ in an atom depends on the maximum kinetic energy at that point exactly in the same manner as it would in a free electron gas with the same maximum kinetic energy. Thus, from eqns (6.62) and (6.63)

$$\rho(r) = \frac{8\pi}{3}\frac{[2meV(r)]^{3/2}}{h^3}. \tag{6.64}$$

2. At each point $\mathbf{r}$, the relation between $V(r)$ and the charge density is given by Poisson's equation

$$\nabla^2 V(r) = -4\pi[-e\rho(r)] \tag{6.65}$$

$$= \frac{32\pi^2 e}{3}\frac{[2meV(r)]^{3/2}}{h^3}. \tag{6.65a}$$

If we now write

$$V(r) = \frac{Ze}{r}U(r), \tag{6.66}$$

and make the change of variable

$$U(r) = \varphi(x),\ r = \frac{a_0 x}{4}\left(\frac{9\pi^2}{2Z}\right)^{1/3}, \tag{6.67}$$

eqn (6.65a) assumes the form

$$\frac{d^2\varphi}{dx^2} = x^{-1/2}\varphi^{3/2}. \tag{6.68}$$

This is the Fermi–Thomas equation. From our discussion in Section 3, the boundary conditions on $\varphi(x)$ are

1. $\varphi(x) \to 1$ as $x \to 0$,
2. $\varphi(x) \to 0$ as $x \to \infty$.

Equation (6.68) has a highly nonlinear form and cannot be solved analytically. However, numerical solutions have been given and are widely used. This equation has a universal form valid for all elements, the unit length for an element Z being $(a_0/4)(9\pi^2/2Z)^{1/3}$, i.e. $0.8853\ a_0 Z^{-1/3}$.

In eqn (6.66), the form of $U(r)$ is to be determined by solving eqn (6.68) with the boundary conditions. The question arises: is there a way of explicitly building the exchange potential into it? The answer is to write down $U(r)$ as in Section 3 and then to subject the resulting expression for the total energy to a variation principle. The result is the Thomas–Fermi–Dirac equation (Lenz, 1932).

The kinetic energy of the electrons per unit volume at a point **r** is

$$\frac{1}{v}\int_0^{k_F(r)} \frac{\hbar^2 k^2}{2m}\,\mathrm{dn} = \int_0^{k_F(r)} \frac{\hbar^2 k^2}{2m} 2\frac{1}{(2\pi)^3} 4\pi k^2\,\mathrm{d}k = \frac{3}{5}\frac{\hbar^2}{2m}\rho(r)\,[3\pi^2\rho(r)]^{2/3}.$$

The total kinetic energy of the system is, therefore,

$$T = \int \mathrm{d}\mathbf{r}\,\frac{3}{5}\frac{\hbar^2\pi^2}{2m}\rho(r)\left[\frac{3\rho(r)}{\pi}\right]^{2/3} =$$

$$= \frac{3}{5}e^2 a_0 \int \mathrm{d}\mathbf{r}\,\frac{\pi^2}{2}\rho(r)\left[\frac{3\rho(r)}{\pi}\right]^{2/3}. \tag{6.69}$$

And the total potential energy is, from Section 3,

$$W = \int\left[-\frac{Ze^2}{r} + \frac{e^2}{2}\int \mathrm{d}\mathbf{r}'\,\frac{\rho(r')}{|\mathbf{r}-\mathbf{r}'|} - \frac{3}{4}e^2\left(\frac{3\rho(r)}{\pi}\right)^{1/3}\right]\rho(r)\,\mathrm{d}\mathbf{r}. \tag{6.70}$$

The factor $\frac{1}{2}$ is introduced in the two electronic interaction terms to avoid counting the electrons twice.

Under equilibrium conditions, the density $\rho(r)$ adjusts itself so that $T+W$ is a minimum. Thus, letting

$$\delta(T+W) = 0,$$

the variation being with respect to $\rho(r)$ as a function of position, we get

$$-\frac{Ze^2}{r} + e^2\int \mathrm{d}\mathbf{r}'\,\frac{\rho(r')}{|\mathbf{r}-\mathbf{r}'|} + \frac{e^2 a_0}{2}\pi^2\left(\frac{3\rho(r)}{\pi}\right)^{2/3} - e^2\left(\frac{3\rho(r)}{\pi}\right)^{1/3} = 0. \tag{6.71}$$

Writing

$$-eV(r) = -\frac{Ze^2}{r} + e^2\int \mathrm{d}\mathbf{r}'\,\frac{\rho(r')}{|\mathbf{r}-\mathbf{r}'|}, \tag{6.72}$$

we can see that $V(r)$ is the external potential experienced by a test charge at the point **r**. Equation (6.71) is a quadratic equation in the quantity $(3\rho(r)/\pi)^{1/3}$. Writing $z = a_0(3\rho(r)/\pi)^{1/3}$, the solution is

$$z = \frac{1}{\pi}\left(\frac{1}{\pi} \pm \sqrt{(2y)}\right), \tag{6.73}$$

where

$$y = \frac{1}{2\pi^2} + \frac{Va_0}{e}. \tag{6.74}$$

In eqn (6.73), we choose the positive sign so as to avoid the possibility of a negative probability density $\rho(r)$. The function y represents a renormalized potential to which we now apply Poisson's equation

$$\frac{1}{r}\frac{\mathrm{d}^2}{\mathrm{d}r^2}(ry) = -4\pi(-e\rho(r))\frac{a_0}{e} = \frac{4}{3a_0^2\pi}\left(\frac{1}{\pi} + \sqrt{2y}\right)^3. \tag{6.75}$$

Again we write

$$r = \frac{a_0 x}{4}\left(\frac{9\pi^2}{2Z}\right)^{1/3} \quad \text{and} \quad ry = Za_0\varphi$$

to get

$$\frac{\mathrm{d}^2\varphi}{\mathrm{d}x^2} = x\left[\left(\frac{\varphi}{x}\right)^{1/2} + \alpha\right]^3, \tag{6.76}$$

where $\alpha = [3/(32\pi^2 Z^2)]^{1/3}$. Equation (6.76) is the Thomas–Fermi–Dirac (TFD) equation. By comparing with eqn (6.68) we can see that α is the contribution due to exchange.

The quantity $0.8853\ a_0 Z^{-1/3}$ serves as a natural unit of length for both the methods. It was the unit used by Herman and Skillman in their computation of numerical orbitals. In fact they observed good agreement between the function $U(r)$ used in their calculation for Zn and the function $\varphi(x)$ over the region $0 \leqslant x \leqslant 10$.

7. Relativistic approaches to the many-electron atom—theory and experiment

The work of Listengarten (1961, 1962) referred to in the last section involved an application of the Thomas–Fermi–Dirac potential to treat screening in Auger calculations. His approach was based on relativistic quantum theory with retarded interaction between the electrons. In fact the treatment of retardation mentioned in the last section was developed by him. He was interested in *KLL* transitions in heavy atoms and his choice of the TFD model was guided by the consideration that in the inner shells of such atoms the statistical model would provide a fair description of the physical situation. Interestingly, the statistical model does overlook shell structure. Listengarten obtained the TFD function $\varphi(x)$ by interpolation from the tables of Metropolis and Reitz (1951) and of Thomas (1954) for the Z-values of interest. To compute the bound-state orbitals, Listengarten used the Thomas–Fermi model. The continuum orbitals were obtained with the inclusion of exchange.

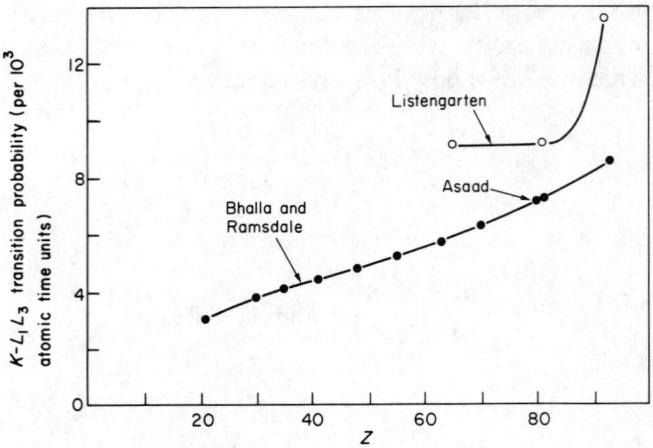

Fig. 6.6. Relativistic KL_1L_3 transition probabilities calculated by different authors (from Bhalla and Ramsdale, 1970a, courtesy of Institute of Physics).

Listengarten's work represents the earliest numerical approach to screening so far as relativistic Auger calculations are concerned. Subsequently, he interpolated his work to include all odd-Z elements between $Z = 65$ and $Z = 92$. He found the total probability for KLL transitions to be more than double the corresponding non-relativistic values. The greatest enhancement occurred for

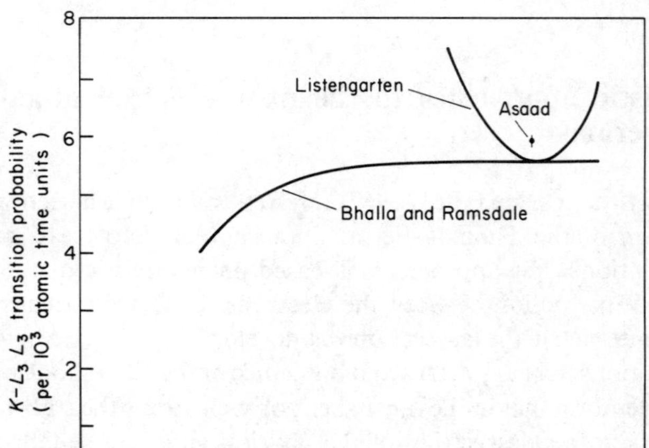

Fig. 6.7. Relativistic KL_3L_3 transition probabilities calculated by different authors (from Bhalla and Ramsdale, 1970b, courtesy of Springer Verlag).

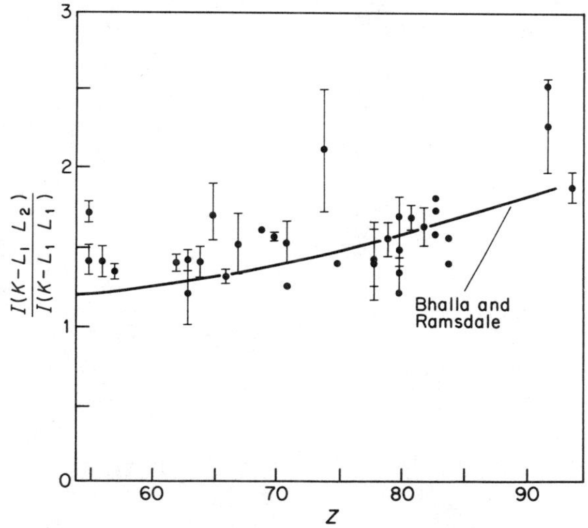

Fig. 6.8. Comparison of relativistic Hartree–Fock–Slater (RHFS) results of Bhalla and Ramsdale (1970b) for the relative intensity of the KL_1L_2 line with experiment. All relative intensities in this and the following five figures are computed with respect to the KL_1L_1 line (from Bhalla and Ramsdale, 1970b, courtesy of Springer Verlag).

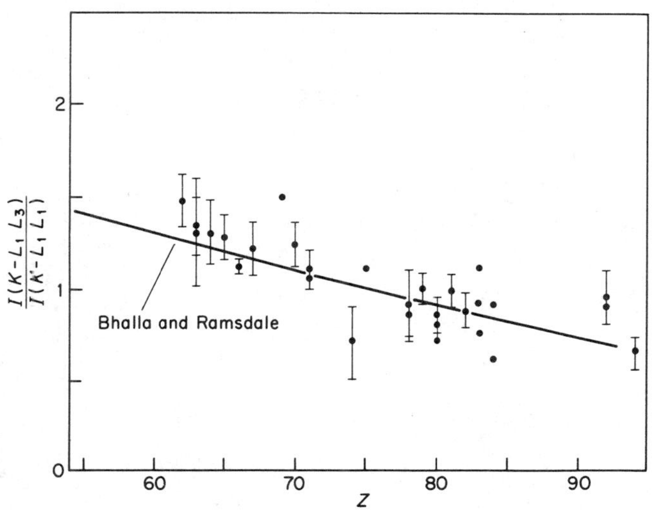

Fig. 6.9. Comparison of RHFS results of Bhalla and Ramsdale (1970b) for relative intensity of the KL_1L_3 line with experiment (from Bhalla and Ramsdale, 1970b, courtesy of Springer Verlag).

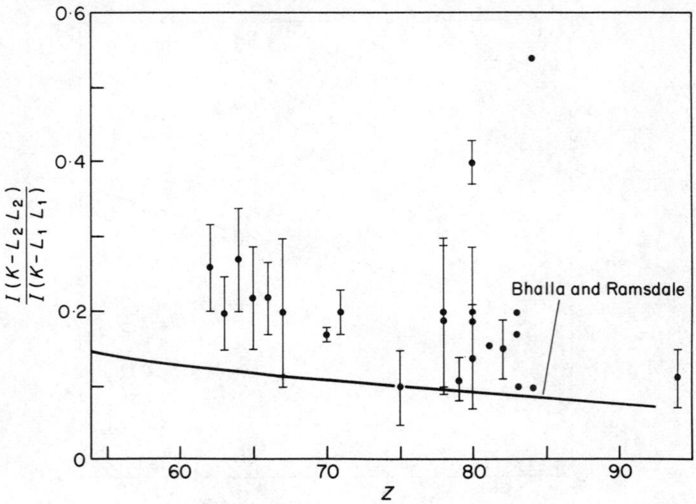

Fig. 6.10. RHFS results of Bhalla and Ramsdale (1970b) for relative intensity of the KL_2L_2 line compared with experiment (from Bhalla and Ramsdale, 1970b, courtesy of Springer Verlag).

the KL_1L_1 and KL_1L_2 lines. At the same time the KL_2L_2, KL_2L_3 and KL_3L_3 lines appeared to be significantly reduced in intensity. These general trends are in agreement with experiment. However, for four of the six lines, i.e. all but the lines KL_2L_3 and KL_3L_3, Listengarten's results disagree by more than 20%

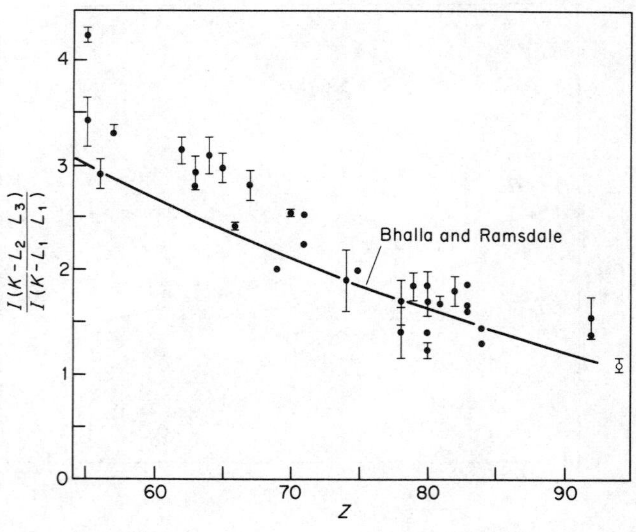

Fig. 6.11. RHFS results of Bhalla and Ramsdale (1970b) for relative intensity of the KL_2L_3 line compared with experiment (from Bhalla and Ramsdale 1970b, courtesy of Springer Verlag).

with the relativistic Hartree–Fock–Slater values obtained by Bhalla and Ramsdale (1970a,b).

Figures 6.6 and 6.7 show KL_1L_3 and KL_3L_3 transition probabilities obtained by Bhalla and Ramsdale (1970a,b), Asaad (1959) and Listengarten (1961, 1962). The results obtained by Bhalla and Ramsdale range from $Z=21$ to $Z=93$. Listengarten's results refer to $Z=65$, 81 and 92, while Asaad's result is for $Z=80$. In both figures, Listengarten's results show significant local structure, whereas Bhalla's results exhibit a smooth Z-dependence. On the other hand, the relative intensities obtained by the former do not exhibit large local structures, presumably because each line has this structure in some measure. The agreement between Asaad's results and those of Bhalla and Ramsdale is much closer.

The relativistic Hartree–Fock–Slater results of Bhalla and Ramsdale are significant because they incorporate all important effects, including that of finite nuclear size. Because of the strong sensitivity of Auger transition rates to the choice of wave functions, they provide a test for the accuracy of the screening model as well as the coupling scheme. These authors used *jj* coupling. Figures 6.8–6.12 show the intensities of the *KLL* lines relative to the KL_1L_1 line. Detailed references to the experimental results are to be found in the work of Ramsdale (1969). The Z-dependence of the theoretical results follows experimental trends quite well. It will be seen that only the KL_1L_2 line increases in intensity with increasing Z. There is a corresponding suppression of the KL_1L_3 line belonging to the same vacancy configuration. Other *KLL* lines

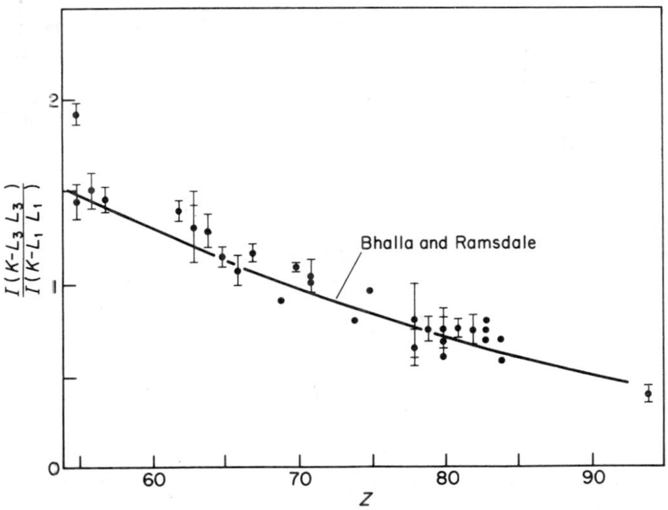

Fig. 6.12. RHFS results of Bhalla and Ramsdale (1970b) for relative intensity of the KL_3L_3 line compared with experiment (from Bhalla and Ramsdale, 1970b, courtesy of Springer Verlag).

Table 8. Comparison of relative intensities of the *KLM* group with theory ($KL_1L_1 = 100$). Hartree–Fock–Slater results from Bhalla (1970b).

Transition	$Z=47$			$Z=49$	
	Expt. (Johnson and Foster, 1953)	Theory[a]	Theory (Asaad and Burhop, 1958)	Expt. (Krisciokaitis and Haynes, 1967)	Theory[a]
KL_1M_1	82 ± 20	33.5	29.2	42 ± 4	34.5
$KL_1M_{2,3}$	80 ± 20	49	50	62 ± 6	49.5
KL_2M_1	20 ± 10	16	15.6	20 ± 2	16.5
$KL_1M_{4,5}$					
$KL_2M_{2,3}$	160 ± 30	81	146	86 ± 9	80
KL_3M_1					
$KL_3M_{2,3}$	170 ± 30	117	121	128 ± 13	116
$KL_2M_{4,5}$	…				
$KL_3M_{4,5}$	55 ± 20	11	43.2	20 ± 2	11.2

[a] Bhalla, Rosner and Ramsdale (1970).

Table 9. Comparison of relative intensities of *KLM* group with theory—relativistic Hartree–Fock–Slater results from Bhalla (1970b)

Transition	$Z=80$		$Z=81$	
	Expt. (Nall *et al.*, 1960)	Theory (Bhalla *et al.*, 1970)	Expt. (Sujkowski, 1961)	Theory (Bhalla *et al.*, 1970)
KL_1M_1	35	40	94 ± 15	85
KL_1M_2	43	34		
KL_1M_3				
$KL_1M_{4,5}$	75[a]	51[a]	66 ± 10	53
KL_2M_1				
KL_2M_2				
KL_2M_3	75[b]	31	40 ± 5	31
$KL_2M_{4,5}$	89	4.6	13 ± 3	6.2
KL_3M_1	16	14		
KL_3M_2	40[c]	29	124 ± 10[e]	75
KL_3M_3	69[d]	31		
$KL_3M_{4,5}$				

[a] Without the $KL_1M_{4,5}$ lines.
[b] With the $KL_1M_{4,5}$ lines.
[c] Includes contributions from the KL_1N group.
[d] Includes contributions from several other groups.
[e] With the contribution of the KL_2N group.

exhibit a decreasing relative intensity as Z increases. Tables 8 and 9 exhibit theoretical and experimental KLM line intensities for the elements $Z = 47$, 49 and $Z = 80$, 81 respectively. These intensities are relative to the KL_1L_1 line. It will be seen that particularly for $Z = 47$, the agreement of the relativistic Hartree–Fock–Slater (RHFS) results with experiment is far from being satisfactory. This may be ascribed to the fact that the jj coupling used by Bhalla (1970b) does not describe the situation for these intermediate atomic numbers. This is particularly true of the $KL_3M_{4,5}$ transition, where the non-relativistic Hartree result of Asaad and Burhop (1958) using intermediate coupling (NRIC) fares much better. For the KL_1M_1, $KL_1M_{2,3}$ and KL_2M_1 lines the two sets do not differ much, whereas for the group $KL_1M_{4,5}$, $KL_2M_{2,3}$ and KL_3M_1 the intermediate coupling result is quite good. There are some notable discrepancies. For example, the experimental intensity of the $KL_1M_{2,3}$ line is larger than both the NRIC and RHFS values by $(10 \pm 10)\%$. For $Z = 52$, the experimental intensity of this same line happens to be lower by $(20 \pm 9)\%$. For heavier atoms, the RHFS results show much better agreement with experiment. This shows the importance of coupling schemes in such calculations.

Chapter Seven
Some Recent Advances In Auger Theory

When the decay of an inner shell vacancy via electron emission was first observed by Auger (1925), it was looked upon as a subsidiary phenomenon. Characteristic X-rays were the principal object of study and Auger yields were of interest in as much as they provided an estimate of fluorescence yields. Gradually, however, Auger studies came into their own because of their intrinsic importance to atomic physics as well as their possible applications. As Carlson (1973) has aptly remarked, the study of a nuisance effect has burgeoned into the field of Auger electron spectroscopy. Quite recently, there has been a corresponding progress in the theoretical understanding of Auger transitions. In this chapter we shall outline some of these developments.

1. Satellite spectra

When an atom has two or more vacancies in the inner shells, it may decay into a final state with three or more vacancies via radiationless transitions. The resulting electron spectrum contains satellite Auger lines. The initial multiple ionization may be caused by bombardment with photons, electrons, protons or heavy ions. A possible mechanism for double ionization in a pair of inner shells is the shake process. An example is afforded by either nuclear electron capture or photo-ionization in an inner shell which occurs so fast that the remaining electrons in the atom experience a sudden change in the nuclear Coulomb field. The resulting excited state of the system relaxes either by emitting another inner or outer shell electron into the continuum (shake-off) or by sending it "up" to an outer shell (shake up). The condition for this description to be valid is that the velocities of the interacting and emitted particles are large compared to that of the bound outer shell electrons. This is called the sudden approximation (Åberg, 1969; Åberg, 1972). Once a double-vacancy state has come about, its decay may give rise to Auger satellites. For example, an atom with an initial *KL* vacancy configuration may decay into a final *LLL* vacancy configuration giving rise to *KL–LLL* Auger satellite lines. Satellites are usually shifted from the normal or *diagram* lines (e.g. *KLL*) towards lower energies.

The first detailed observation and analysis of Auger satellite lines were made by Körber and Mehlhorn (1966) during their investigation of the *KLL* Auger spectrum of neon. Ionization was caused by electron impact and the resulting Auger spectrum containing diagram as well as *KL–LLL* satellite lines was analysed with an electrostatic spectrometer having an energy resolution of 0.12%. Assuming pure Russell–Saunders coupling, the initial double vacancy states were $KL_1(^1S, ^3S)$ and $KL_{2,3}(^1P, ^3P)$. The possible final states were $L_{2,3}L_{2,3}L_{2,3}(^4S, ^2D, ^2P)$, $L_1L_{2,3}L_{2,3}(^4P, ^2D, ^2S, ^2P)$ and $L_1L_1L_{2,3}(^2P)$. Under the usual selection rules $\Delta L = \Delta S = \Delta J = \Delta\pi = 0$, not all transitions are allowed. Of the allowed lines, the following ten were actually measured by Körber and Mehlhorn.

Initial state	Final state	
$KL_{2,3}(^3P)$	$L_{2,3}L_{2,3}L_{2,3}(^2D \text{ and } ^2P)$ $L_1L_{2,3}L_{2,3}(^2D \text{ and } ^2P)$ $L_1L_1L_{2,3}(^2P)$	5 lines
$KL_{2,3}(^1P)$	$L_{2,3}L_{2,3}L_{2,3}(^2D \text{ and } ^2P)$ $L_1L_{2,3}L_{2,3}(^2P)$ $L_1L_1L_{2,3}(^2P)$	4 lines
$KL_1(^3S)$	$L_1L_{2,3}L_{2,3}(^2D)$	1 line

To compute the energies of these lines, Körber and Mehlhorn made the simple assumption that the *L*-shell binding energy of a neon atom with a 1*s* vacancy is nearly equal to that of a sodium atom ionized in the 3*s* shell. An improved measurement and a more comprehensive analysis of the *KL–LLL* Auger satellite lines of neon were given by Krause *et al.* (1971). By using X-rays at 10 keV as well as monochromatic electrons between 3 and 10 keV to excite the diagram and satellite lines; they showed that the character of the spectrum was independent of the mode of excitation (Krause *et al.*, 1970). Other satellite spectra of noble gases have also been studied, the more important ones being the $L_{2,3}M$–*MMM* lines of argon (Volz and Rudd, 1970; Werme *et al.*, 1973), $L_{2,3}M$–*MMM*, $L_{2,3}N$–*MMN* and $M_{4,5}N$–*NNN* lines of krypton (Werme *et al.*, 1972; Mehlhorn *et al.*, 1972). The Auger spectra resulting from an initial $M_{4,5}$, i.e. 3*d*, vacancy in krypton and an initial $N_{4,5}$ vacancy in xenon are particularly rich in satellite lines. Some of these have not yet been positively identified. Werme *et al.* (1972) used Moore's tabulation of atomic energy levels to identify some of them. Reliable calculations could of course help such identification. The recent work of

McGuire (1974a, b) on Auger transitions in atoms with partially filled shells has a special relevance in this context.

Hypersatellite Auger lines are produced when a configuration with multiple vacancies in an inner shell but no other vacancies elsewhere decays through the radiationless emission of electrons. Such lines were observed by Briand *et al.* (1970, 1973) in atoms with a completely vacant K shell. Hypersatellites may be observed in metallic atoms with double-vacancy initial states so long as the decay does not involve a valence band (Löfgren and Wallden, 1973).

Finally, initial states with multiple vacancies in both the innermost shell and in the outer shells may give rise to what are called satellites of hypersatellite lines. Such lines were observed by Richard *et al.* (1972) in calcium bombarded with 30 MeV oxygen ions.

The categorization of all such lines as satellites presumes that they are, after all, subsidiary objects of low intensity. This is more or less true so long as the initial excitation of the atomic system is caused by photons or electrons. The situation becomes different with ion–atom collisions where multiple vacancy states are copiously produced due to the high energies involved. Here satellites and hypersatellites may have relatively high intensities, fluorescence yields being rather low ($\omega < 0.01$). A study of these spectra may become quite important to the understanding of the main features of ion–atom collisions.

The first theoretical treatment of KL–LLL Auger satellites in neon was given by Schmidt (1973) who used LS coupling with configuration interaction. As with diagram lines, the transitions are caused by the Coulomb interaction between electrons in the non-relativistic limit. Furthermore, Schmidt used the frozen atom approximation which assumes that the single-electron wave functions are the same in the initial and final states. The initial state configuration has two holes in the K and L shells respectively and a hole in the continuum with orbital angular momentum l. These three holes couple to give the initial wave function $\varphi_i(SLM_S M_L)$. In the final state there are three holes in the L shells which couple to give the wave function $\varphi_f(S'L'M_{S'}M_{L'})$. This is the picture in the absence of configuration interaction.

With the inclusion of configuration interaction, two possible final state configurations, namely $(1s)^2(2s)^2(2p)^3$ and $(1s)^2(2p)^5$, both having a term ^2P, can be connected by the Coulomb interaction. Therefore we get an off-diagonal term in the Hamiltonian,

$$H_{21} = H_{12} = \langle (2p)^{-3}\,{}^2PM_S M_L \left| \sum \frac{e^2}{r_{ij}} \right| (2s)^{-2}(2p)^{-1}\,{}^2PM_S M_L \rangle =$$

$$= -\sqrt{(2/3)}\, G^1(2s, 2p). \qquad (7.1)$$

In eqn (7.1) we have used vacancies to denote the configurations. Schmidt neglected the interaction between initial state configurations on the grounds that the continuum electron interacts only weakly with the bound electrons. The diagonal matrix elements were

$$H_{11} = \langle (2s)^{-2}(2p)^{-1}\,{}^2PM_S M_L \left| \sum \frac{e^2}{r_{ij}} \right| (2s)^{-2}(2p)^{-1}\,{}^2PM_S M_L \rangle =$$

$$= 2E(2s) + E(2p) + F^0(2s,2s) + 2F^0(2s,2p) - \tfrac{1}{3}G^1(2s,2p), \qquad (7.2)$$

and

$$H_{22} = \langle (2p)^{-3}\,{}^2PM_S M_L \left| \sum \frac{e^2}{r_{ij}} \right| (2p)^{-3}\,{}^2PM_S M_L \rangle =$$

$$= 3E(2p) + 3F^0(2p,2p). \qquad (7.3)$$

On diagonalizing this matrix, one gets the energy eigenvalues and the mixing coefficients. The resulting final-state eigenfunctions for the mixed configuration were

$$\Psi_{1f} = 0.9904\,\varphi_1((2s)^{-2}(2p)^{-1}\,{}^2P) - 0.1380\,\varphi_2((2p)^{-3}\,{}^2P) \qquad (7.4)$$

and

$$\Psi_{2f} = 0.1380\,\varphi_1((2s)^{-2}(2p)^{-1}\,{}^2P) + 0.9904\,\varphi_2((2p)^{-3}\,{}^2P). \qquad (7.5)$$

To obtain theoretical intensities of the satellite lines Schmidt used radial integrals from two different sources: (a) integrals given by McGuire (1969a) and (b) integrals derived from experimental relative intensities of normal *KLL* lines (Krause *et al.*, 1971). The expression for intensity used by him was

$$I(i \to f) = I_0\, Q(i) P(i \to f) / \sum_f P(i \to f), \qquad (7.6)$$

where I_0 is the total intensity of the *KL–LLL* spectrum, $Q(i)$ the relative probability for the production of the initial state i and $P(i \to f)$ the transition probability from the initial to the final state. In Table 10, his results using the radial integrals (b) with configuration mixing are compared with the experimental values of intensity obtained by (A) Krause *et al.* (1971) and (B) Körber and Mehlhorn. Agreement is good with (A), where the resolution was better than (B) by a factor of 2.5.

Schmidt found that the intensities of satellite lines with initial states $(1s)^{-1}(2s)^{-1}\,{}^1S$ and $(1s)^{-1}(2s)^{-1}\,{}^3S$ could not be readily compared with experiment because of their low intensity. He also encountered some discrepancies between theory and experiment with regard to the predicted energies of

Table 10. Intensities of *KL–LLL* Auger satellite lines of neon. All intensities are normalized to the total intensity of the *KL–LLL* spectrum (from Schmidt, 1973)

Line	Initial state	Final state	Calculated intensity using radial integrals (b), with configuration interaction (normalized to experimental total intensity)	Experiment (Krause *et al.*, 1971)	Experiment (Körber and Mehlhorn, 1966)
1	$(1s)(2s)^2(2p)^{5\,1}P$	Ψ_{1f}	1.09	0.9 (2)	1.6 (4)
2	$(1s)(2s)^2(2p)^{5\,1}P$	$(1s)^2(2s)(2p)^{4\,2}S$	2.4×10^{-4}	0.12 (5)	—
3	$(1s)(2s)^2(2p)^{5\,1}P$	$(1s)^2(2s)(2p)^{4\,2}P$	2.37	2.3 (1)	1.8 (3)
4	$(1s)(2s)^2(2p)^{5\,1}P$	$(1s)^2(2s)(2p)^{4\,2}D$	1.2×10^{-3}	~0.1	—
5	$(1s)(2s)^2(2p)^{5\,1}P$	Ψ_{2f}	2.49	2.4 (3)	3.3 (3)
6	$(1s)(2s)^2(2p)^{5\,1}P$	$(1s)^2(2s)^2(2p^3)^2D$	4.65	4.8 (1)	4.0 (3)
7	$(1s)(2s)^2(2p)^{5\,3}P$	Ψ_{1f}	1.38	1.2 (1)	1.7 (4)
8	$(1s)(2s)^2(2p)^{5\,3}P$	$(1s)^2(2s)(2p)^{4\,2}S$	0.33	0.2 (1)	—
9	$(1s)(2s)^2(2p)^{5\,3}P$	$(1s)^2(2s)(2p)^{4\,4}P$	0.71	~0.3	—
10	$(1s)(2s)^2(2p)^{5\,3}P$	$(1s)^2(2s)(2p)^{4\,2}D$	1.30	1.1 (1)	1.3 (3)
11	$(1s)(2s)^2(2p)^{5\,3}P$	$(1s)^2(2s)(2p)^{4\,2}D$	1.67	1.8 (1)	1.5 (3)
12	$(1s)(2s)^2(2p)^{5\,3}P$	Ψ_{2f}	3.15	3.2 (2)	3.3 (3)
13	$(1s)(2s)^2(2p)^{5\,3}P$	$(1s)^2(2s)^2(2p)^{3\,2}D$	5.88	6.6 (2)	6.7 (4)

KL_1–LLL transitions. This seems to arise from the initial assumption (Körber and Mehlhorn, 1966) that the L shell binding energy of a neon ion with a $1s$ vacancy is approximately equal to that of a sodium ion with a $3s$ vacancy. The difference appears to be about 10 eV.

One of the assumptions made by Schmidt in the above work was that the single particle wave functions used for normal Auger transitions are not essentially changed by the additional hole in the L shell. This is what permits one to use the same radial integrals to calculate transition probabilities for the satellites. Obviously, this is a rather strong assumption. Also, the initial K-shell vacancy is a special case and a more general approach to the problem of multiple-vacancy transitions is needed.

2. Auger transitions in partially filled shells

McGuire (1974a,b) has recently given a generalized treatment of Auger transitions involving partially filled shells. In the initial state, he considers an inner shell with orbital angular momentum l_1 having $m + 1$ holes which couple to give quantum numbers α_1, L_1, S_1 and J_1. The symbol α_1 stands for all other quantum numbers besides L_1, S_1 and J_1 that may be necessary to label the initial state of this shell. There is also a hole in the continuum having energy ε and quantum numbers l_2 and j_2. Apart from these holes, the atom in its initial state has partially filled outer shells as follows: an l_3 shell with n holes described by quantum numbers α_3, L_3, S_3; an l_4 shell with p holes and quantum numbers α_3, L_3, S_3; an l_4 shell with p holes and quantum numbers α_4, L_4, S_4; and other open shells with quantum numbers L_R, S_R. All these holes are coupled together to yield the initial state of the system with total angular momentum J. The final state has an l_1 shell with m holes and quantum numbers β_1, P_1, Q_1, an l_3 shell with $n + 1$ holes and quantum numbers β_3, P_3, Q_3, an l_4 shell with $p + 1$ holes and quantum numbers β_4, P_4, Q_4 and other open shells with quantum numbers P_R, Q_R. The symbol β_i stands for quantum numbers other than angular momenta characterizing a shell i, P_i stands for the total orbital angular momentum of the shell and Q_i for its total spin. Evidently, the partially filled shells labelled 1, 3 and 4 are the ones which participate in the Auger transition. The remaining incomplete shells act as spectators and we can see that the transition probability should contain a factor $\delta_{L_R, P_R} \delta_{S_R, Q_R}$ which comes from the orthonormalization of the wave function $u(L_R, S_R)$ describing these shells.

McGuire has used the jj coupling scheme for the initial state. The wave function for this state has to be obtained with the following constraints: (a) L_1 and S_1 must couple to J_1, (b) J_1 couples to j_2 to yield J_{12} and (c) because the interaction e^2/r_{12} preserves both L and S, we are interested in a wave function which is characterized by quantum numbers P and Q in the Russell–

Saunders limit. Thus we write, using the *SLJM* representation,

$$\Psi_i(J_1 j_2(J_{12})JM) = \sum_{\substack{A_2\, B_2\, A_4 \\ B_4\, C_4}} \sum_{\substack{\text{all} \\ m,s}} C^{L_3\ L_4\ A_2}_{M_{L_3}\, M_{L_4}\, M_{A_2}} C^{A_2\ L_R\ A_4}_{M_{A_2}\, M_{L_R}\, M_{A_4}} C^{S_3\ S_4\ B_2}_{M_{S_3}\, M_{S_4}\, M_{B_2}} C^{B_2\ S_R\ B_4}_{M_{B_2}\, M_{S_R}\, M_{B_4}} \cdot$$

$$\cdot C^{A_4\ B_4\ C_4}_{M_{A_4}\, M_{B_4}\, M_{C_4}} C^{L_1\ S_1\ J_1}_{M_{L_1}\, M_{S_1}\, M_1} C^{l_2\ s_2\ j_2}_{m_{l_2} m_{s_2} m_2} C^{J_1\ j_2\ J_{12}}_{M_1\ m_2\ M_{12}} \cdot$$

$$\cdot C^{J_{12}\ C_4\ J}_{M_{12} M_{C_4} M}\, u(l_1^{m+1}\,\alpha_1 L_1 S_1)\, u(l_3^n \alpha_3 L_3 S_3) \cdot$$

$$\cdot u(l_4^p \alpha_4 L_4 S_4)\, u(L_R, S_R)\, \Psi^{m_{l_2}}_{\eta_2 l_2} \chi^{m_{s_2}}_{s_2}. \qquad (7.7)$$

Here we have coupled the orbital angular momenta of the holes in shells 3, 4 and R to A_4, their spins to B_4 and then A_4 and B_4 to the total "core" angular momentum C_4. We now couple J_1 and j_2 to J_{12} and finally J_{12} and C_4 to get the total angular momentum J. Eventually we shall have to uncouple j_2 from J_1 so that we can project this wave function onto the quantum numbers P and Q via 9-*j* symbols. This will be implicit in the final expression for the matrix element.

$\Psi^{m_{l_2}}_{\eta_2 l_2}$ is the space part of the wave function of the hole in the continuum and $\chi^{m_{s_2}}_{s_2}$ its spin part. The functions $u(l_1^{m+1}\,\alpha_1 L_1 S_1)$, $u(l_3^n \alpha_3 L_3 S_3)$ and $u(l_4^p \alpha_4 L_4 S_4)$ are the wave functions corresponding to the shells 1, 3 and 4. Each of them represents a number of equivalent holes. One of the holes with orbital angular momentum l_1 participates in the Auger process and we can write down $u(l_1^{m+1}\alpha_1 L_1 S_1)$ in an antisymmetrized form

$$u(l_1^{m+1}\,\alpha_1 L_1 S_1) = \sqrt{(m+1)} \sum_{\alpha_1'\,\lambda_1\,\sigma_1} (l_1^{m+1}\,\alpha_1 L_1 S_1\{|l_1^m \alpha_1' \lambda_1 \sigma_1) \times$$

$$\times\, u(l_1^m \alpha_1' \lambda_1 \sigma_1; l_1 s_1 \alpha_1 L_1 S_1). \qquad (7.8)$$

The quantities $(l_1^{m+1}\alpha_1 L_1 S_1\{|l_1^m \alpha_1' \lambda_1 \sigma_1)$ are coefficients of fractional parentage. The functions $u(l_1^m \alpha_1' \lambda_1 \sigma_1; l_1 s_1 \alpha_1 L_1 S_1)$ are direct products of the type $u(l_1^m \alpha_1' \lambda_1 \sigma_1) \otimes \Psi(N_1 l_1 s_1)$ which are antisymmetric in the m holes of the parent configuration represented by the first factor but not in the last $(m+1)$th hole. (Racah, 1943; De-Shalit and Talmi, 1963). The factor $\sqrt{(m+1)}$ arises from the normalization procedure for such antisymmetric functions (McGuire, 1969).

The final state wave function is simply

$$\Psi_f'(P, Q, J, M) = \sum_{\substack{A_1, A_3 \\ B_1, B_3}} \sum_{\substack{\text{all} \\ m's}} C^{P_3\ P_4\ A_1}_{M_{P_3}\, M_{P_4}\, M_{A_1}} C^{A_1\ P_1\ A_3}_{M_{A_1}\, M_{P_1}\, M_{A_3}} C^{A_3\ P_R\ P}_{M_{A_3}\, M_{P_R}\, M_P} \cdot$$

$$\cdot C^{Q_3\ Q_4\ B_1}_{M_{Q_3}\, M_{Q_4}\, M_{B_1}} C^{B_1\ Q_1\ B_3}_{M_{B_1}\, M_{Q_1}\, M_{B_3}} C^{B_3\ Q_R\ Q}_{M_{B_3}\, M_{Q_R}\, M_Q} \cdot$$

$$\cdot C^{P\ \ Q\ \ J}_{M_P\, M_Q\, M}\, u(l_1^m \beta_1 P_1 Q_1)\, u(l_3^{n+1}\,\beta_3 P_3 Q_3) \cdot$$

$$\cdot u(l_4^{p+1}\,\beta_4 P_4 Q_4)\, u(P_R, Q_R). \qquad (7.9)$$

Again we can express the functions $u(l_3^{n+1}\beta_3 P_3 Q_3)$ and $u(l_4^{p+1}\beta_4 P_4 Q_4)$ as linear combinations of functions which are antisymmetric only in the parent configurations:

$$u(l_3^{n+1}\beta_3 P_3 Q_3) = \sqrt{(n+1)} \sum_{\beta_3' \lambda_3 \sigma_3} (l_3^{n+1}\beta_3 P_3 Q_3\{|l_3^n \beta_3' \lambda_3 \sigma_3) \cdot$$
$$\cdot\, u(l_3^n \beta_3' \lambda_3 \sigma_3; l_3 s_3 \beta_3 P_3 Q_3) \tag{7.10}$$

and

$$u(l_4^{p+1}\beta_4 P_4 Q_4) = \sqrt{(p+1)} \sum_{\beta_4' \lambda_4 \sigma_4} (l_4^{p+1}\beta_4 P_4 Q_4\{|l_4^p \beta_4' \lambda_4 \sigma_4) \cdot$$
$$\cdot\, u(l_4^p \beta_4' \lambda_4 \sigma_4; l_4 s_4 \beta_4 P_4 Q_4). \tag{7.11}$$

The Auger transition rate is given by

$$W_{if} = \frac{2\pi}{\hbar}\left|\left\langle \Psi_f \left| \frac{e^2}{r_{12}} \right| \Psi_i \right\rangle\right|^2, \tag{7.12}$$

where we have to sum over final states and average over initial states. The statistical weight of the initial state is

$$\omega_i = (2J_1+1)(2L_3+1)(2S_3+1)(2L_4+1)(2S_4+1)(2L_R+1)(2S_R+1). \tag{7.13}$$

We note that the active part of the matrix element will contain the two-electron components of Ψ_i and Ψ_f. We write this as

$$M = \int d\mathbf{r}_1\, d\mathbf{r}_2\, \Psi^{*m_{s1}}_{N_1 l_1 m_{l_1}} \Psi^{*m_{l_2}}_{\eta_2 l_2} \chi^{\dagger m_{s_2}}_{s_2} \frac{e^2}{r_{12}} [\Psi^{m_{s_3}}_{N_3 l_3 m l_3}(1)\, \Psi^{m_{s4}}_{N_4 l_4 m l_4}(2) -$$
$$-\Psi^{m_{s3}}_{N_3 l_3 m_{l_3}}(2)\, \Psi^{m_{s4}}_{N_4 l_4 m_{l_4}}(1)]$$
$$= \sum_{\delta} \prod_i (2l_i+1) \left\{ \sum_k R^k(N_1 l_1, \eta_2 l_2; N_3, l_3, N_4 l_4) \begin{pmatrix} l_1 & k & l_3 \\ 0 & 0 & 0 \end{pmatrix} \cdot \right.$$
$$\cdot\, (-1)^{l_2+l_3+l_1+l_4} \begin{pmatrix} l_2 & k & l_4 \\ 0 & 0 & 0 \end{pmatrix} (-1)^{m_{l_3}+m_{l_2}} \delta_{m_{s_1}, m_{s_3}} \delta_{m_{s_2}, m_{s_4}} \begin{pmatrix} l_1 & k & l_3 \\ m_{l_1} & \delta & -m_{l_3} \end{pmatrix} \cdot$$
$$\cdot \begin{pmatrix} l_4 & k & l_2 \\ m_{l_4} & \delta & -m_{l_2} \end{pmatrix} - \sum_{k'} R^{k'}(N_1 l_1, \eta_2 l_2; N_4 l_4, N_3 l_3) \begin{pmatrix} l_1 & k' & l_4 \\ 0 & 0 & 0 \end{pmatrix} \cdot$$
$$\cdot\, (-1)^{l_2+l_4+l_1+l_3} \begin{pmatrix} l_2 & k' & l_3 \\ 0 & 0 & 0 \end{pmatrix} (-1)^{m_{l_4}+m_{l_2}} \delta_{m_{s_1}, m_{s_4}} \delta_{m_{s_2}, m_{s_3}} \begin{pmatrix} l_1 & k' & l_4 \\ m_{l_1} & \delta & -m_{l_4} \end{pmatrix} \cdot$$
$$\cdot \begin{pmatrix} l_3 & k' & l_2 \\ m_{l_3} & \delta & -m_{l_2} \end{pmatrix}, \tag{7.14}$$

For the purpose of obtaining the matrix element in eqn (7.12) in terms of M, we decouple the active electrons from the core in the form $C^{g_1\ g\ P}_{m_g\ m_g m_P}\,\Psi(L_c S_c \to g_1)$ $\Psi(l_1 l_2 \to g)$, etc. and do the necessary recoupling at the end. Of course $\mathbf{l}_1 + \mathbf{l}_2 = \mathbf{g} = \mathbf{l}_3 + \mathbf{l}_4$. As usual, we note that

$$\frac{1}{r_{12}} = \sum_{k=0}^{\infty} \frac{r_<^k P_k(\cos\theta_{12})}{r_>^{k+1}}, \tag{7.15}$$

where θ_{12} is the angle between $\mathbf{r}_1$ and $\mathbf{r}_2$. The Legendre function P_k is expressed in terms of spherical harmonics. We now write explicitly

$$u(m_1^m \alpha_1' \lambda_1 \sigma_1; l_1 s_1 \alpha_1 L_1 S_1) = \sum_{\mu_{\lambda_1} \mu_{\sigma_1}} C^{\lambda_1\ l_1\ L_1}_{\mu_{\lambda_1} m_{l_1} M_{L_1}} C^{\sigma_1\ s_1\ S_1}_{\mu_{\sigma_1} m_{s_1} M_{s_1}}\, u(l_1^m \alpha_1' \lambda_1 \sigma_1)\, \Psi^{m_{s_1}}_{N_1\, l_1\, m_{l_1}}. \tag{7.16}$$

Since the parent $u(l_1^m \alpha_1' \lambda_1 \sigma_1)$ represents the "core" hole configuration which is not affected by the Auger transition,

$$\int u(l_1^m \alpha_1' \lambda_1 \sigma_1)\, u^*(l_1^m \beta_1 P_1 Q_1)\, \mathrm{d}\tau_m = \delta_{\lambda_1,\, P_1} \delta_{\sigma_1,\, Q_1} \delta_{\alpha_1',\, \beta_1}.$$

Thus only a single term in the sum over $(\alpha_1' \lambda_1 \sigma_1)$ in eqn (7.8) contributes to the Auger matrix element, namely, that corresponding to the quantum numbers β_1, P_1 and Q_1. The wave function occurring in this term is

$$u(l_1^m \beta_1 P_1 Q_1; l_1 s_1 \alpha_1 L_1 S_1) = \sum_{M_{P_1} M_{Q_1}} C^{P_1\ l_1\ L_1}_{M_{P_1}\ m_{l_1}\ M_{L_1}} C^{Q_1\ s_1\ S_1}_{M_{Q_1}\ m_{s_1}\ M_{S_1}} \cdot u(l_1^m \beta_1 P_1 Q_1)\, \Psi^{m_{s_1}}_{N_1\, l_1\, m_{l_1}}. \tag{7.17}$$

Similarly, the relevant wave functions occurring on the right-hand side of eqns (7.9) and (7.10) are

$$u(l_3^n \alpha_3 L_3 S_3; l_3 s_3 \beta_3 P_3 Q_3) = \sum_{M_{L_3} M_{S_3}} C^{L_3\ l_3\ P_3}_{M_{L_3}\ m_{l_3}\ M_{P_3}} C^{S_3\ s_3\ Q_3}_{M_{S_3}\ m_{s_3}\ M_{Q_3}} \cdot u(l_3^n \alpha_3 L_3 S_3)\, \Psi^{m_{s_3}}_{N_3\, l_3\, m_{l_3}} \tag{7.18}$$

and

$$u(l_4^p \alpha_4 L_4 S_4; l_4 s_4 \beta_4 P_4 Q_4) = \sum_{M_{L_4} M_{S_4}} C^{L_4\ l_4\ P_4}_{M_{L_4}\ m_{l_4}\ M_{P_4}} C^{S_4\ s_4\ Q_4}_{M_{S_4}\ m_{s_4}\ M_{Q_4}} \cdot u(l_4^p \alpha_4 L_4 S_4)\, \Psi^{m_{s_4}}_{N_4\, l_4\, m_{l_4}}. \tag{7.19}$$

Collecting all these results and combining the Clebsch–Gordan coefficients into 6-j and 9-j symbols, one obtains

$$W_{if}=\frac{2\pi}{\hbar}\frac{(2L_1+1)(2S_1+1)(2P_3+1)(2Q_3+1)(2P_4+1)(2Q_4+1)(2j_2+1)(2J_{12}+1)}{(2L_R+1)(2S_R+1)(2L_3+1)(2S_3+1)(2L_4+1)(2S_4+1)}$$

$$\cdot (2P+1)(2Q+1)(2J+1)\prod_{i=1}^{4}(2l_i+1)\sum_{A_i B_i C_4}(2C_4+1)\prod_{i=1}^{4}(2A_i+1)\cdot$$

$$\cdot (2B_i+1)\delta_{L_R,\,P_R}\,\delta_{S_R,\,Q_R}|M'|^2F^2, \tag{7.20}$$

where

$$F=[(m+1)(n+1)(p+1)]^{1/2}(l_1^{m+1}\alpha_1 L_1 S_1\{|l_1^m\beta_1 P_1 Q_1)\cdot$$

$$\cdot (l_3^{n+1}\beta_3 P_3 Q_3\{|l_3^n\alpha_3 L_3 S_3)(l_4^{p+1}\beta_4 P_4 Q_4\{|l_4^p\alpha_4 L_4 S_4) \tag{7.21}$$

The matrix element

$$M'=\sum_{\substack{f\ g\ d\\ f_1\ g_1\ d_1\\ b}}(-1)^{2f-f_1-g-g_1+d+d_1}I(fg)(2f+1)(2g+1)(2f_1+1)(2g_1+1)(2d+1)\times$$

$$\cdot (2d_1+1)(2b+1)\begin{Bmatrix}A_4 & g_1 & P_1\\ l_1 & L_1 & d\end{Bmatrix}\begin{Bmatrix}B_4 & f_1 & Q_1\\ \frac{1}{2} & S_1 & d_1\end{Bmatrix}\begin{Bmatrix}l_2 & l_1 & g\\ g_1 & P & d\end{Bmatrix}\begin{Bmatrix}\frac{1}{2} & \frac{1}{2} & f\\ f_1 & Q & d_1\end{Bmatrix}\cdot$$

$$\cdot\begin{Bmatrix}J & J_{12} & C_4\\ J_1 & b & j_2\end{Bmatrix}(-1)^{\theta}\begin{Bmatrix}l_2 & \frac{1}{2} & j_2\\ P & Q & J\\ \mathrm{d} & d_1 & b\end{Bmatrix}\begin{Bmatrix}L_1 & d & A_4\\ S_1 & d_1 & B_4\\ J_1 & b & C_4\end{Bmatrix}\begin{Bmatrix}A_1 & P_4 & P_3\\ A_2 & L_4 & L_3\\ g & l_4 & l_3\end{Bmatrix}\cdot$$

$$\begin{Bmatrix}P & A_3 & P_R\\ g & A_1 & A_2\\ g_1 & P_1 & A_4\end{Bmatrix}\begin{Bmatrix}B_1 & Q_4 & Q_3\\ B_2 & S_4 & S_3\\ f & \frac{1}{2} & \frac{1}{2}\end{Bmatrix}\begin{Bmatrix}Q & B_3 & Q_R\\ f & B_1 & B_2\\ f_1 & Q_1 & B_4\end{Bmatrix}. \tag{7.22}$$

The phase θ is given by

$$\theta=L_4+P_4-P_3-L_3+P_R-P_1-L_1+2J_{12}+A_1-A_4+$$

$$+S_4+Q_4-Q_3-S_3+Q_R-Q_1-S_1+B_1-B_4. \tag{7.23}$$

Radial functions enter eqn (16) via the expression

$$I(fg)=\sum_k D(k)\begin{Bmatrix}l_1 & l_3 & k\\ l_4 & l_2 & g\end{Bmatrix}+(-1)^{f-g}\sum_{k'}E(k')\begin{Bmatrix}l_1 & l_4 & k'\\ l_3 & l_2 & g\end{Bmatrix}, \tag{7.24}$$

$$D(k) = R^k(N_1 l_1, \eta_2 l_2; N_3 l_3, N_4 l_4) \begin{pmatrix} l_1 & k & l_3 \\ 0 & 0 & 0 \end{pmatrix} \begin{pmatrix} l_2 & k & l_4 \\ 0 & 0 & 0 \end{pmatrix}, \tag{7.25}$$

$$E(k) = R^k(N_1 l_1, \eta_2 l_2; N_4 l_4, N_3 l_3) \begin{pmatrix} l_1 & k & l_4 \\ 0 & 0 & 0 \end{pmatrix} \begin{pmatrix} l_1 & k & l_3 \\ 0 & 0 & 0 \end{pmatrix}. \tag{7.26}$$

D(k) and $E(k)$ are the direct and exchange radial matrix elements, $R^k(N_1 l_1, \eta_2, l_2; N_3 l_3, N_4 l_4)$ is the general Slater integral and N_1, N_3, N_4 are the principal quantum numbers associated with l_1, l_3, l_4.‡ As has been pointed out by McGuire, the radial matrix elements do not, in general, uncouple from the structure, i.e. the complete matrix element is not simply the radial matrix element times a structure factor. The coupling of angular momenta leads to summation over a number of dummy indices, namely, A_i, B_i, C_4, d, d_1, f, f_1, g, g_1 and b. Equation (7.20) embodies a general result applicable to the case where even the passive shells may be open. Of course, most of the earlier work assumes a simpler situation and the results obtained there follow as special cases of eqn (7.20).

Consider, for example, diagram Auger lines, i.e. lines resulting from transitions in atoms with a single initial vacancy in one of the inner shells and two final vacancies in the ground state configuration. Here $P_1 = Q_1 = 0$, i.e. the shell 1 becomes complete in the final state. The shells 3 and 4 are complete in the initial state, so that $L_3 = S_3 = 0$ and $L_4 = S_4 = 0$. The passive shells are also complete, i.e. $P_R = Q_R = 0$. This implies that $Q_3 = Q_4 = \frac{1}{2}$, $L_1 = l_1$, $S_1 = \frac{1}{2}$, $A_1 = A_3 = g$, $B_1 = B_3 = f$, $A_2 = B_2 = 0$, $A_4 = g_1 = P_R = 0$, $B_4 = f_1 = Q_R = 0$, $C_4 = 0$. Also, $d = l_1$, $d_1 = \frac{1}{2}$, $b = J_1$ and $J = J_{12}$, the vector sum of J_1 and j_2 as well as of P and Q. We finally get

$$W_{if} = (2j_2+1)(2J_{12}+1)(2P+1)(2Q+1)\prod_{i=1}^{4}(2l_i+1)\begin{Bmatrix} \frac{1}{2} & \frac{1}{2} & Q \\ l_1 & l_2 & P \\ J_1 & j_2 & J_{12} \end{Bmatrix}^2 |M''|^2,$$

where

$$M'' = \sum_k D(k)\begin{Bmatrix} l_1 & l_3 & k \\ l_4 & l_2 & P \end{Bmatrix} + (-1)^{P+Q}\sum_{k'} E(k')\begin{Bmatrix} l_1 & l_4 & k' \\ l_3 & l_2 & g \end{Bmatrix}. \tag{7.27}$$

This is the result which was obtained by El Ibayari *et al.* (1972) using the mixed coupling scheme. In the final state of this scheme, we assume *LS* coupling, i.e. l_3 and l_4 add up to give P, s_3 and s_4 give Q and, of course, P and Q couple to give J. To go to the *jj* coupling case, we can couple l_3 and s_3 to j_3, l_4 and s_4

‡ η_2 is the Sommerfield parameter for the Auger electron.

to j_4 and then j_3 and j_4 to J. The resulting final state can be obtained via the transformation

$$T: M'' \to \sum_{PQ} \sqrt{((2j_3+1)(2j_4+1)(2P+1)(2Q+1))} \begin{Bmatrix} S_4 & S_3 & Q \\ l_4 & l_3 & P \\ j_4 & j_3 & J \end{Bmatrix} M''. \tag{7.28}$$

We then obtain (McGuire, 1974a)

$$W_{if} = (2J+1)(2j_2+1)(2j_3+1)(2j_4+1) \prod_i (2l_i+1) \cdot$$

$$\cdot \left| \sum_k D(k) \begin{Bmatrix} j_3 & j_4 & J \\ j_2 & j_1 & k \end{Bmatrix} \begin{Bmatrix} \frac{1}{2} & l_4 & j_4 \\ k & j_2 & l_2 \end{Bmatrix} \begin{Bmatrix} \frac{1}{2} & l_3 & j_3 \\ k & j_2 & l_2 \end{Bmatrix} - (-1)^{J-2j_4} \sum_{k'} E(k') \times \right.$$

$$\left. \cdot \begin{Bmatrix} j_3 & j_4 & J \\ j_1 & j_2 & k' \end{Bmatrix} \begin{Bmatrix} \frac{1}{2} & l_4 & j_4 \\ k' & j_1 & l_1 \end{Bmatrix} \begin{Bmatrix} \frac{1}{2} & l_3 & j_3 \\ k' & j_2 & l_2 \end{Bmatrix} \right|^2 . \tag{7.29}$$

This is the result obtained by Asaad (1963) in the limit of jj coupling. In the limit of LS coupling, the transition rate is found by summing over j_2, J_{12} and J. For diagram lines this gives the well-known result

$$W_{if} = \frac{1}{2(2l_1+1)} \prod_{i=1}^{4} (2l_i+1) \sum_{fg} (2f+1)(2g+1)|I(fg)|^2. \tag{7.30}$$

McGuire (1974a) has also discussed the case of equivalent initial and final state holes. For example, an initial inner shell vacancy may be transferred from one subshell to another within the same shell so that only one of the participating shells experiences a change in the number of its vacancies. Using the language of this section, the initial state has an l_1 shell with $m+1$ holes and quantum numbers $\alpha_1 L_1 S_1 J_1$, a continuum hole with quantum numbers $l_2 s_2 j_2$, an l_3 shell with n holes and quantum numbers $\alpha_3 L_3 S_3$, and passive incomplete shells with quantum numbers L_R, S_R. The total angular momentum of this state is J. In the final state, the l_1 shell has $m+1$ holes and quantum numbers $\beta_1 P_1 Q_1$, the l_3 shell has $n+1$ holes and quantum numbers $\beta_3 P_3 Q_3$, and passive shells with quantum numbers P_R, Q_R. The basic difference is that we have to sum over m-hole parent states of the l_1 shell which are not observed. The expression F in eqn (7.21) is modified to

$$F' = \sum_{\alpha' \lambda' \sigma'} (-1)^{\lambda'+\sigma'} \sqrt{((m+1)(n+1)(m+1))} (l_1^{m+1} \alpha_1 L_1 S_1 \{| l_1^m \alpha' \lambda' \sigma') \cdot$$

$$\cdot (l_1^{m+1} \beta_1 P_1 Q_1 \{| l_1^m \alpha' \lambda' \sigma')(l_3^{n+1} \beta_3 P_3 Q_3 \{| l_3^n \alpha_3 L_3 S_3). \tag{7.31}$$

The transition rate

$$W_{if} = \frac{(2L_1+1)(2S_1+1)(2P_1+1)(2Q_1+1)\,(2P_3+1)(2Q_3+1)(2j_2+1)(2J_{12}+1)}{(2L_3+1)(2S_3+1)(2L_R+1)(2S_R+1)} \cdot$$

$$\cdot\,(2P+1)(2Q+1)(2J+1)\prod_{i=1}^{4}(2l_i+1)\sum_{A_i\,B_i\,C}(2C+1)\prod_{i=1}^{2}(2A_i+1)\cdot$$

$$\cdot\,(2B_i+1)\,\delta_{L_R,\,P_R}\,\delta_{S_R,\,Q_R}|F'M|^2, \tag{7.32}$$

where

$$M = \sum_{dd_1b}(2d+1)(2d_1+1)(2b+1)\begin{Bmatrix} J & J_{12} & C \\ J_1 & b & j_2 \end{Bmatrix}\begin{Bmatrix} l_2 & s_2 & j_2 \\ P & Q & J \\ d & d_1 & b \end{Bmatrix}\begin{Bmatrix} L_1 & d & A_1 \\ S_1 & d_1 & B_1 \\ J_1 & b & C \end{Bmatrix}\cdot$$

$$\cdot\sum_{gx}(-1)^{2g}(2x+1)(2g+1)\begin{Bmatrix} L_1 & l_2 & x \\ P & A_1 & d \end{Bmatrix}\begin{Bmatrix} P & A_1 & x \\ L_3 & A_2 & L_R \end{Bmatrix}\begin{Bmatrix} L_3 & A_2 & x \\ P_1 & l_3 & P_3 \end{Bmatrix}\cdot$$

$$\cdot\begin{Bmatrix} P_1 & l_3 & x \\ g & P' & l_1 \end{Bmatrix}\begin{Bmatrix} g & \lambda' & x \\ L_1 & l_2 & l_1 \end{Bmatrix}\sum_{f,y}(-1)^f(2y+1)(2f+1)\begin{Bmatrix} S_1 & s_2 & y \\ Q & B_1 & d_1 \end{Bmatrix}\cdot$$

$$\cdot\begin{Bmatrix} Q & B_1 & y \\ S_3 & B_2 & S_R \end{Bmatrix}\begin{Bmatrix} S_3 & B_2 & y \\ Q_1 & s_3 & Q_3 \end{Bmatrix}\begin{Bmatrix} Q_1 & s_3 & y \\ f & \sigma' & S_1 \end{Bmatrix}. \tag{7.33}$$

In the limiting case $\lambda' = \sigma' = L_3 = S_3 = L_R = S_R = 0$, this result reduces to eqn (7.27).

McGuire has made calculations on the observed satellite spectra $(3d)^{-1}(4p)^{-1} \rightarrow (4p)^{-3}$ and $(3d)^{-1}(4p)^{-1} \rightarrow (4s)^{-1}(4p)^{-2}$ in krypton using the alternative assumptions of (a) a statistical distribution of shake-off and (b) Auger decay of $3p$ vacancies as mechanisms for populating the $(3d)^{-1}(4p)^{-1}$ configuration (McGuire, 1974c). He found that the $(3d)^{-1}(4p)^{-1} \rightarrow (4p)^{-3}$ calculations with an initial $(3d)^{-1}(4p)^{-1}$ population distribution resulting from Auger decay of $3p$ vacancies were in excellent agreement with experiment, while the statistically distributed shake off hypothesis led to relatively poor agreement. For the transition $(3d)^{-1}(4p)^{-1} \rightarrow (4s)^{-1}(4p)^{-2}$, neither calculation was in good agreement with experiment, probably due to neglect of configuration interaction. The summed theoretical intensity for the initial $(3d)^{-1}(4p)^{-1}$ configuration populated by Auger decay of $3p$ vacancies gives better agreement than under the shake-off hypothesis.

Multiple inner-shell vacancies give rise to satellite spectra. We naturally ask: do outer-shell vacancies always behave as a passive structure? In the event that different outer-shell vacancy configurations give the same Auger energy, we can sum over the final state variables P and Q in eqn (7.20) and obtain an expression which is independent of P_R and Q_R, i.e. of the passive vacancy structure. However, the so-called passive shells may affect the Auger spectrum in two different ways.

1. Various distributions of outer-shell vacancies may shift the Auger energy in a continuous manner, thereby giving rise to a continuous spectrum.
2. Outer-shell vacancies caused by autoionization may affect the energetics of the system, provided that they have a lifetime comparable to that of the inner-shell vacancies. An experiment by Thomson *et al.* (1971) indicates that lifetimes of autoionizing states may not be significantly shorter.

3. Effects of electron correlation

One of the problems in Auger transitions in light elements is the poor agreement between theory and experiment with regard to relative intensities. In Section 4.5 we have seen how mixing between the $(2s)^{-2}$ and $(2p)^{-2}$ final vacancy configurations improves the relative intensities of the KLL lines of neon (Bhalla, 1973). An interesting point in this calculation is that interaction between these two configurations should directly affect the intensities of only the $KL_1L_1(^1S)$ and $KL_2L_2(^1S)$ lines belonging to these configurations. Actually, however, considerable improvement occurs in the intensities of the other lines over the Hartree–Fock–Slater results. Moreover, the intensity of the KL_2L_2 line relative to the KL_1L_1 line is still poor, being 0.99 as compared to the average experimental value of 1.52.

The fact that mixing between the two ^1S states, namely $L_1L_1(^1S)$ and $L_2L_2(^1S)$, indirectly affects the intensities of lines associated with other final vacancy states indicates the importance of electron correlations in these transitions. Obviously, the self-consistent-field potential does not completely describe the situation. The correlation energy E_{corr} of a system is defined by

$$E_{\text{corr}} = E - E_{\text{HF}}, \tag{7.34}$$

where E is the exact non-relativistic energy and E_{HF} the restricted Hartree–Fock value of energy. Since apart from exchange, the latter is based on an independent particle description of the atom, the difference E_{corr} is indeed a measure of correlation effects. A physically interesting, though not necessarily unique, way of looking at these effects is to incorporate them in a reasonable ansatz for the configuration-space wave function of the two participating electrons (Chattarji and Talukdar, 1973). Such calculations at least indicate that the inclusion of correlation improves the KLL Auger results in the right direction. A more rigorous and complete procedure is provided by the many-body perturbation theory of Brueckner (1955) and Goldstone (1957). Details of this theory may be found in the work of these authors. We shall briefly outline their method and describe its application by Kelly (1971, 1974) to Auger transitions.

Consider a system of n electrons with the Hamiltonian

$$H = \sum_{i=1}^{n} -\frac{\hbar^2}{2m}\nabla_i^2 + \sum_{i<j} v_{ij}, \tag{7.35}$$

where the first term represents the kinetic energy of the system and v_{ij} the interaction between particles i and j. Let V_i be a single-particle potential acting on the particle i in a reasonable independent particle model of the system. The Hartree–Fock potential V_{HF} is a typical example. We write

$$H_0 = H_{\text{HF}} = \sum_{i=1}^{n}\left(-\frac{\hbar^2}{2m}\nabla_i^2 + V_i\right), \tag{7.36a}$$

$$H_1 = \sum_{i<j} v_{ij} - \sum_{i} V_i, \tag{7.36b}$$

and

$$H = H_0 + H_1. \tag{7.36c}$$

Let Ψ_m be a series of one-particle eigenstates of H_0, so that

$$H_0 \Psi_m = E_m \Psi_m.$$

The bound states of this system are assumed to be discrete.

Let $a_m^\dagger$ and a_m be creation and annihilation operators for the state Ψ_m which satisfy the anticommutation relations

$$\{a_m, a_n^\dagger\}_+ = \delta_{mn}, \tag{7.37a}$$

$$\{a_m, a_n\}_+ = \{a_m^\dagger, a_n^\dagger\}_+ = 0, \tag{7.37b}$$

where $\{\ \}_+$ denotes the anticommutator. Then, using the formalism of second quantization, we can write

$$H_0 = \sum_n E_n a_n^\dagger a_n, \tag{7.38}$$

$$H_1 = \sum \langle pq|v|mn\rangle a_p^\dagger a_q^\dagger a_m a_n - \sum \langle p|V|m\rangle a_p^\dagger a_m. \tag{7.39}$$

Here $|m\rangle$ is a single-particle state of the type $\Psi_m(\mathbf{r}_1, \tau_1)$ and $|mn\rangle$ is an unsymmetrized two-particle state of the type $\Psi_m(\mathbf{r}_1, \tau_1)\Psi_n(\mathbf{r}_1, \tau_2)$. The symbol τ stands for the spin. The two-particle matrix element $\langle pq|v|mn\rangle$ implies the transition $\Psi_m \to \Psi_p$ and $\Psi_n \to \Psi_q$. Thus $\langle pq|v|mn\rangle$ is distinct from $\langle pq|v|nm\rangle$, but not from $\langle qp|v|nm\rangle$. The two-particle interaction in eqn (7.39) is summed over all distinct matrix elements, so that antisymmetry is automatically ensured by the anticommutation relations (7.37).

The basic idea is to construct the eigenstates of H from those of H_0 by using a time-dependent perturbation theory. It is assumed that H_0 has a non-degenerate ground state Φ_0 formed from the lowest Z of the single-particle states Ψ_n. Of these, the states which are occupied are called unexcited states. All higher states Ψ_n, are called excited states. An unoccupied unexcited state is called a "hole" and an occupied excited state is a particle. The Pauli exclusion principle is invoked by regarding an unexcited state as being automatically occupied, unless a hole is explicitly introduced. Thus the ground state corresponds formally to the "vacuum". Excitations are described by specifying departures from this vacuum state, i.e. the holes in unexcited states and the particles in excited states. Because occupied unexcited states do not explicitly enter the theory, there is an asymmetry between particles and holes. Thus, for an unexcited state Ψ_m, a_m represents a hole creation operator (but not a particle annihilation operator) and $a_m^\dagger$ a hole annihilation operator (but not a particle creation operator).

The lowest eigenstate Ψ_0 of H is derived from Φ_0 by adiabatically switching on the interaction H_1 over the time-interval $-\infty \leqslant t \leqslant 0$. For this purpose one writes

$$H_1(t) = \mathrm{e}^{iH_0 t/\hbar} H_1 \,\mathrm{e}^{-iH_0 t/\hbar} \mathrm{e}^{\alpha t}. \tag{7.40}$$

Then

$$\Psi_0 = \lim_{\alpha\to 0} \frac{U_\alpha \Phi_0}{\langle \Phi_0|U_\alpha|\Phi_0\rangle}, \tag{7.41a}$$

where

$$U_\alpha = \sum_{n=0}^{\infty} (-i\hbar)^n \int_{t_2}^{0} \mathrm{d}t_1 \int_{t_3}^{t_1} \mathrm{d}t_2 \int_{t_4}^{t_2} \cdots \int_{t_n}^{t_{n-2}} \mathrm{d}t_{n-1} \int_{t_{n+1}}^{t_{n-1}} \mathrm{d}t_n \, H_1(t_1)\, H_1(t_2) \ldots H_1(t_n). \tag{7.41b}$$

The adiabatic theorem states that

$$H\Psi_0 = (E_0 + \Delta E)\,\Psi_0, \tag{7.42a}$$

where

$$H_0\,\Phi_0 = E_0\,\Phi_0 \tag{7.42b}$$

and

$$\Delta E = \langle \Phi_0|H_1|\Psi_0\rangle = \lim_{\alpha\to 0} \frac{\langle \Phi_0|H_1\, U_\alpha|\Phi_0\rangle}{\langle \Phi_0|U_\alpha|\Phi_0\rangle}. \tag{7.42c}$$

The quantity ΔE represents the perturbation energy for the lowest state Ψ_0 of the perturbed system. To obtain the time-dependent form $H_1(t)$ of the perturbation, we substitute in eqn (7.39) the expression $a_n(t) = a_n e^{-iE_n t/\hbar}$, etc. and multiply by $e^{\alpha t}$.

Equation (7.41b) is the well-known perturbation series, so that Feynman graph techniques may be applied to obtain the correlation energy ΔE to any order n. Each graph represents a term in the series of H. Consider, for example, the graph in Fig. 7.1(a). A line in the direction of increasing time,

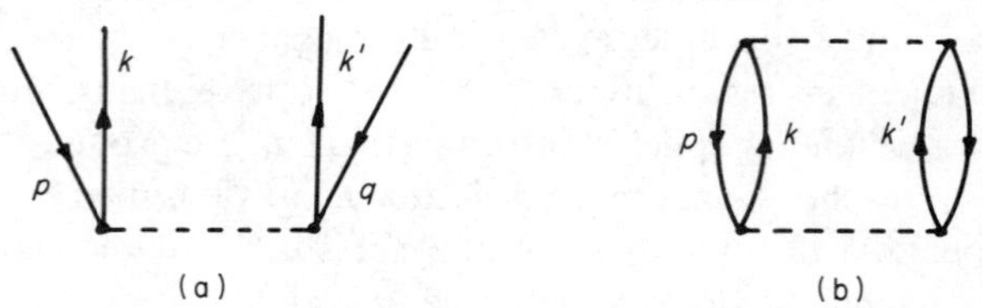

Fig. 7.1. (a) First order energy diagram for normal Auger transition in Brueckner–Goldstone many-body perturbation theory. This diagram represents the situation in a pure Hartree–Fock formalism, i.e. without correlation. (b) Second order energy diagram representing direct correlation.

i.e. a line directed upward, represents a particle in an excited state. A line in the opposite direction, i.e. one directed downward denotes a hole in an unexcited state. Thus, in this graph, we have two excited electrons k and k' in the final state and two unexcited bound electrons p and q in the initial state (i.e. two bound vacancies p and q in the final state). Such a diagram does not involve any correlation and represents the contribution to ΔE from the term $n = 1$ in eqn (7.41b) for the system just described. The corresponding matrix-element is $\langle kk'|v|pq\rangle$ and we have to sum over all k, k' to get the total contribution. It is easy to see that this represents the situation in a pure Hartree–Fock formalism.

The first non-trivial correlation diagram appears in the second order and is shown in Fig. 7.1(b). The corresponding contribution to ΔE is (Chase *et al.*, 1971)

$$\Delta E^{(2)}(p,q) = \sum_{kk'} \frac{|\langle kk'|v|pq\rangle|^2}{D} \tag{7.43}$$

where

$$D = \varepsilon_p + \varepsilon_q - \varepsilon_k - \varepsilon_{k'}. \tag{7.44}$$

The quantities ε represent single-particle energies. If both k and k' are continuum, then obviously $\varepsilon_p + \varepsilon_q \neq \varepsilon_k + \varepsilon_{k'}$ and the expression (7.43) is non-singular. However, if the excited state $|k\rangle$ represents an inner-shell vacancy, which would normally be occupied in the ground state of the atom, then D

may vanish for some particular excited state $|k'\rangle$. In this case, for a given k, we must write eqn (7.43) in the form

$$\Delta E^{(2)}(p,q) = \lim_{\eta\to 0} \sum_{k'} \frac{|\langle kk'|v|pq\rangle|^2}{D+i\eta}. \tag{7.45}$$

Now we use the well-known result (Dirac, 1958)

$$\lim_{\eta\to 0} \frac{1}{D+i\eta} = P\left(\frac{1}{D}\right) - i\pi\delta(D), \tag{7.46}$$

where P represents a principal-value integration over continuum states and $\delta(D)$ the Dirac delta function. Using $k_0 = [2m(\varepsilon_p + \varepsilon_q - \varepsilon_k)]^{1/2}/\hbar$ and $\varepsilon_{k'} = \hbar^2/k'^2/2m$,
we get

$$\Delta E^{(2)}(p,q) = -\frac{2m}{\hbar^2}\{P\sum_{k'} |\langle kk'|v|pq\rangle|^2\, [(k'-k_0)(k'+k_0)]^{-1} - -\pi i/k_0^2 |\langle kk_0|v|pq\rangle|^2\}. \tag{7.47}$$

The imaginary term on the right-hand side of eqn (7.47) represents a decay rate $2m/\hbar^3 k_0^2 |\langle kk_0|v|pq\rangle|^2$ corresponding to the Auger transition $pq \to kk_0$.‡

Kelly (1974) has used this theory to obtain *KLL* Auger transition rates for neon. In eqn (7.47) the Auger term results from integration of the delta-function term in eqn (7.46), so that one now needs to consider only that part of the energy diagram which is either above or below the singularity. Figure 7.2 shows the diagrams corresponding to the second-order perturbation theory. All these diagrams have the same external hole and particle lines as the first order Auger matrix element in Fig. 7.1(a) and therefore constitute correlation corrections to this matrix element. Some exchange diagrams are not explicitly shown, e.g. the matrix element $\langle 1sk'|v|pr\rangle$ in Fig. 7.2(a) may be replaced by $\langle k'1s|v|pr\rangle$ to get the exchange term. Kelly (1974) calculated single particle orbitals for the excited state in the field of the $Z-1$ other electrons by using, instead of the Hartree–Fock potential V_{HF}, the Silverstone–Yin–Huzinaga–Arnau (SYHA) potential (Silverstone and Yin, 1968; Huzinaga and Arnau, 1970). This is written as

$$V_{\mathrm{SH}} = V_{\mathrm{HF}} + (1-P)\Omega(1-P), \tag{7.48}$$

where Ω is an arbitrary Hermitian operator and P a projection operator

$$P = \sum_{i=1}^{Z} |\Psi_i\rangle\langle\Psi_i|. \tag{7.49}$$

For $i \leqslant Z$,

$$(1-P)\Psi_i = 0,$$

‡ The continuum state $|k_0\rangle$ is normalized to one particle per unit energy interval.

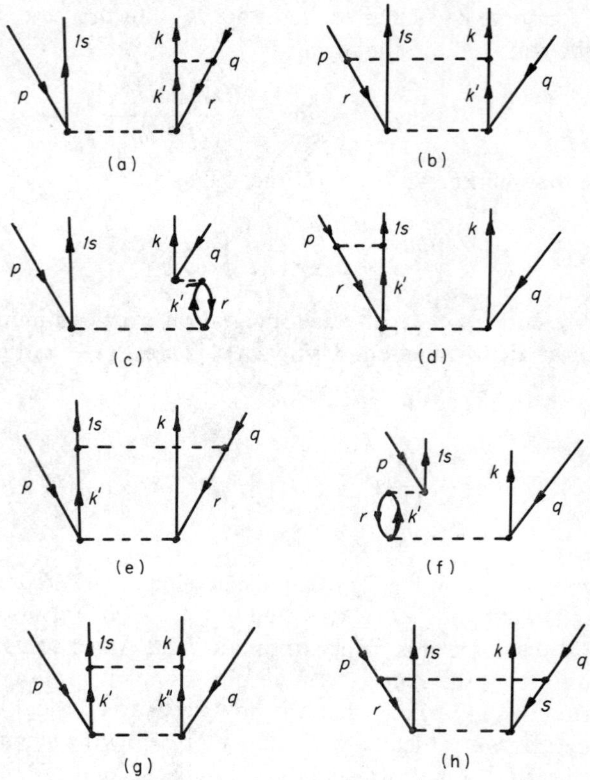

Fig. 7.2. Diagrams for correlation corrections to the Auger transition rate (from Kelly, 1974).

so that

$$V_{\mathrm{SH}}\,\Psi_i = V_{\mathrm{HF}}\,\Psi_i, \tag{7.50}$$

On the other hand, for $j > Z$,

$$V_{\mathrm{SH}}\,\Psi_j = (V_{\mathrm{HF}} + \Omega - P\Omega)\,\Psi_j. \tag{7.51}$$

Since they belong to the same Hermitian potential, the Ψ_i are all orthogonal. The operator Ω depends on the physical problem under consideration. For $l = 0$ orbitals, it is given by

$$\Omega(l=0) = J^0_{1s} - (2J^0_{2s} - K^0_{2s}), \tag{7.52}$$

where

$$J^k_{Nl}\,P_i(r) \equiv \int_0^\infty \mathrm{d}r'\,\frac{r^k_<}{r^{k+1}_>}\,P^*_{Nl}(r')\,P_{Nl}(r')\,P_i(r), \tag{7.53}$$

$$K^k_{Nl} P_i(r) \equiv \int_0^\infty dr' \frac{r^k_<}{r^{k+1}_>} P^*_{Nl}(r') P_i(r') P_{Nl}(r). \tag{7.54}$$

In other words, J^k_{Nl} and K^k_{Nl} are direct and exchange integral operators defined in terms of the Hartree–Fock radial functions $P_{Nl}(r)$.

The results obtained by Kelly (1974) for transition rates of the *KLL* Auger lines of neon are compared with the experimental values of Mehlhorn *et al.* (1968) in Table 11. The total Auger transition rate comes out to be 8.055×10^{-3}

Table 11. Auger transition rates for Neon with and without correlations. Theoretical results are compared with experimental values of Mehlhorn *et al.* (1968). (From Kelly, 1974)

	Theory			
Transition	Hartree–Fock result	With correlation in lowest order	With higher order corrections	Experiment (Mehlhorn *et al.*, 1968)
$KL_1L_1(^1S)$	0.9508	0.4879	0.4902	0.35 ± 0.07
$KL_1L_{2,3}(^1P)$	2.0335	1.3956	1.3670	0.96 ± 0.19
$KL_1L_{2,3}(^3P)$	0.7888	0.5015	0.4922	0.35 ± 0.07
$KL_{2,3}L_{2,3}(^1S)$	0.4560	0.8643	0.7707	0.55 ± 0.11
$KL_{2,3}L_{2,3}(^1D)$	5.6849	8.4476	4.9349	3.28 ± 0.066
Total rate	9.9140	8.4476	8.0550	5.49 ± 0.51

atomic units, whereas the experimental value is $(5.49 \pm 0.51) \times 10^{-3}$ a.u. The result obtained by Walters and Bhalla (1971a) in a Hartree–Fock–Slater calculation was 10.0×10^{-3}. Column 4 of this table shows the results with higher order corrections included (i.e. beyond the second order), whereas column 3 contains results with lowest order correlations, i.e. those in Fig. 7.2. Both of these are in much better agreement with experiment than the Hartree–Fock values obtained by Kelly in the lowest order, i.e. from Fig. 7.1(a). It will be seen that the individual rates are much more affected by correlation than the total Auger rate. This is because the most intense line, i.e. the ^1D line belonging to the $(2p)^{-2}$ configuration, is least affected by the inclusion of correlation.

Relative intensities of the *KLL* lines computed relative to the $KL_1L_1(^1S)$ line are shown in Table 12. The relative intensity of the ^1S line belonging to the $(2p)^{-2}$ configuration shows much better agreement with experiment than

Table 12. Relative intensities of KLL Auger lines of neon with respect to KL_1L_1 line. Theoretical results with and without correlations are compared with experiment (from Kelly, 1974)

Type of result	$KL_1L_{2,3}(^1P)$	$KL_1L_{2,3}(^3P)$	$KL_{2,3}L_{2,3}(^1S)$	$KL_{2,3}L_{2,3}(^1D)$
Hartree–Fock result obtained by Kelly (1974)	2.139	0.830	0.480	5.979
Correlation calculation by Kelly (1974) with higher order corrections	2.789	1.004	1.572	10.067
Bhalla (1973) with configuration interaction between $KL_1L_1(^1S)$ and $KL_{2,3}L_{2,3}(^1S)$	2.98	0.99	0.99	8.3
Experimental relative intensities by Körber and Mehlhorn (1966)	3.06 ± 0.07	1.00 ± 0.05	1.67 ± 0.08	13.1 ± 0.6
Experimental relative intensities by Stalherm (1968)	2.73 ± 0.04	1.00 ± 0.02	1.55 ± 0.03	9.31 ± 0.15
Experimental relative intensities by Krause *et al.* (1970) (excitation via electron impact)	2.87	1.06 ± 0.05	1.5 ± 0.1	10.00 ± 0.18

Table 13. KLL Auger line energies of neon in electronvolts (eV). Theoretical results of Kelly (1974) with and without correlation are compared with experiment (from Kelly, 1974)

	Theory		
Transition	Hartree–Fock results of Kelly (1974)	With correlation and relativistic corrections (Kelly, 1974)	Experiment (Körber and Mehlhorn, 1966)
$KL_1L_1(^1S)$	746.99	748.15	748.0 ± 0.1
$KL_1L_{2,3}$	770.86	771.71	771.4 ± 0.1
$KL_1L_{2,3}(^3P)$	783.01	782.45	782.0 ± 0.1
$KL_{2,3}L_{2,3}(^1S)$	800.97	801.27	800.4 ± 0.1
$KL_{2,3}L_{2,3}(^1D)$	806.04	804.51	804.15 ± 0.4

obtained by Bhalla (1973) with configuration interaction. The ^1D line belonging to the same configuration also agrees very well, particularly with the results of Krause *et al.* (1970).

Finally, the results obtained by Kelly for KLL Auger energies in this same calculation are shown in Table 13. They are in excellent agreement with the experimental values of Körber and Mehlhorn (1966). These energies are also corrected for relativistic effects.

Chapter Eight

Auger Electron Spectroscopy and its Application to Surface Science

1. Introduction

The earlier chapters have tried to build a theoretical framework for the understanding of the Auger process and to review the various methods for making calculations. The bias so far has been predominantly theoretical. It happens, however, that the Auger effect is far from being an esoteric phenomenon of interest only to the pure scientist. Auger spectra are being increasingly used as a tool in the investigation of solid surfaces. The necessity of preparing solid surfaces of known chemical purity in materials science and communication engineering cannot be overemphasized. Obviously, this calls for sensitive techniques for the detection of impurities. The use of Auger electrons for surface chemical analysis was first suggested by Lander (1953). Recent developments in Auger electron spectroscopy (AES) have led to its emergence as one of the most effective techniques in this kind of work. This is one more example of the interaction between pure science and technology. In this chapter, an account will be given of AES and its applications to surface studies.

The importance of surface science results from the fact that solids interact with their surroundings through their surfaces. Many important physical and chemical processes in solids such as catalysis, optical reflection, adhesion, corrosion, thermionic and photoelectric emission, etc. depend critically on the nature and condition of the surface. The term surface is used here to denote the top atomic layer or the top layer along with a few atomic layers beneath. The processes just mentioned depend on the structure and chemical composition of this region. Scientific interest in surfaces also stems from the fact that quite often they have properties with no counterpart in bulk solid state physics. With the development of new techniques for observing these surface properties and of theoretical models that translate observations into quantitative measures of the condition of the outermost layers of a solid, the study of surfaces is rapidly becoming a quantitative science.

A complete description of a solid surface involves answering the following questions:

1. What atomic species are present on the surface?

2. How are they arranged?
3. How are their valence electrons distributed?
4. What are their atomic motions?

These questions may be answered only by fitting together the results of many experiments in a consistent fashion. In order that this may be done, we must have a stable environment in which reproducible surfaces of atomically clean single crystals can be prepared and maintained for a sufficient time to permit detailed experimental study. Modern vacuum technology provides the surface scientist with such a controlled environment. Pressures in the 10^{-10} torr range are now easily attainable. The need for such high vacua can be seen from the following. Gas molecules collide with surfaces at rates that would deposit roughly one atomic monolayer (i.e. about 10^{15} atoms/sq. cm) per second at 10^{-6} torr, if all the molecules stick to the surface. Actually, sticking probabilities are usually much less than unity, particularly after the first monolayer has been adsorbed. Thus at pressures less than 10^{-8} torr there is usually sufficient time for obtaining data before the surface is appreciably altered. At a pressure of 10^{-10} torr, it takes about two hours for a monolayer to condense on a surface.

Various experimental techniques have been developed to examine surfaces. Some of these techniques are suitable for the study of a particular aspect of surface structure, while others may be more versatile. The general requirements of any surface analytical method are that it should be simple, non-destructive and of high sensitivity to the properties of the surface region rather than to bulk properties. It should give accurate quantitative information as far as possible and require no special specimen preparation. At present no single technique meets all these requirements and one or more of them has to be combined to get a comprehensive picture of the surface. The importance of AES lies in the fact that it nearly satisfies all the above conditions.

For a macroscopic specimen, an obvious way to explore the surface is by scattering experiments. The ultimate in sensitivity would of course be to image each individual atom, which can be done, for example, by means of the atom-probe field-ion microscope (Müller *et al.*, 1968). However, for a bulk sample, this would imply a colossal amount of information collected over a long interval of time, the measurement for each atom being the average of a large number of independent readings. Quite often what is actually needed is a statistical profile of the surface and that is exactly what a scattering experiment provides. For example, in determining the chemical composition of a surface, it is necessary to know the nuclear charges of the species present. This can be done by performing a scattering experiment in which the region of impingement of the projectile beam is large enough to provide a good sampling of the surface but still small by macroscopic standards. In principle, the projectile may be any incident particle having energy greater than the binding energy of a core electron of the element in question.‡ In practice, the most convenient and

‡ Since a core electron more nearly experiences the actual nuclear charge ($Z_{eff} \approx Z$), ionization of an inner shell is implied.

least destructive probes are X-ray photons and electrons (Duke and Park, 1972). For a complete description of the surface properties, all available techniques must be used so that both X-rays and electrons play important roles in surface analysis. However, electrons do have an edge over X-rays because of the following reasons (Grivet, 1972).

1. *Penetrating power* Penetration of the primary beam into the solid is limited basically by inelastic events. Electrons interact more closely with matter and have a vastly smaller penetrating power than X-rays. For example, an aluminium sheet 3 cm thick absorbs only 90% of the beam-intensity of hard X-rays ($\lambda = 0.2$ Å), while 5 μ of the same material can halt 100 keV electrons completely. Hence, use of a low-energy electron beam ensures that only a few surface layers are sampled.

2. *Monochromatism* By using suitable electron optics it is much easier to obtain a monochromatic electron beam than with X-rays. The latter generally consist of lines superimposed on an intense continuous background. To monochromatize such a beam, we must use either filters or selective reflection at a crystal, with a marked reduction in beam intensity.

3. *Power and intensity* Fine and intense electron beams are easily obtained, whereas this is quite difficult for X-rays owing to the generally low efficiency of X-ray tubes (~1/1000) and the difficulty in collimating X-ray beams. The low selective reflection coefficient of X-rays ($\sim 10^{-4}$) also implies long exposures.

Once a suitable probe has been decided upon, a method has to be devised for identifying the chemical species occurring in the surface region, with the help of this probe. The simplest approach in principle would be to directly measure the ionization energy, i.e. the energy required to create a hole in the core electronic state, which is closely related to the nuclear charge of the species involved. In practice, it may be easier to measure another quantity, namely the energy of the electronic transition involved in the subsequent recombination of this hole with a valence electron. Such recombination can take place either via radiative transitions or through the emission of an Auger electron. The different possible sets of primary and secondary particles participating in these processes are shown in Fig. 8.1. In Fig. 8.1(a) we have a core electron from a surface layer knocked out by an X-ray photon. This is basically what happens in ESCA (acronym for "electron spectroscopy for chemical analysis"), a technique developed by Siegbahn *et al.* (1967a). In this method, an X-ray beam of discrete energy is directed onto a solid specimen and photoelectrons emitted by the surface are collected and their energy analysed. Figure 8.1(b) shows ionization by electron impact. Figures 8.1(a) and (b) thus complete the possible primary ionization events with X-ray photons and electrons. Figure 8.1(c) represents the recombination of a hole, created by such an ionization event, with a valence electron via radiative transition, i.e. X-ray emission. If the primary particle causing the ionization is an X-ray photon, there is what is called X-ray fluorescence, i.e. the situation in Fig. 8.1(a) followed by that in Fig. 8.1(c). Because both the primary and secondary photons have a large penetration depth, X-ray fluorescence is not suitable for surface studies. Alternatively, we may have the event in 8.1(b) followed by that in 8.1(c), i.e. ionization by electron impact followed by emission of

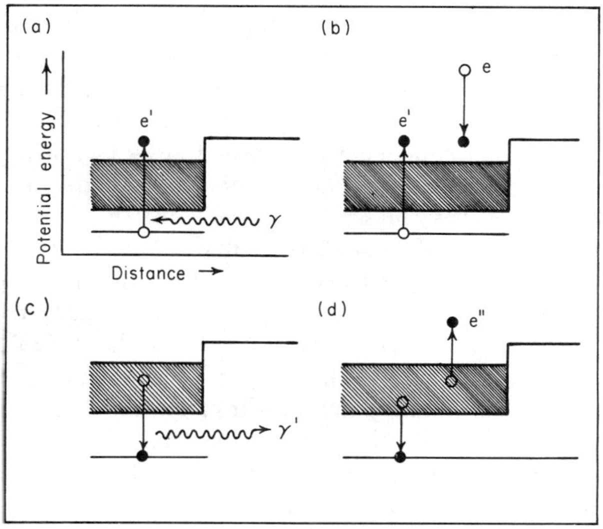

Fig. 8.1. Possible modes of core-level excitation and recombination in solid surfaces. The exciting species is either an X-ray photon (a) or an electron (b). In the recombination process, the emitted species is either an X-ray photon (c) or an electron (d). Atomic species can be identified from the energies of the ejected core electrons e', or from the characteristic X-rays γ' or Auger electrons e'' that result from recombination of the core hole. Only X-ray recombination following X-ray excitation is insensitive to surface region, because both the incident and emitted photons penetrate into the bulk solid easily (from Duke and Park, 1972, courtesy of *Physics Today*).

characteristic X-rays. This is the basis for the technique known as electron probe microanalysis (Castaing, 1951, 1960; Reuter, 1971). Spectral analysis of the X-ray signal is used here for the qualitative and quantitative determination of the sample composition. The method has a spatial resolution of better than 1 mμ and a relative sensitivity to within an error of about $\pm 5\%$. Unfortunately, the sampling depth ranges from 200 to 20 000 Å depending on operating conditions, so that, strictly speaking, this too is not a surface technique.

Finally, we come to the combination of events given by Figs 8.1(b) and 8.1(d) taken in succession, i.e. ionization by electron impact followed by the non-radiative emission of an electron. This is the familar Auger process described in the earlier chapters. The corresponding technique is Auger electron spectroscopy. If the primary particle is an energy-analysed electron in the range 1–3 keV, Auger electrons with energies usually less than 1 keV will be generated at depths up to the limit of penetration of the primary beam, which would be about 600–1000 Å. Auger electron spectroscopy is based on the fact that each element has a well-defined hierarchy of characteristic Auger lines—well defined with regard to both energies and relative intensities—which can serve as a kind of chemical signature for the element. In contrast to the atomic Auger spectra discussed in the earlier chapters, the lines arising from solid state layers exhibit some energy broadening due to environmental effects and appear as low peaks against a general inelastic background. Even then, with suitable instrumentation, Auger electrons from a particular transition can be recognized by their energy. It is this property which is utilized in AES. The fact that the method should have the desired surface sensitivity can be seen from the following. Let us focus our attention on a particular Auger peak associated with an element present in the solid. As the electrons generated within the solid travel towards the surface, they will lose energy to excitations of the valence electron fluid or plasma and undergo inelastic collisions with other electrons. Therefore, only the Auger electrons coming from the few outermost layers of the solid will be detected as part of the Auger peak in question. Auger electrons generated below these layers will lose enough energy to be scattered out of the peak into the inelastic background and will not be detected as Auger electrons. Typically, the mean escape depth of Auger electrons may extend from two to four atomic layers (4–8 Å) (Palmberg and Rhodin, 1968). Thus the Auger electrons actually detected come from the surface region.

It should be emphasized here that the results obtained by any one technique will not generally suffice for a complete description of the surface properties and the data obtained by any given method have to be supplemented with other measurements. Even in the determination of the chemical composition of a surface, other methods besides AES have been used, namely electron probe microanalysis, ion microprobe analysis, ion scattering, ESCA, field ion microscopy (FIM), field electron microscopy (FEM) and ion neutralization spectroscopy. Some of these are superior to AES with regard to one or more of the analytical requirements discussed earlier. For example, the electron probe is capable of very high absolute sensitivity ($\sim 10^{-14}$ g) and is particularly suitable where a sample can be irradiated by a focused electron beam of energy up to 30 keV. With the appropriate choice of the electron energy, it can even be adapted to the analysis of thin films deposited on a substrate so that the X-ray emission is actually confined to the film layer. However, de-excitation via X-ray emission, which is the physical process used in electron probe microanalysis, has a low probability, particularly for the lighter elements and for low primary electron energies. Below 2 keV Auger emission becomes a more efficient process and remains relatively constant with Z. Similarly, the ion microprobe is a quantitative device of great sensitivity (a few parts in 10^9), where a focused beam of rare-gas ions strikes the specimen and sputters atoms from the surface into a precision mass spectrometer. This method is destructive and cannot be readily combined with other techniques.

A word may be said here about ESCA, which is a powerful method for chemical analysis. As we have already indicated, in this method an X-ray beam of discrete energy is directed onto the specimen and electrons emitted from the surface are collected and energy analysed. Both Auger and photo-electrons are emitted, but attention is primarily focused on the photo-electrons because they have a simpler energy spectrum with relatively narrow and well-defined peaks. The emitting species can be more easily identified from the energies of the photoelectrons. Unfortunately, early workers overestimated the inelastic-collision damping length of the electrons and therefore presumed that they were investigating bulk rather than surface properties of the solid. As a result, not much effort was made towards employing ESCA as a surface method. Actually, the depth of sampling is determined, as in AES, by the escape depth of ejected electrons. It follows that ESCA is a surface sensitive technique. In fact surface sensitivity may turn out to be its most important property for practical use (Bank *et al.*, 1972). Its main drawback is that the cross-section for photo-electron emission is small compared to that for Auger emission with electron excitation, so that longer counting times are required. It is also relatively difficult to produce a well-focused X-ray beam, so that spatial resolution is likely to be poorer. However, rapid progress is now being made in ESCA technology. With the development of high transmission electron spectrometers and the technique of increasing intensity by pre-retarding the electrons before they enter the spectrometer (Helmer and Weichert, 1968), the time needed for scanning a spectrum has been reduced by a factor of 10 or 100. High absolute energy resolution is also being achieved.

Along with the chemical composition of the surface, we of course need to know the structure of the surface layers, i.e. the arrangement of atoms in these layers. Because of the advantages of using electrons in surface investigations, elastic electron diffraction naturally suggests itself as a possible tool. Two different geometries have been used: glancing-incidence reflection high-energy electron diffraction (RHEED) (Sewell and Cohen, 1967) and the near-normal-incidence reflection of 10–500 eV electrons called "low-energy electron diffraction" (LEED) (Estrup and McRae, 1971; Prutton, 1971). The basic principle underlying these techniques is the Bragg law and they can be simply understood from Fig. 8.2. Another simple way of representing the diffraction process is to look at the physical and reciprocal lattices.‡ In the limiting case when the separation a between the layers parallel to the surface goes to zero, so that $1/a \to \infty$, and we are left with the two-dimensional reciprocal "Bravais net" of the surface. The corresponding positions of the diffraction maxima are shown in Figs 8.3(a) and (b).

‡ i.e. a lattice characterized by a separation $1/a$, the inverse of the actual separation.

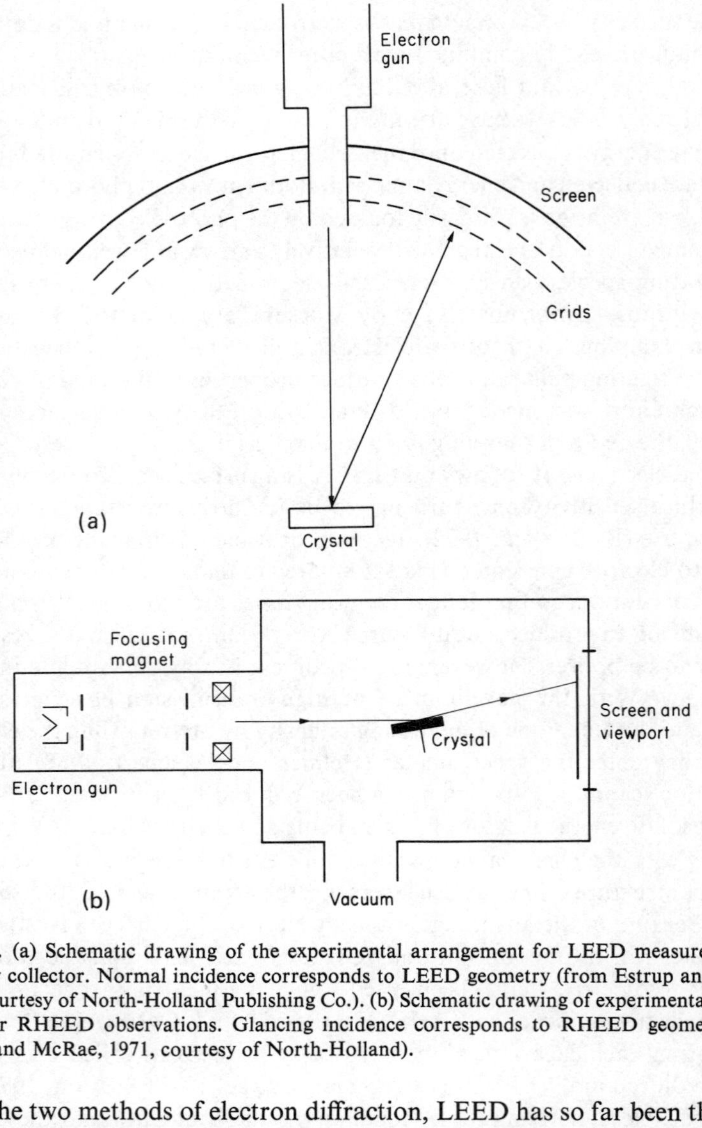

Fig. 8.2. (a) Schematic drawing of the experimental arrangement for LEED measurements by Faraday collector. Normal incidence corresponds to LEED geometry (from Estrup and McRae, 1971, courtesy of North-Holland Publishing Co.). (b) Schematic drawing of experimental arrangement for RHEED observations. Glancing incidence corresponds to RHEED geometry (from Estrup and McRae, 1971, courtesy of North-Holland).

Of the two methods of electron diffraction, LEED has so far been the more popular and productive. It is a true surface technique, sensitive only to the condition of the topmost layers of the solid, because an electron with an energy of about 100 eV moving normal to the surface will most likely lose 10–20 eV before it has penetrated more than 5–10 Å into the solid. With the kind of energy resolution now available, this guarantees that the elastically diffracted electrons coming from the top layers can be clearly distinguished from the "energy loss" electrons. As a surface-structure technique, LEED is obviously applicable only to single-crystal specimens. The question naturally arises: can

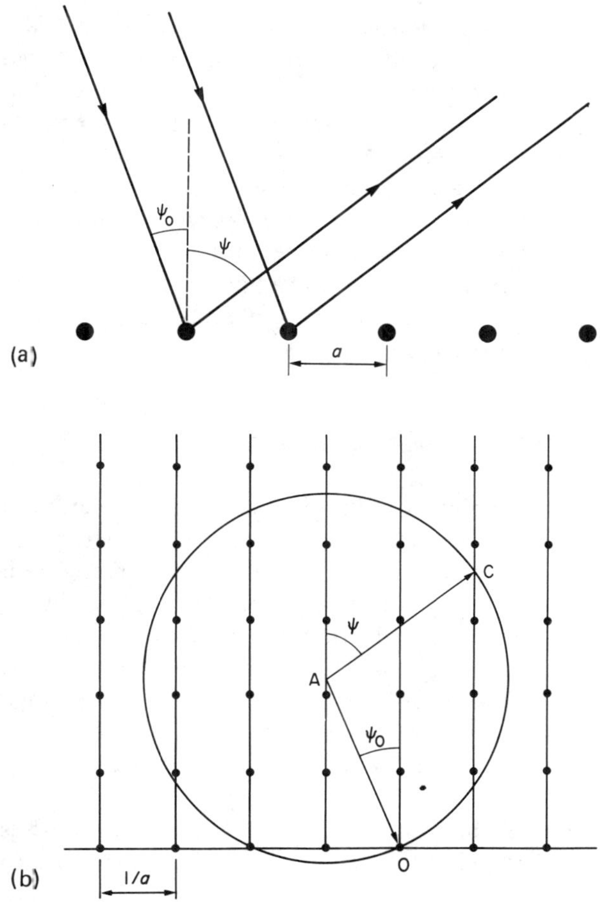

Fig. 8.3. (a) Geometrical construction of the Bragg diffraction condition $a(\sin \Psi_0 - \sin \Psi) = n\lambda$, a being the lattice constant and n the order of diffraction. Angles of incidence and diffraction, Ψ_0 and Ψ, are measured relative to the normal (from Estrup and McRae, 1971, courtesy of North-Holland). (b) Ewald sphere diagram of simple cubic lattice. Radius of the sphere is $1/\lambda$. AO represents the primary beam and AC one of the possible diffracted beams. If the intersection C coincides with a reciprocal lattice point, the Bragg condition is satisfied (from Estrup and McRae, 1971, courtesy of North Holland).

LEED be used to develop a surface crystallography as precise as the X-ray crystallography of the bulk? This was the hope expressed by Davisson and Germer (1927) in their classic paper on electron diffraction. Their expectation has so far been only partially fulfilled by later developments. For example, LEED patterns were used by Farnsworth *et al.* (1958) to show that surface atoms of Silicon and Gemanium are displaced from their normal lattice positions by surface forces. In fact this was the discovery that led to the growth of LEED during the 1960s'. LEED patterns can also exhibit the rearrangement of surface structures caused by the presence of adsorbed species. Unfortunately, electrons also interact inelastically with the solid so as to undergo multiple scattering. This complicates the interpretation of LEED patterns and the unambiguous assignment of atomic positions within the unit mesh.

Quite apart from the importance of LEED in the determination of the structure of solid surfaces, it has a bearing on any account of Auger electron spectroscopy, because the same method of primary ionization is used in both these techniques and the experimental arrangement for LEED can be conveniently adapted for Auger studies. Later in this chapter we shall see how the LEED-Auger analyser utilizes some of the basic features of a LEED unit to advantage in the analysis of Auger spectra.

2. Auger electron energies—general features

Auger spectroscopy explores the electronic energy levels in atoms and solids. Because of the discrete nature of most electronic energy levels, Auger electrons are best characterized by their energies. The latter can therefore be used to identify the chemical species emitting the Auger spectra. Such identification is made by measuring the energy distribution, $N(E)$, of secondary electrons, where E is the electron energy. In actual practice, experimental $N(E)$ curves contain much more information than just that concerning the Auger process. In the later sections of this chapter we shall see how the necessary information is extracted from these raw data. For the present let us review the basic features of Auger spectra.

By slightly modifying the notation of eqn (5.77) we can write Bergstrom and Hill's prescription for the kinetic energy of a $W_0X_pY_q$ Auger electron emitted by an atom,

$$E(Z) = E_{W_o}(Z) - E_{X_p}(Z) - E_{Y_q}(Z) - \Delta E_{X_pY_q} \quad (8.1)$$

$$= E_{W_o}(Z) - E_{X_p}(Z) - E_{Y_q}(Z) - \Delta Z_{X_p}[E_{Y_q}(Z+1) - E_{Y_q}(Z)]. \quad (8.1a)$$

Here $E_{X_p}(Z)$ is the binding energy of the X_p shell for atomic number Z and ΔZ_{X_p} the "effective incremental charge" for the same atom with a vacancy in the X_p shell ($0 < \Delta Z_{X_p} < 1$). With the atom already ionized in this shell, the Y_q electron experiences an effective nuclear charge greater than Z and the last term in eqn (8.1a), namely $\Delta Z_{X_q}[E_{Y_q}(Z+1) - E_{Y_q}(Z)]$, corrects for this fact. Values of ΔZ can be obtained by fitting eqn (8.1a) to experimentally observed binding energies. It has been found that ΔZ changes only slowly with Z so that it is possible to use a value, once obtained, over a wide range of elements (Burhop and Asaad, 1972).

It will be noticed that although eqns (5.76) and (8.1) are formally symmetric in X_p and Y_q, eqns (5.77) and (8.1a) are not. Equation (8.1a) treats the Auger process as an external conversion of the X-ray photon produced by the transition $X_p \rightarrow W_0$ as a result of which an electron is ejected from the shell Y_q. Since the two holes in the shells X_p and Y_q are actually indistinguishable, it is desirable to have an empirical energy equation which is symmetric in X_p and Y_q. Such an alternative expression was suggested by Geffrion *et al.* (quoted by Erman and Sujkowski, 1961):

$$\left(\frac{1}{\Delta E_{X_pY_q}}\right)^n = \left(\frac{1}{E_{Y_q}(Z+1) - E_{Y_q}(Z)}\right)^n + \left(\frac{1}{E_{X_p}(Z+1) - E_{X_p}(Z)}\right)^n, \quad (8.2)$$

where n is an empirically determined integer. $E(Z)$ can be found by substituting this value of $\Delta E_{X_pY_q}$ in eqn (8.1). By using $n = 1$ and 2, comparison has been made with measured values for the *KLL* spectrum of caesium, but the agreement is no better than that given by eqn (8.1a).

More recently, Chung and Jenkins (1970) developed an approximate formula in which they took the effective binding energy for a doubly ionized shell to be given by the average of the binding energies corresponding to atomic numbers Z and $Z + 1$. Using this ansatz in eqn (5.75), we get

$$E(Z) = E_{W_0}(Z) - \tfrac{1}{2}[E_{X_p}(Z) + E_{X_p}(Z+1) + E_{Y_q}(Z) + E_{Y_q}(Z+1)]. \quad (8.3)$$

This formula is obviously symmetric in X_p and Y_q. It has a typical accuracy of the order of 5 eV.

Because of the importance of Auger electron energies in surface chemical analysis, it would be good to have accurate first principle predictions by solving the Schrödinger equation for the initial and final states. Such attempts have so far been confined mainly to *KLL* transitions (Siegbahn *et al.*, 1967a) and may require semi-empirical estimates of some quantities (Mehlhorn and Asaad, 1966). First principle Hartree–Fock calculations have been carried out by Larkins (1971a, b) on the *KLL* spectrum of neon and the *LMM* spectrum of argon with reasonable success. The accuracy of these calculations is not, however, appreciably better than that of the empirical estimates given above. These expressions provide a fair starting point, particularly in view of the instrumental and environmental uncertainties inherent in experimental methods available so far.

KLL Auger spectra have been the subject of particularly close experimental and theoretical study. As we know, the nine lines of this spectrum given by intermediate coupling merge into six lines for high Z values (*jj* coupling) and five lines for low Z values (*LS* coupling). With the analysing voltages and resolutions of the experimental equipment we shall discuss in this chapter, only the *KLL* spectrum of elements up to silicon ($Z = 14$) are conveniently measured, the highest *KLL* energy for silicon being about 1.6 keV. At these energies, the competing process of X-ray emission is negligible. The escape depth of the Auger electrons is also small, so that the method is surface sensitive.

Table 14. *KLL* Auger energies in electron-volts (eV) for elements with $6 \leqslant Z \leqslant 14$ (from Siegbahn *et al.*, 1967a).

Configuration and spectra / Elements	$2s^0 2p^6$	$2s^1 2p^5$				$2s^2 2p^4$			
	$^1S_0$	$^1P_1$	$^3P_0$	$^3P_1$	$^3P_2$	$^1S_0$	$^1D_2$	$^3P_0$	$^3P_2$
	KL_1L_1	KL_1L_2	KL_1L_2	KL_1L_3	KL_1L_3	KL_2L_2	KL_2L_3	KL_3L_3	KL_3L_3
C 6	243	252	258	258	258	265	266	267	267
N 7	356	362	369	369	369	373	375	377	377
O 8	474	486	495	495	495	504	507	509	509
F 9	610	627	638	638	638	650	654	657	657
Ne 10	761	781	794	794	794	808	813	816	816
Na 11	928	952	967	967	967	984	989	993	993
Mg 12	1105	1135	1151	1151	1151	1172	1179	1183	1183
Al 13	1301	1336	1354	1354	1354	1379	1387	1392	1392
Si 14	1516	1554	1574	1574	1574	1602	1611	1616	1617

The energy resolution is usually such that not all six lines are observed. Table 14, taken from Siegbahn *et al.* (1967a), lists the Auger energies of the KLL group from carbon ($Z = 6$) to silicon ($Z = 14$) based upon precision measurements and semi-empirical expressions. Even for silicon, the $KL_3L_3(^3P_{0,2})$ line is so close to the $KL_2L_3(^1D_2)$ line that it is not observed. The $KL_2L_2(^1S_0)$ and $KL_2L_3(^1D_2)$ lines are also close together, so with an instrumental line width of 0.5% they cannot be resolved.‡ Therefore only four components of the KLL spectrum are resolved. Figure 8.4(a) shows the KLL spectrum for silicon as obtained with the retarding field analyser which will be described later in this chapter.

As one goes to higher atomic numbers, the Auger spectra become more complex. Fortunately, some of the peaks of the M, N groups, etc. are quite weak, so that over the available energy range the spectra are not as crowded as one might expect. The stronger peaks belonging to the lighter surface species are generally more prominent than the weaker peaks belonging to heavier elements.

When atoms combine to form a solid, the basic features of the resulting changes can be understood in a qualitative way by considering the perturbations in the energy levels of an atom which are caused by the presence of its neighbours in the crystal lattice array. The energy of an ion in a lattice is of the order of several electron volts. Since this is a hundred to a thousand times smaller than the binding energy of the inner (i.e. X-ray) electronic levels of an atom, the inner levels are only slightly affected by the presence of the neighbouring atoms. They are at best somewhat shifted in energy, but remain quasi-atomic in character and the atomic notation provides an adequate description. The outer, or optical, levels are, however, very considerably affected because of their low binding energy. They may be thought of as being shifted variously in magnitude and constituting a very great number of levels characteristic of the crystal as a whole rather than of the individual atoms composing it. The energy range over which these valence levels occur is of the order of only a few electron volts, so that they form an effectively continuous "valence band" within the crystal. The situation in this valence band is best described by the density of states function. Figure 8.4(b) shows the energy level structure in a silicon ($Z = 14$) crystal. The binding energies for the X-ray levels K, L and M, etc. are shown on the left and the level designations on the right. The density of states for the valence band is also drawn in. A hole in the valence band is usually assigned the letter V.

Where valence levels are involved particularly in the conduction band of metals, the two holes in the conduction band may not interact very strongly owing to screening by the electron gas comprised of the free electrons in this

‡ Instrumental line width (ILW) is defined as the ratio $\Delta E/E \times 100\%$, where ΔE is the full width at half maximum (FWHM).

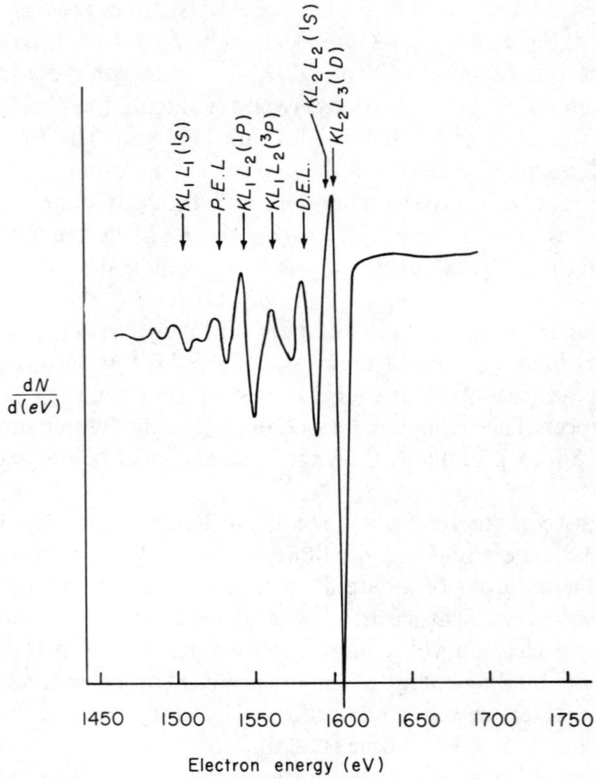

Fig. 8.4. (a) Derivative of *KLL* spectrum of silicon obtained with retarding field analyzer (from Taylor, 1971, courtesy of Wiley Interscience).

band. The Chung and Jenkins formula then assumes the form

$$E_{WXV}(Z) = E_W(Z) - \tfrac{1}{2}[E_X(Z) + E_X(Z+1)] - E_V \text{ (Metal)}, \qquad (8.3a)$$

or

$$E_{WVV}(Z) = E_W - 2\, E_V \text{ (Metal)}, \qquad (8.3b)$$

depending on whether we have a transition of the *WXV* or the *WVV* type. Equations (8.3a and b) were given by Coad and Rivière (1971). In these equations, E_V (Metal) is the most prominent conduction band energy measured relative to the vacuum level.

In writing an expression for the Auger energy we have so far neglected the work done by the ejected electron in overcoming the potential barrier at the surface, i.e. the work function φ. Taking this into account and using the

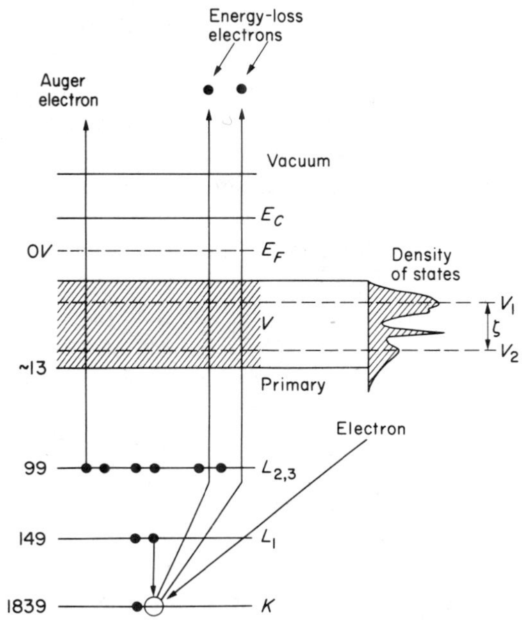

Fig. 8.4. (b) Electronic energy levels in singly ionized silicon atom. Energy levels are listed on the left (in eV) with zero of energy at the Fermi level E_F. E_c is the bottom of the condition band. A $KL_1L_{2,3}$ Auger process is shown after primary electron ionization (from Chang, 1974, courtesy of Plenum Press).

Chung–Jenkins ansatz, the energy of the *WXY* Auger electron becomes

$$E_{WXY}(Z) = E_W(Z) - \tfrac{1}{2}[E_X(Z) + E_X(Z+1) + E_Y(Z) + E_Y(Z+1)] - \varphi. \tag{8.4}$$

We now remember the fact that the energy of the electron passing through the analyser is now measured relative to the Fermi level of the analyser material. This introduces an additional term $-(\varphi_A - \varphi)$, the work function of the analyser *A* measured relative to the material under investigation. We then have

$$E_{WXY}(Z) = E_W(Z) - \tfrac{1}{2}[E_X(Z) + E_X(Z+1) + E_Y(Z) + E_Y(Z+1)] - \varphi_A. \tag{8.5}$$

Analogous equations may be written for transitions involving valence band electrons.

3. Auger electron spectroscopy as a surface technique—historical background

The history of Auger electron spectroscopy as a method of surface chemical analysis dates back to Lander (1953). Using a 90° spherical electrostatic analyser, he identified certain peaks in the secondary electron energy spectra of various materials bombarded by low-energy electrons as Auger peaks and suggested that this could be of use in surface analysis. His work was followed up to a certain extent by Harrower (1956), who tried to use the "energy loss peaks" due to primary electrons that have lost discrete amounts of energy to the electron gas through plasma excitation, to supplement the characteristic X-ray tables at low energies. Actual use of the energy analysis of Auger electrons in surface studies started only recently. Interest was renewed as a result of work carried out by two separate groups.

1. Tharp and Scheibner (1967) showed that the Auger peaks could be detected by means of a three-grid LEED arrangement used as a retarding field analyser. The LEED primary beam bombarded the surface under investigation and the energy distribution of the secondary electrons was obtained by the technique of electrical differentiation of the current-versus-potential plot of the analyser.
2. The idea of electrical differentiation of the retarding field plot had been suggested earlier by Leder and Simpson (1958). They used it in a convenient combination with LEED and showed, by monitoring the Auger peaks, that changes in surface composition could occur and yet go undetected by LEED. Differentiation played a particularly important role because of the following circumstances. When a solid is irradiated by an electron beam, the resulting Auger electrons are superimposed on a large and varying background of secondary electrons.

Before we can use the Auger peaks as an analytical tool, they must be separated from this background. Differentiation of the secondary current sharply accentuates the Auger peaks in contrast to the background. A major contribution in this direction was made by Harris (1968a, 1968b). He first achieved marked enhancement in the detectability of the Auger peaks through electronic differentiation of the measured energy distribution function, and convincingly demonstrated the application of this technique in a large number of cases. Harris used a cylindrical electrostatic sector analyser in his work. However, Weber and Peria (1967) showed that a three-grid LEED system could also be used quite effectively to obtain this derivative. This led to the progressive use of AES, both to complement LEED measurements and also as an independent non-destructive technique of surface analysis.

Palmberg *et al.* (1969) have shown that the cylindrical mirror analyser can be used to measure Auger spectra so as to increase the sensitivity by one or two orders of magnitude relative to the other analysers previously used. This increased sensitivity can be traded, if desired, for much greater scan speeds.

Regardless of the type of analyser used, the emergence of AES as the most important technique in surface chemical analysis in recent years can be ascribed to two reasons.

1. When a solid is irradiated with a beam of electrons having energy in the 100 eV–3 keV range, most of the resulting Auger electrons originate within the

first few atomic layers of the surface. The depth of excitation depends of course on the energy of the incident beam and the angle of incidence. But the Auger electrons leaving the surface for the most part originate from a much shallower surface region because of their lower energy. Thus the technique is very sensitive to surface conditions. Most of the possible surface contaminants are composed of atoms of low atomic number, for example, adsorbed gases such as carbon monoxide, hydrocarbons or water, or small amounts of material reaching the surface through surface segregation from the bulk of the solid or by diffusion. The Auger spectra of such light elements are quite easily detected.

2. The other secondary mode resulting from the irradiation, namely X-ray fluorescence, drops sharply as the atomic number Z decreases. In fact it becomes negligible for values of E_W (i.e. the binding energy corresponding to the initial vacancy) less than 500 eV.‡ The Auger current, however, remains relatively constant with Z provided that one chooses to study the lower energy transitions for the heavier elements. Thus X-ray fluorescence becomes progressively less competitive for low Z values. The K-shell fluorescence yield W_K for silicon ($Z = 14$) is about 0.05, whereas for oxygen ($Z = 8$) it is only about 0.005. For high-Z materials, the sensitivity of AES is still useful for transitions involving primary vacancies in the outer shells. Even though the spectra become more complicated, for any given element there are some prominent lines which can be sought if presence of that element is suspected in the surface.

As for the source of primary excitation, we have already enumerated some advantages of using electrons rather than X-rays. Electron beams can be produced relatively easily and inexpensively with the electron gun. The cross-section for ionization of a given level is also greater with electrons than with X-rays of comparable energy.

4. The secondary electron spectrum

To observe the Auger electrons emitted from the topmost layers of a solid, it is necessary to measure the energy distribution $N(E)$ of the secondary electrons as a function of the energy E. A typical $N(E)$ curve obtained by irradiating a solid with primary electrons of energy E_P is shown in Fig. 8.5. This curve shows three major features.

I. A sharp peak at the primary energy E_P containing electrons elastically reflected without energy loss.

II. A region with relatively few electrons, sometimes called the region of rediffused primary electrons, lying between the features (I) and (III).

III. A somewhat broader peak at energy $E = 0$ with a half-width of at least 10 eV.

The elastic peak at E_P represents the primary electrons used in electron diffraction work—a LEED or RHEED set-up could be used. This peak is

‡ Obviously, E_W for a given shell W falls off with decreasing Z.

useful as a fiduciary energy for calibrating the energy scale, for aligning the Auger analyser and for measuring the analyser resolution. The small peaks close to the elastic peak are produced by incident electrons which have undergone well-defined energy losses. The most common source of such energy loss is the excitation of collective plasma oscillations of the electrons in the solid. This gives rise to a series of plasma peaks of decreasing magnitude spaced by energy $\hbar\omega_P$, where ω_P is the frequency of plasma oscillation (Pines, 1955).

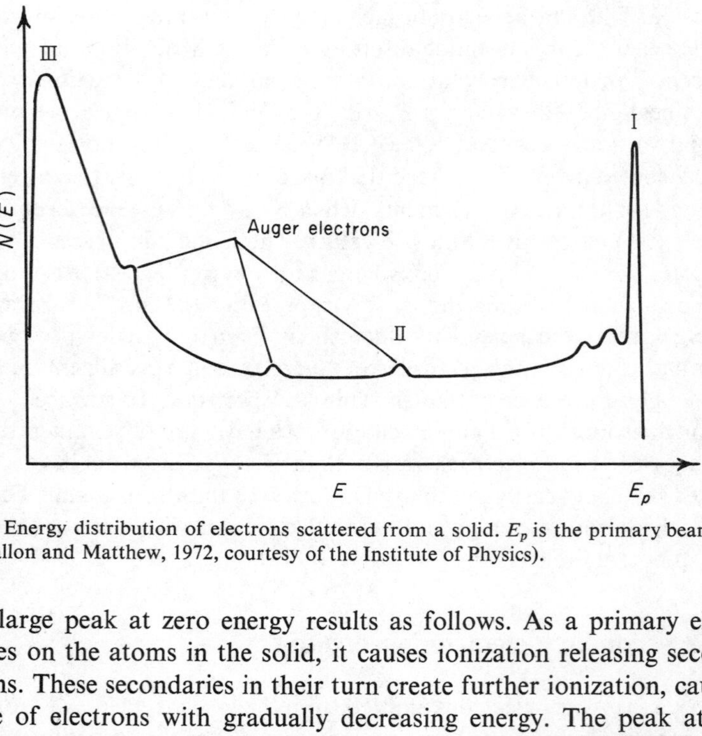

Fig. 8.5. Energy distribution of electrons scattered from a solid. E_p is the primary beam energy (from Gallon and Matthew, 1972, courtesy of the Institute of Physics).

The large peak at zero energy results as follows. As a primary electron impinges on the atoms in the solid, it causes ionization releasing secondary electrons. These secondaries in their turn create further ionization, causing a cascade of electrons with gradually decreasing energy. The peak at $E=0$ would shoot up to a great height but for the fact that the secondary electrons registered by the analyser must overcome the surface work function, which the less energetic electrons cannot do.

Region (II) in Fig. 8.5 contains inelastically scattered primary electrons and the more energetic secondary electrons. This is the region where most of the Auger peaks occur. In the N(E) versus E plot, these peaks are just resolved. Being superimposed on a variable background where the total secondary current is much larger than the Auger current, the Auger peaks are only barely discernible and need to be accentuated before any quantitative information can be gained from them. This is accomplished by differentiating the energy distribution so as to get the $dN(E)/dE$ curve shown in Fig. 8.6. Here each Auger peak is represented by a double peak which stands out clearly from the background. This differential method has led to the high sensitivity and practical utility of AES as a technique for surface analysis.

An important point in the analysis of the secondary spectrum is to distinguish the Auger peaks from the energy loss peaks. This is done by remembering that for a given element the Auger energies are quite well defined and independent of the primary energy. On the other hand, for the energy loss spectra, it is the

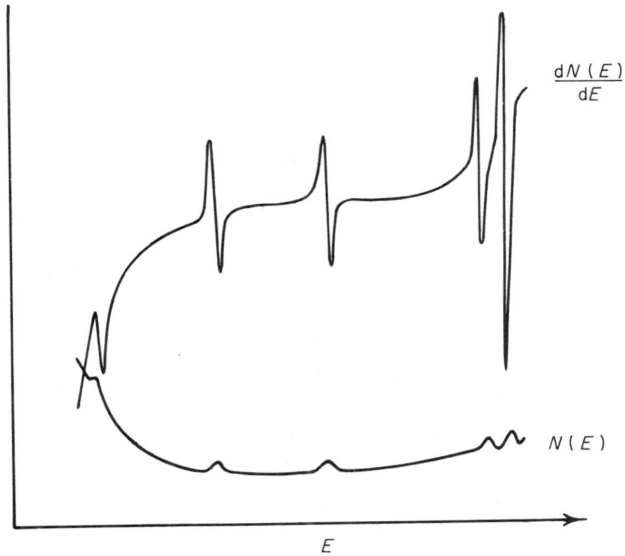

Fig. 8.6. The differential of the energy distribution of electrons shown in Fig. 8.5. (from Gallon and Matthew, 1972, courtesy of the Institute of Physics).

energy loss which is well defined with respect to the primary energy. Thus a shift in the primary energy causes a corresponding shift of the energy loss spectra.

A. Energy loss peaks

These peaks are due to primary electrons that have lost discrete amounts of energy by exciting collective oscillations of the valence electrons in the solid—the plasma oscillations. Alternatively, plasma gain may occur when the plasma loses energy to the primary electrons. A physical picture of these excitations was given by Pines and Bohm (1951, 1952, 1953) where the electron gas displays both collective and individual particle aspects. In a collective oscillation, the system oscillates as a whole—each individual electron suffering a small periodic perturbation of its velocity and position due to the combined Coulomb interaction of all the other particles. Because of the long range of this interaction, a very large number of electrons contribute to the potential at a given point. The energy of the plasma oscillations, or plasmons, can be calculated by considering the motion of a homogeneous mixture of electrons and positive cores. Classically, this is done as follows (Chang, 1974). Let the electron gas momentarily expand radially outward from some origin, $\xi(r)$ being the radial displacement at r. The number of electrons leaving a sphere of radius r is $4\pi nr^2\xi(r)$, where n is the electron density. This gives rise to a radial electric field

$$E(r)=\frac{4\pi ner^2\,\xi(r)}{r^2}=4\pi ne\xi(r),$$

and a force

$$F=-4\pi ne^2\,\xi(r)$$

on the electrons, where e is the electronic charge. Equating this to the force of restitution $m\ddot{\xi}(r)$,

$$\ddot{\xi}+\frac{4\pi ne^2}{m}\xi=0. \tag{8.6}$$

This implies a harmonic motion with the oscillation frequency

$$\omega_P=\left(\frac{4\pi ne^2}{m}\right)^{1/2}, \tag{8.7}$$

and energy

$$\varepsilon_P=\hbar\omega_P. \tag{8.8}$$

In metals, ω_P is of the order of 10^{16} radians/s, and $\hbar\omega_P$ is about 15 eV. In a crystal, m is replaced by the effective mass m^* and the band structure of the material enters into the calculation. This classical description can now be translated into the quantum mechanical domain.

It was shown by Bohm and Pines (1953) that an electron gas embedded in a uniform background of positively charged particles in a unit volume can be described, in a second-quantized formulation, by the Hamiltonian

$$H=\sum_k c_{k\sigma}^*\,c_{k\sigma}\frac{\hbar^2k^2}{2m}+2\pi e^2\sum_k\frac{\rho_k\,\rho_{-k}}{k^2}, \tag{8.9}$$

where $c_{k\sigma}$ and $c_{k\sigma}^*$ are the fermion annihilation and creation operators, k being the wave number and σ the two-valued spin. ρ_k is the transition density operator

$$\rho_K=\sum_k c_{k\sigma}^*\,c_{k+K\,\sigma}. \qquad \begin{array}{l} k=m\pi,\ m \text{ an integer} \\ \sigma=+\tfrac{1}{2}\text{ or }-\tfrac{1}{2}. \end{array} \tag{8.10}$$

Now, ρ_k consists of a collective part q_k and an individual electron part due to random motion of the electrons. Defining the displacement operator for frequency ω

$$\xi_{K,\omega}=\sum_k \frac{c_{k\sigma}^* c_{k+K\,\sigma}}{\omega-\left(\hbar\mathbf{k}\cdot\mathbf{K}-\dfrac{\hbar K^2}{2m}\right)} \tag{8.11}$$

and

$$q_K=\tfrac{1}{2}[(\xi_{K,\omega}-\xi_{K,-\omega})/\omega], \tag{8.12}$$

it can be shown that the $\xi_{K,\omega}$ satisfy

$$\dot{\xi}_{K,\omega}+i\omega\xi_{K,\omega}=0 \tag{8.13}$$

and

$$\ddot{q}_K+\omega^2 q_K=0. \tag{8.14}$$

It happens that $\omega=\omega_P$, so that eqn (8.14) is the quantum analogue of eqn (8.6). Obviously, q_K is a generalized coordinate operator. In the usual coordinate representation, it has the form

$$q_K=\sum_i \frac{1}{\omega^2-(\mathbf{K}\cdot\mathbf{P}_i/m-\hbar K^2/2m)^2}\exp(-i\mathbf{K}\cdot\mathbf{X}_i), \tag{8.15}$$

i.e. it is a collective normal mode of the system resulting from a superposition of plane waves.

At the surface of a solid, the electric field is weaker and the ideal "surface plasmon" has a smaller energy (Pines, 1964)

$$\hbar\omega_S=\frac{1}{\sqrt{2}}\hbar\omega_P. \tag{8.16}$$

If the surface is covered by a layer with dielectric constant ε, the surface plasmon energy becomes

$$\hbar\omega_{ms}=\frac{1}{\sqrt{1+\varepsilon}}\hbar\omega_P. \tag{8.17}$$

5. Auger electron spectra of atoms in solids

Auger electron spectroscopy of atoms in a solid consists of three stages, namely:

1. Production of the initially ionized state of atoms in the solid under investigation.

2. Decay of this excited state by Auger emission.
3. Escape and collection of these Auger electrons.

All these stages are influenced by the solid-state environment of the atom so that the resulting Auger peaks contain information about this environment as well as the atom itself. To understand the qualitative and quantitative methods employed in AES for the study of solid surfaces, it is necessary to consider the basic features of Auger electron emission from atoms in condensed matter and the way they are influenced by the environment.

The physical situation is, of course, much more complicated than in free atoms. Even in a homogeneous solid, the secondary spectrum contains, in addition to Auger peaks, plasma loss peaks associated with both the elastic peak and the more pronounced Auger peaks. Furthermore, there may be ionization loss peaks due to an incident electron losing an amount of energy slightly greater than the binding energy of the electron which it knocks out. Interband transitions may also occur. The task facing the experimentalist is to correctly classify the peaks, separate the Auger peaks from the rest and then identify the associated chemical species. For quantitative analysis, a detailed characterization of the Auger signal in terms of the various parameters involved is necessary. Because of the complications introduced by the solid-state environment, much of the theoretical formulation is still semi-empirical. Theoretical models are quite often based on simplifying assumptions which try to approximate the experimental conditions without making the treatment too complex or intractable. However, interesting physical weights can and have been gained from some of these models and reliable quantitative techniques are being rapidly developed.

The information content of the Auger spectra themselves consists of the energies of the peaks, their shapes, widths and, finally, their relative intensities. Let us take a look at these features for atoms in solids, keeping in mind the free-atom situation as a frame of reference. Except where impurities are specifically mentioned, a homogeneous material with a well-defined surface will be implied.

A. Energy, line width and line shape

The first question is of course the choice of a suitable energy range for the primary beam. The energies at which prominent Auger peaks for different elements are observed can be seen in Fig. 8.7. This chart does not include minor peaks which cannot be useful in AES anyway. It will be seen that elements from $Z=3$ to $Z=14$ are comparatively easily identified by *KLL* Auger transitions. As we shall presently see, ionization cross-sections decrease rapidly with increasing binding energy. The resulting decrease in intensity makes *KLL* Auger peaks increasingly difficult to detect for elements above $Z=14$. Radiative transitions also become increasingly competitive in this

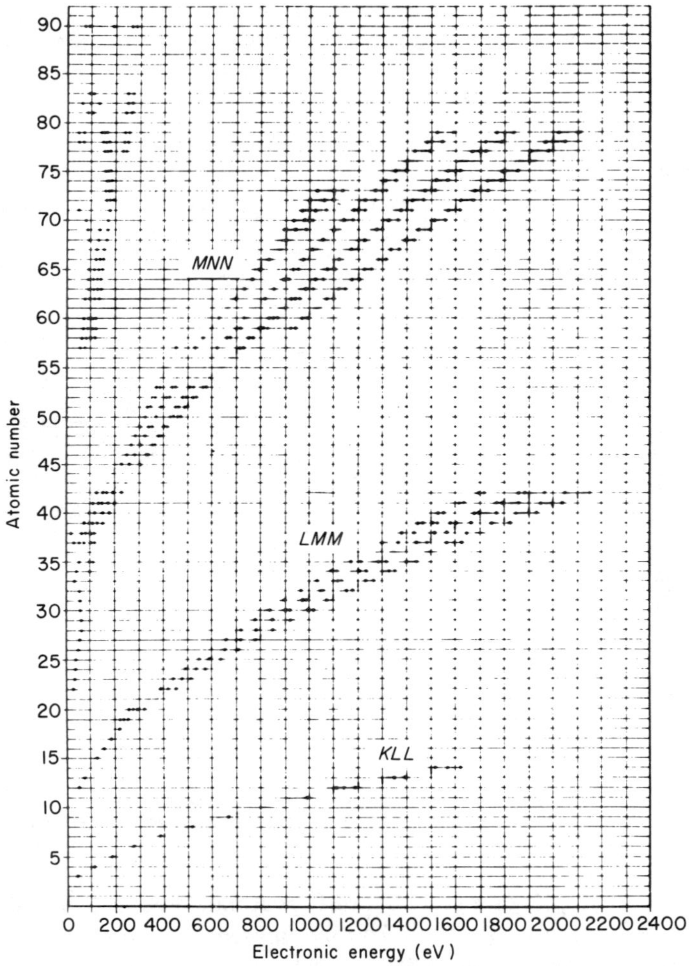

Fig. 8.7. Experimentally observed prominent Auger peaks from elements. Only major peaks are included (from Palmberg, 1972, courtesy of North-Holland Publishing Co.).

region. *LMM* transitions can, however, be observed more easily for $Z \geqslant 14$. The energy-spread for any given atomic number implies that the spectra are now more complex, corresponding to the larger numbers of possible transitions. Similarly, *MNN* peaks can be useful in identifying elements in the region $40 \leqslant Z \leqslant 79$.

The most important point about Fig. 8.7 is that the characteristic Auger spectra lie practically within the range 0–2 keV. For the lighter elements, the range is 0–1.5 keV. Since primary electrons with energy 2 keV are known to have a range of about 500 Å and most of the observed Auger electrons originate within the first few layers of the material, we can reasonably assume that the primary electrons do not suffer much attenuation in this region. Hence, a primary beam of two keV should be adequate for the purposes of Auger electron spectroscopy. The resulting Auger electrons have energies in the range 10–1000 eV. This is a range conveniently accessible to the simple energy analysers described later in this chapter.

Some interesting physical insight can be gained into this question from Fig. 8.8 (Bauer, 1972). Here we have the binding energies E_W of the $K, L_{1,2,3}$

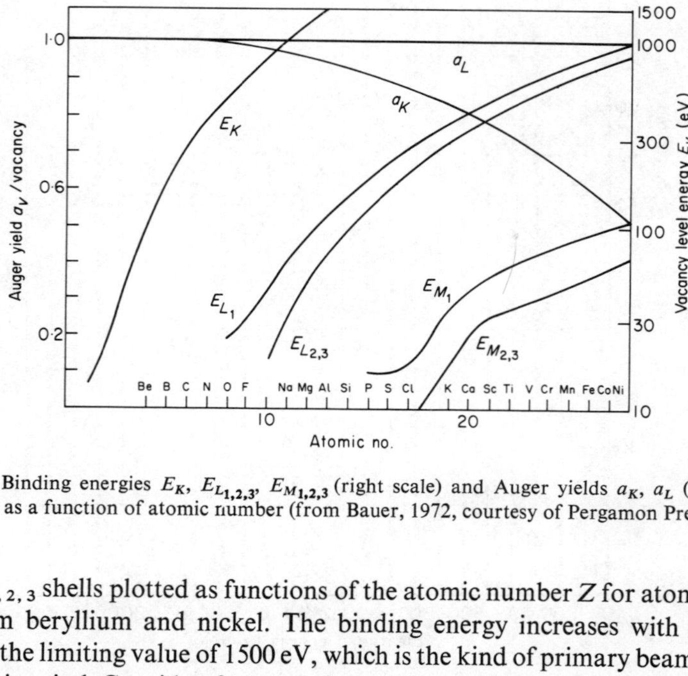

Fig. 8.8. Binding energies E_K, $E_{L_{1,2,3}}$, $E_{M_{1,2,3}}$ (right scale) and Auger yields a_K, a_L (left scale) expressed as a function of atomic number (from Bauer, 1972, courtesy of Pergamon Press).

and $M_{1,2,3}$ shells plotted as functions of the atomic number Z for atoms ranging from beryllium and nickel. The binding energy increases with Z till it reaches the limiting value of 1500 eV, which is the kind of primary beam energy we have in mind. Consider, for example, a KLL transition. As a rough approximation, we take $E_{KLL} \approx E_K$, the K-shell binding energy. Obviously, E_{KLL} cannot be greater than the primary beam energy. Since $E_K \approx 1500$ eV for Al, this is the last element in which we could expect a KLL transition, or, for that matter, any K-Auger transition. Similar considerations hold for the M- and N-series. The physical situation is really more complicated, because even if an Auger transition is energetically permitted, line intensities must be taken into account before we can have a complete picture. Nevertheless, this gives us some idea as to why Auger transitions resulting from a particular inner-shell vacancy are more suitable for the study of a particular region of the periodic table. Table 8.4 exhibits the limiting Z-values corresponding to primary vacancies in the K, L, M and N shells (Bauer, 1972).

Next is the question of line widths, for free atoms these are determined primarily by the lifetime of the vacancies involved, i.e. by the transition rates. According to the uncertainty principle

$$\Gamma_W = \sum_{X, Y} \Gamma_{WXY} = \frac{\hbar}{\tau_W}, \qquad (8.18)$$

where τ_W is the lifetime of the vacancy W and Γ_W represents the total width corresponding to all Auger transitions resulting from this vacancy. Typically, τ_W is of the order of 10^{-16} s, so that $\Gamma_W \approx 0.1$ eV. For example, Γ_K for magnesium is about 0.5 eV, while the L_2 and L_3 levels have a width of less than 0.05 eV, corresponding to the low transition probability for Auger transitions to these levels. On the other hand, the L_1 level is filled more rapidly through Coster–Kronig transitions and is broader, having a width of about 2 eV. The widths of weakly bound shells are generally smaller unless Coster–Kronig transitions are allowed. This indicates that chemical analysis should be more convenient if we use transitions involving primary vacancies in such shells.

The whole question of line widths and line shapes becomes considerably more involved as we go to atoms in solids. Unlike the sharp and narrow peaks looming prominently above the background which are customary for a gaseous target, the peaks are now broadened, shifted and modified in shape due to various environmental factors. Leaving instrumental broadening aside for the moment, the major changes are:

1. Valence band effects.
2. Chemical shifts.
3. Energy loss processes.

Lifetime broadening of course continues to play a role. Because this is not a peculiarity of the solid-state environment, we shall not, however, discuss it further.

1. Valence band effects

As we have already pointed out, the valence band in a solid has a complex structure. For example, the changes which take place in going from (a) the aluminium atom to (b) aluminium metal or (c) alumina (Al_2O_3) are illustrated in Fig. 8.9 (Bauer, 1972). The discrete energy levels of the outer electrons broaden into valence energy bands which are partially (b) or completely filled (c). The probability of finding an electron at a given energy E within the band is determined by the density of states function $\rho(E)$ and of course on the chemical environment, i.e. the nature of the bonding, etc. The fact that $\rho(E)$ depends to a considerable degree on the chemical environment can be seen by comparing (b) and (c). For a given $\rho(E)$, although both of the valence electrons participating in an Auger process of the type WVV may originate at any level

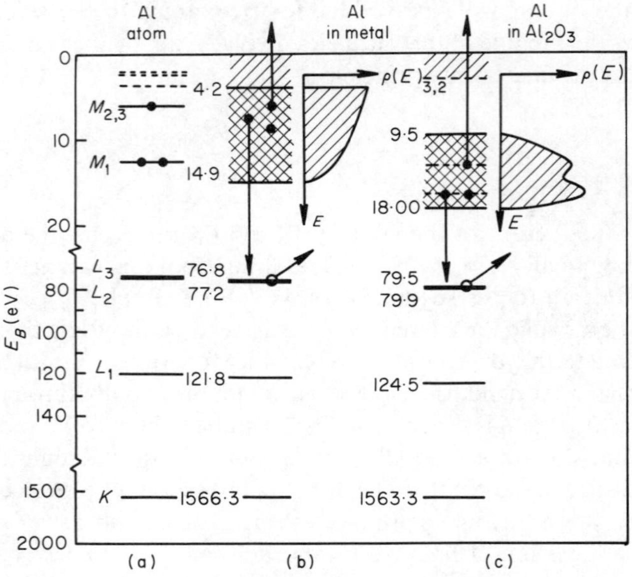

Fig. 8.9. Schematic energy level diagram of aluminium in three forms: (a) Al atom, (b) Al in metal and (c) Al in Al_2O_3. Cross-hatched areas represent filled bands, hatched areas empty bands. $\rho(E)$ is the density of filled states (from Bauer, 1972, courtesy of Pergamon Press).

within the band, certain energy-values are preferred because of the structure in $\rho(E)$.

Even if we neglect the level width of the inner shells, the broadening of the valence band gives rise to a corresponding broadening of Auger peaks. Consider, for example, a valence band of width ζ. The energy E_1 for the $L_{2,3}V_1V_1$ transition, i.e. a transition involving one extremity V_1 of the valence band, is given by

$$E_1 = E_{L_{2,3}} - 2E_{V_1}, \tag{8.19}$$

ignoring work function and screening effects. Similarly, for the $L_{2,3}V_2V_2$ transition involving the other extremity of the same band,

$$E_2 = E_{L_{2,3}} - 2E_{V_2}. \tag{8.20}$$

Thus the resulting width of the $L_{2,3}VV$ Auger peak is

$$E_1 - E_2 = 2(E_{V_2} - E_{V_1}) = 2\zeta, \tag{8.21}$$

which is twice the width of the valence band (Fig. 8.4b).

In addition to this broadening, the structure in $\rho(E)$ should be reflected in the Auger peaks so long as the participating electrons come from the valence band. In other words, the effect becomes pronounced for the *KLL* spectra of second-row elements in the periodic table, the *LMM* spectra of third-row elements, e.g. silicon, etc. The simplest example is, of course, the *KLL* spectrum of carbon. However, because transitions like $L_{2,3}V_AB_B$ are also possible, A and B being states anywhere within the band V, the structure of the Auger peak does not simply reflect the density distribution within the band. Actually the Auger line shape turns out to be a weighted fold of the density of states. Using suitable deconvolution techniques, which take into account the fact that Auger electrons suffer energy losses on their way out of the material and are described by an escape probability function, information on $\rho(E)$ can be derived from the Auger line shape (Hagstrum, 1966; Amelio, 1970; Mularie and Peria, 1971).

To see how this is done, let us look a little more closely at an Auger transition involving two valence band electrons. The process may be visualized as follows.

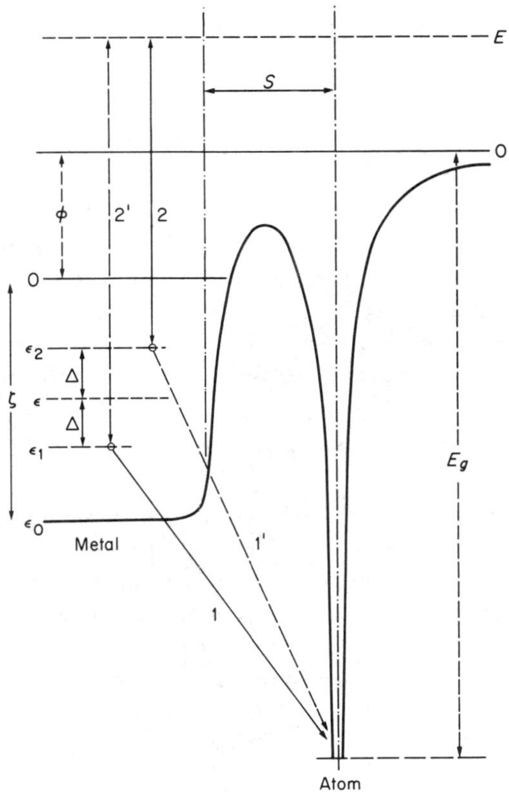

Fig. 8.10. Energy-level diagram of a metal with an ion just outside its surface. The electron transition pairs (1, 2) and (1′, 2′) illustrate Auger-type transitions. Final electron energy is measured from zero at the vacuum level. The energies of initial states in the filled band. ε, are measured positively down into the band from the top of the band (Fermi level for a metal, φ eV below the vacuum level) (from Hagstrum, 1966, courtesy of the American Institute of Physics).

One of the electrons makes a transition to an inner "atomic" shell by losing an amount of energy. The other electron in the valence band gains the same amount of energy and makes a transition to the vacuum above the metal (Fig. 8.10). Let us focus our attention on a particular value of the continuum energy E. Let $-E_g$ be the energy of the "atomic" ground state, energies being measured from the zero of the vacuum level. Let $\varepsilon = \frac{1}{2}(E - E_g)$, i.e. the median between these two energy states. Consider any two valence band electrons with energies $\varepsilon + \Delta$ and $\varepsilon - \Delta$, Δ being some energy less than half the bandwidth ξ. Obviously these two electrons can make the type of "up" and "down" transitions described above without violating the conservation of energy. In fact, for different values of Δ, we get all possible electron-pairs which satisfy this condition. For a given Δ, the probability of finding two such electrons in the band is $\rho(\varepsilon + \Delta)\rho(\varepsilon - \Delta)$. The corresponding probability of having an Auger transition is $|M(\varepsilon,\Delta)|^2 \rho(\varepsilon + \Delta)\rho(\varepsilon - \Delta)$, where M is the Auger matrix element for these electrons. The total probability of producing an Auger electron with energy E is then

$$f(\varepsilon) = \int_0^{\xi} |M(\varepsilon, \Delta)|^2 \rho(\varepsilon + \Delta)\rho(\varepsilon - \Delta)\,\mathrm{d}\Delta. \qquad (8.22)$$

It is customary to write the integrand in the form $g(\varepsilon + \Delta)g(\varepsilon - \Delta)$, where $g(\varepsilon)$ is called the transition density. Obviously, it is essentially a product of the matrix element and the density of states. The function $f(\varepsilon)$ is called the self-convolution, convolution square or fold of the function. So far we have not considered the potential barrier at the surface. Let $P(E)$ be the probability of an electron with energy E escaping over this barrier. Then the external Auger distribution of energy is

$$X(E) = P(E)f(\varepsilon).$$

To obtain $\rho(\varepsilon)$ from a knowledge of $f(\varepsilon)$ is the goal of a deconvolution technique. If we assume that the transition matrix element M depends only weakly on the energy, we can at once see that peaks in $\rho(\varepsilon)$ will be mirrored as peaks in $f(\varepsilon)$. In fact the distribution of $2s$ band electrons in metallic beryllium was inferred from these considerations by Musket and Fortner (1971). They wrote the total density of states for $2s$ and $2p$ band electrons as a sum of s and p terms:

$$\rho_T(\varepsilon) = \rho_s(\varepsilon) + \rho_p(\varepsilon). \qquad (8.23)$$

With the partial density $\rho_p(\varepsilon)$ determined from X-ray emission experiments (Sagawa, 1968) and $\rho_T(\varepsilon)$ obtained from calculations (Loucks and Cutler, 1964), ρ_s would be determined (Fig. 8.11). The convolution $f(\varepsilon)$ is given by

$$f(\varepsilon) = f_{ss}(\varepsilon) + f_{sp}(\varepsilon) + f_{pp}(\varepsilon). \qquad (8.24)$$

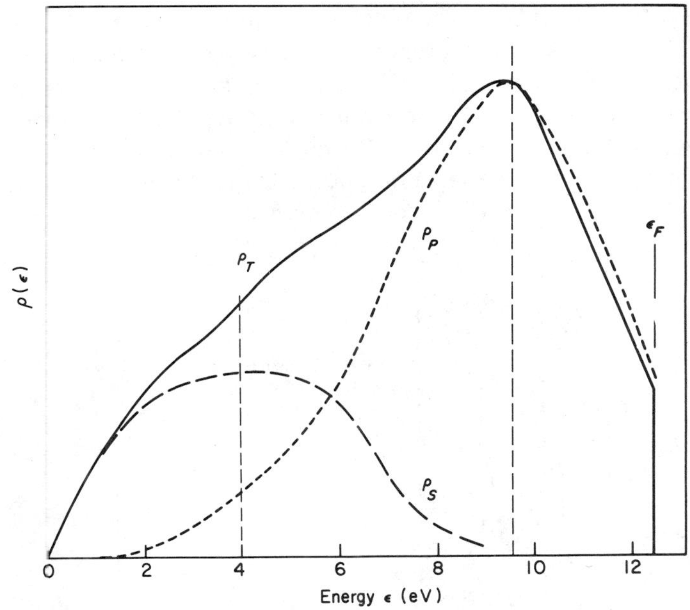

Fig. 8.11. Electron density of states for clean beryllium: $\rho_T(\varepsilon)$ is the calculated total density of states; $\rho_p(\varepsilon)$ is the partial density of states for electrons with p symmetry as determined from X-ray emission experiments; and $\rho_p(\varepsilon)$ is the partial density of states for electrons with s symmetry and is given by the difference between ρ_T and ρ_p (from Musket and Fortner, 1971, courtesy of the American Institute of Physics).

Here f_{ss} stands for a transition where both the participating electrons have $l = 0$, f_{sp} for one s and one p electron, and f_{pp} for two p electrons.

2. *Chemical shifts*

From Fig. 8.9 we can see that although the level widths of the inner shells are not changed significantly, their energies are shifted by a few electron volts. The shifts for Al_2O_3 are greater than those for metallic aluminium. Thus even the inner levels, which retain a quasi-atomic character, undergo a chemical shift depending on the environment. For the outer shells or valence bands, these shifts are considerably greater. This is generally true for different types of chemical effects. With a high-resolution spectrometer, we could expect these shifts to be reflected in the Auger spectra. Figure 8.12 shows such shifts in Al_2O_3 and silicon oxide. Writing the energy of the Auger transition in the form

$$E_{WXY} = E_W - E_X - E_Y,$$

the corresponding shift of the peak becomes

$$\Delta E_{WXY} = \Delta E_W - \Delta E_X - \Delta E_Y. \tag{8.25}$$

Here ΔE_W, ΔE_X and ΔE_Y are the shifts of the individual levels involved. Equation (8.25) can be useful in a situation where we wish to detect the presence of impurities in the surface region. For example, the adsorption of an electronegative gas like oxygen onto a clean metal surface leads to a charge transfer from the metal atoms to the adsorbed atoms. As the electron density around a metal atom decreases, the remaining electrons in this atom "see" a greater

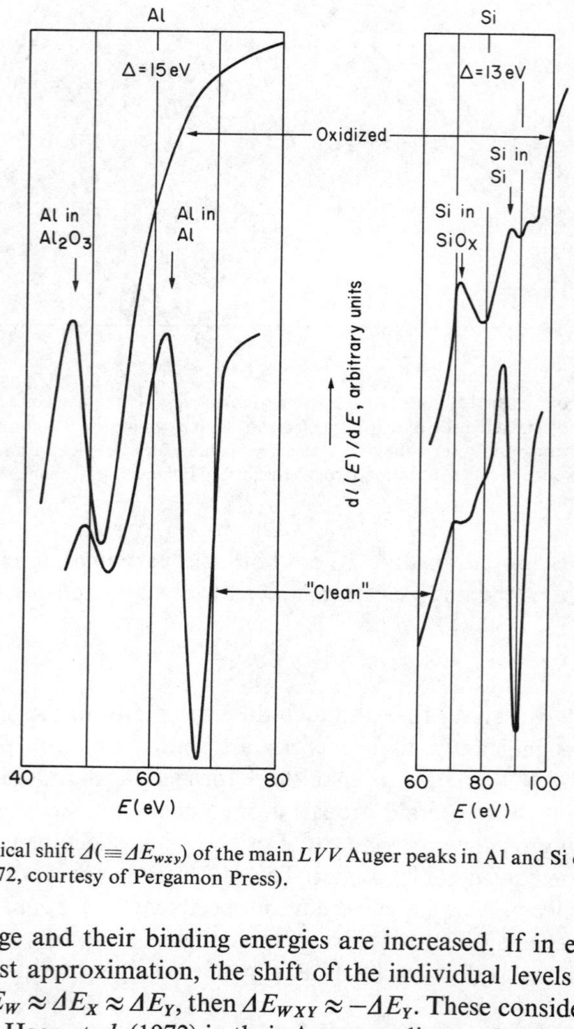

Fig. 8.12. Chemical shift $\Delta(\equiv \Delta E_{wxy})$ of the main LVV Auger peaks in Al and Si due to oxidation (from Bauer, 1972, courtesy of Pergamon Press).

nuclear charge and their binding energies are increased. If in eqn (8.25) we take, as a first approximation, the shift of the individual levels to be nearly equal, i.e. $\Delta E_W \approx \Delta E_X \approx \Delta E_Y$, then $\Delta E_{WXY} \approx -\Delta E_Y$. These considerations were employed by Haas *et al.* (1972) in their Auger studies on the chemisorption of oxygen on refractory metal surfaces such as Mo(110), Nb(110) and Ta(110).

Chemical effects are often much more complicated than a simple shift of energy levels. In an element characterized by groups of Auger lines in different energy ranges, the lower-energy lines may quite often involve ejection of an electron from the valence band, while the higher-energy lines correspond to the emission of a core electron. Changes in the chemical environment have a greater influence on the former. This may lead to a change in the relative intensity of one group of Auger lines to another, in addition to the energy shift described above.

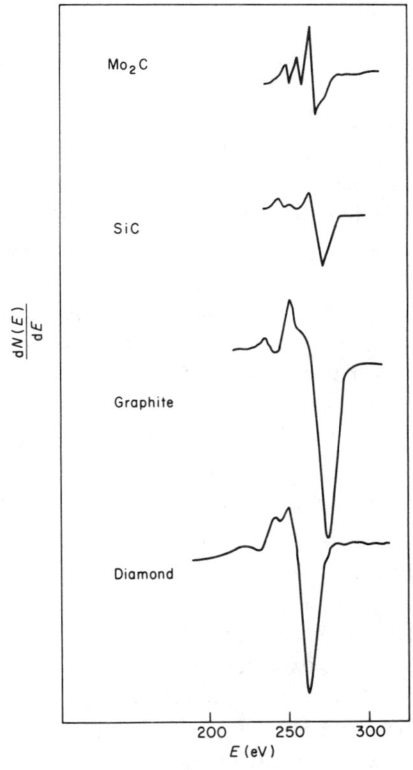

Fig. 8.13. *KLL* carbon Auger spectra from graphite, diamond, silicon carbide and Mo_2C. The diamond spectrum is shifted in energy somewhat because of the charging up of the insulating crystal used (from Haas *et al.*, 1972, courtesy of the Institute of Physics).

The chemical environment also leaves a distinctive imprint on Auger spectra involving valence band transitions. Valence electron wave functions are the most susceptible to changes in bonding. Assuming, therefore, that the escape probability $P(E)$ is a smoothly varying function of energy, the external Auger distribution $X(E)$ should reflect these changes. It was shown by Haas and Grant (1970; see also Haas *et al.*, 1972) that the resulting changes in the carbon *KLL* spectrum could be used as a fingerprint for carbon occurring in the diverse chemical forms of graphite, diamond, silicon carbide and Mo_2C (Fig. 8.13).

In particular, this can be useful in defining the state of surface atoms. For example, following bakeout in vacuum, most materials show the presence of carbon in their Auger spectra. In almost every case, the peak shape is identical to that of graphite. This indicates that the carbon comes from the decomposition of such residual gases as CO, CO_2, hydrocarbons, etc. It also suggests that the surface conditions existing within a vacuum system are akin to those in graphite.

3. *Energy loss processes*

Auger peaks are also broadened by small energy losses suffered by the escaping electrons. For example, an Auger electron may ionize one of the less tightly bound levels with binding energy E_B so as to produce two energy loss electrons, i.e. the original Auger electron which is now scattered out of the Auger peak and the one it has knocked out. This gives rise to a characteristic tailing structure on the low energy side of the Auger peak. There is a limit imposed by the maximum available energy, namely $E_{WXY} - E_B$. This gives rise to a step-like structure, the steps corresponding to the different possible values of the binding energy E_B. Losses may also occur due to interband transitions and the creation of electron-hole pairs.

Associated with the more prominent Auger peaks there can be plasma loss peaks. Because plasmons are characterized by shorter lifetimes, plasma peaks can be distinguished from Auger peaks by their larger width.

Quite apart from the processes described above, broadening of Auger peaks results inevitably because of instrumental limitations. This is best understood by remembering that even if a monoenergetic beam of electrons having energy E is injected into an Auger analyser, a peak width ΔE is recorded due to various aberrations resulting from the experimental set-up. Usually, this width ΔE is measured by the energy-difference between the two points on either side of the peak maximum where the line intensity falls to half its maximum value. The technical name for this width is full width at half maximum (FWHM). In Section 6 we shall discuss some of the factors which contribute to instrumental broadening.

B. Auger yields and intensities

The experimentally measured current from an Auger peak for a pure solid specimen is typically about $10^{-5} i_p$, where i_p is the primary beam current. In absolute terms, this amounts to an Auger current of about 10^{-11} amperes, the primary current being of the order of a micro-ampere. The problem, therefore, is how to maximize the Auger current so that a peak may be easily detected? This is what makes the question of intensity so crucial.

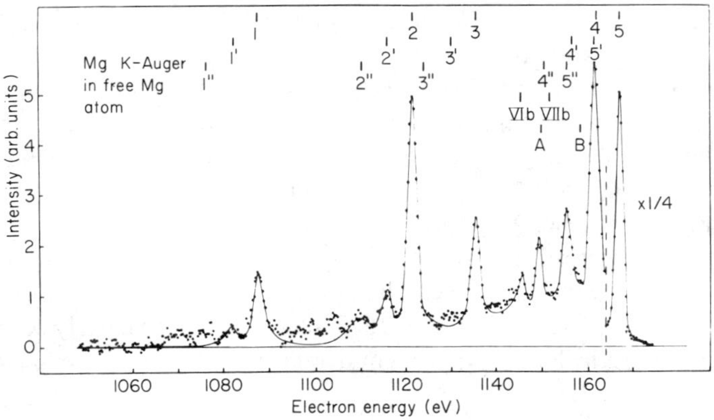

Fig. 8.14. *K* Auger spectrum of the free magnesium atom as observed in magnesium vapour. The lines 1, 2, 3, 4 and 5 are the *KLL* diagram lines in the following order: $KL_1L_1(^1S)$, $KL_1L_2(^1P)$, $KL_1L_{2,3}(^3P)$, $KL_{2,3}L_{2,3}$ (^1S), $KL_{2,3}L_{2,3}(^1D)$. The primed lines are satellites resulting from 3*s*–4*s* shake up and the double-primed lines satellites resulting from 3*s* shake off (from Breuckmann and Schmidt, 1974, courtesy of Springer-Verlag).

In view of the complications introduced by the solid-state environment, Auger yields and intensities in free atoms can act only as a very rough guideline for atoms in a solid. The striking difference between the two can be seen by comparing Figs 8.14 and 8.15. Figure 8.14 depicts the *K*-Auger spectrum of

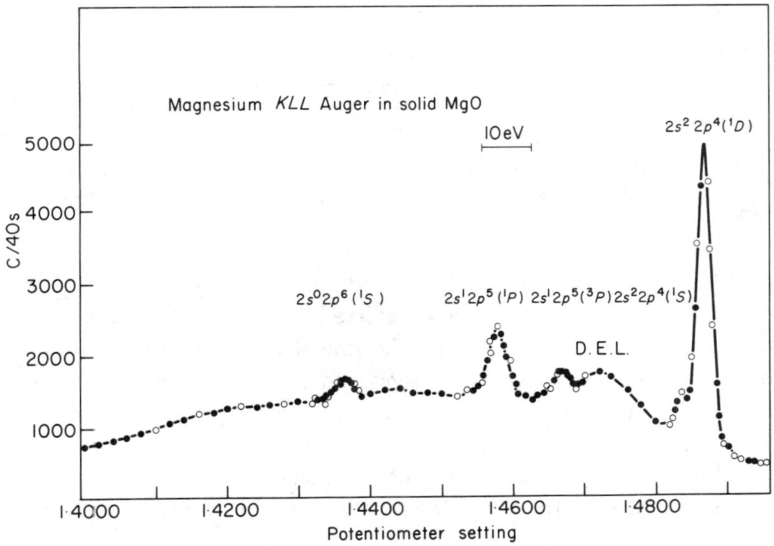

Fig. 8.15. *KLL* Auger spectrum of magnesium in MgO. The oxide was deposited onto a 7 μm thick aluminium backing. The structure marked D.E.L. is due to degraded energy loss electrons (from Fahlman *et al.*, 1966, courtesy of Springer-Verlag).

the free Magnesium atom‡ including satellite lines as analysed by a 127° cylindrical electrostatic analyser with a reciprocal energy resolution of 0.16% (Breuckmann and Schmidt, 1974). The Auger electrons were detected by a Geiger–Müller counter. Figure 8.15 shows the *KLL* Auger spectrum of Mg obtained with a solid target of magnesium oxide (Fahlman *et al.*, 1966). Here the Auger electrons were analysed in a 30 cm double-focusing spectrometer of the current sheet type. Apart from the fact that satellite lines *KM–LLM* and *KL–LLL* labelled by primed and double-primed arabic numerals are clearly visible in Fig. 8.14, it will be seen that even the normal Auger lines exhibit different relative intensities in Fig. 8.15. This is even more clearly seen in Fig. 8.16 where theoretical and experimental relative intensities are shown. For

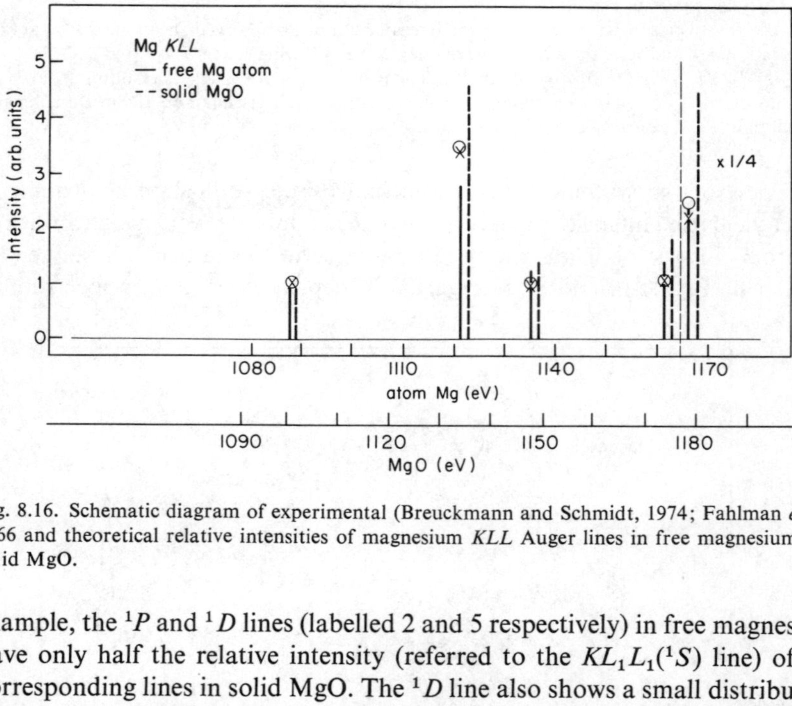

Fig. 8.16. Schematic diagram of experimental (Breuckmann and Schmidt, 1974; Fahlman *et al.*, 1966 and theoretical relative intensities of magnesium *KLL* Auger lines in free magnesium and solid MgO.

example, the ^1P and ^1D lines (labelled 2 and 5 respectively) in free magnesium have only half the relative intensity (referred to the $KL_1L_1(^1S)$ line) of the corresponding lines in solid MgO. The ^1D line also shows a small distribution of energy-loss electrons in the solid target, the peak of this distribution corresponding to an energy loss of 22 eV. The most striking feature of the atomic spectrum is that the peaks are considerably narrower and sharper.§ Thus statistical fluctuations are much more of a problem with solid specimens than with free atoms, particularly if we remember the presence of trace elements and

‡ Magnesium vapour was used as the target.

§ Similar studies have been made with the free sodium atom by Hillig *et al.* (1974) which may be compared with work done on solid-state sodium by Fahlman *et al.* (1966).

impurities which a surface analyst has to contend with. Viewed against a large background and given a low counting rate, fluctuations are more likely to appear as spurious peaks in a solid. This is what makes the noise-to-signal ratio so important in AES for surface studies. It also explains why collector grids (or Faraday cups) in conjunction with electron multipliers are more useful for detection than pulse counters. Modulation of the Auger signal has been used to accentuate the peaks in a manner which we shall describe in the next section. Background nulling techniques have also been reported.

Keeping these differences in mind, let us examine the factors which determine the observed Auger intensity in a material. If the primary ionization of an inner shell W is caused by an incident beam of electrons, then the observed Auger current $i_A(E_{WXY})$ associated with a transition WXY may be taken as a measure of this intensity. It will depend on (a) the cross-section for ionization σ_W and (b) the Auger yield per vacancy a_{WXY}. The cross-section σ_W will depend on the energy of the primary beam and the binding energy E_W of the shell W, but should not depend on the environment as such. The Auger yield a_{WXY}, on the other hand, will depend on the environment so long as at least one of the participating electrons comes from the valence region. There must of course be other parameters, such as the primary beam current i_p, the atom density ρ of the material, etc. In a solid target, geometrical factors such as the angle of incidence will play a role. However, σ_W and a_{WXY} are the dynamical quantities forming a kind of common denominator for the problem.

In order to maximize the Auger current, we must understand the behaviour of the ionization cross-section σ_W as a function of the primary beam energy E_p. Calculations of this type have been made in the Born approximation (Worthington and Tomlin, 1956) as well as in classical electrodynamics (Gryzinski, 1959, 1965; Vriens, 1969). Experimental results for K-shell ionization of metallic aluminium obtained by Hink and Ziegler (1969) with a solid target are considerably higher than the Born approximation values, while the classical results seem to be a good approximation if multiplied by a factor 1.23. In any case the general trend is indicated by the simple choice

$$\sigma_W = \frac{\gamma}{E_W^2} f(U), \tag{8.26}$$

where $U = E_p/E_W$ and E_W is the ionization potential of the shell W. The factor γ is a constant depending on the number of electrons in the shell W. The function f has the form

$$f(U) = \frac{1}{U} \ln U, \tag{8.27}$$

so that f has a maximum at $U \approx 2.72$, i.e. $E_p \approx 2.72\ E_W$.

The important point is that the dependence of σ_W on the primary energy E_p is given by the same function for all species. Moreover, the argument of this function contains E_p only via the quotient E_p/E_W, so that eqn (8.27) is in the nature of a law of corresponding states and the binding energy E_W is a natural unit of energy for the purpose of this equation. Figure 8.17 shows that $f(U)$ has the same general trend in the Born and classical models. Calculations made by Rudge and Schwartz (1966) using Coulomb wave functions for the

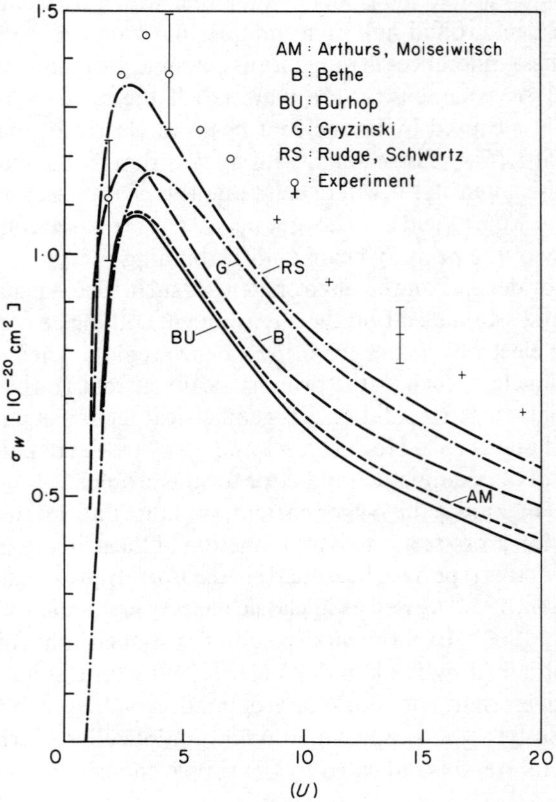

Fig. 8.17. *K*-shell ionization cross-section σ_w of aluminium as a function of relative primary energy $U = E_p/E_W$. The curves are from various theoretical approximations. For details see Hink and Ziegler (1969) (from Bauer, 1972, courtesy of Pergamon Press).

primary electron, also shown in this figure, are closest to the measurements. It will be seen that the theoretically predicted maxima occur almost at the same value of U for different models and that this value is somewhat lower than the experimental result. The curve rises sharply from a zero-value at the threshold $U = 1$, but descends more slowly beyond the maximum. Because of the correspondence in binding energy, different shells of the atoms in a solid exhibit maxima in the same relative range.

Now let us see what conclusions we can draw about the energy of the primary beam from the behaviour of σ_W.

1. To efficiently ionize the material under investigation, the energy E_p should be at least 2.5 times higher than the binding energy E_W. This is the region where σ_W begins to be large.
2. In order to obtain large cross-sections and therefore high Auger intensities, the binding energy E_W should not be large. For cross-sections greater than 10^{-20} cm^2, eqn (8.26) demands that E_W should be less than 1600 eV.
3. σ_W decreases rapidly with increasing nuclear charge Z. For example, $\sigma_K^{\text{Al}} \approx \frac{1}{30}\sigma_K^{\text{carbon}}$. This, taken together with the fact that higher intensities may be expected with lower binding energies, suggests that for heavier atoms we should use L- and M-shell Auger spectra. Because of their lower binding energies, these shells provide larger cross-sections. For aluminium, $E_L = 117$ eV and $\sigma_L \approx 180\,\sigma_K$. It will be seen that these conclusions are in conformity with our earlier discussion on primary beam energies.

The other factor which determines Auger intensities is the Auger yield a_{WXY}. Because of its dependence on the environment, values of a_{WXY} obtained for free atoms are not very useful in relation to condensed matter. However, some general remarks may be made. Figure 8.8 shows, in addition to the inner shell binding energies, calculated total Auger yields for the K and L shells in atoms ranging from Be to Ni. For low atomic numbers these yields are 100% in the energy range considered, i.e. below 1500 eV. In fact this is one major reason why AES should be a sensitive technique in this region. At higher energies, the fluorescence yield ω_W increases at the cost of a_W, so that AES becomes less useful.

Let us now try to see how far these considerations can be woven into a more quantitative picture of the situation in a solid with a plane surface which is chemically homogeneous in the lateral direction. We shall broadly follow the approach of Bauer (1972). In the absence of calculations on the influence of the solid state on the Auger yield a_W, we shall focus our attention on its effect on the other two aspects mentioned at the beginning of this section, namely the primary ionizing beam and the Auger current. It can be easily seen that inelastic and elastic scattering processes should modify both of them.

Since we are looking at a solid with a well-defined surface, geometrical parameters need to be introduced. The most important parameters are the following.

1. If we chose the direction $\hat{n}$ of the normal to the surface to be the z-direction, then z denotes the depth below the surface and dz the thickness of a layer at this depth.
2. The solid angle element dΩ is in direction $\hat{s}$ in which the Auger current i_A is being observed, the angle between $\hat{s}$ and $\hat{z}$ being θ.
3. The direction of the incident beam is $\hat{s}_0$ and the corresponding angle with $\hat{z}$ is θ_0.

In addition to these, we have the density $\rho(z)$ which may vary with depth. Both the primary beam and the emitted Auger current get attenuated in the course of their progress through the solid. We write the primary current incident on an atom at depth z as $i_p(E_p, \hat{s}_0, E, z, C)$, where C stands for the chemical environment and E the attenuated energy of the beam. Then the Auger intensity at this depth should be proportional to

$$\rho(z)\, a(E_{WXY}, C)\, i_p(E_p, \hat{s}_0, E, C)\, \sigma_W(E),$$

where we have now parametrized the total effect of the environment on the Auger yield by C. If we now consider the Auger current which travels towards the detector in the direction $\hat{s}$ through the solid, then this current also gets attenuated due to scattering. At the same time, the current in direction $\hat{s}$ is somewhat enhanced due to scattering from beams travelling in other directions $\hat{s}'$. The overall effect may be incorporated into a coefficient $A(E_{WXY}, \hat{s}, z, C)$, so that we can write

$$\frac{d^2 i_A}{d\Omega\, dz}\, d\Omega\, dz = \rho(z)\, a(E_{WXY}, C) \int_{E_W}^{E_P} i_p(E_p, \hat{s}_0, E, z, C)$$
$$\sigma_W(E) A(E_{WXY}, \hat{s}, z, C) dE\, d\Omega\, dz. \qquad (8.28)$$

This expression represents the contribution of the atoms in the layer dz at depth z to the Auger current emitted into the solid angle dΩ in direction $\hat{s}$.

The attenuation coefficient $A(E_{WXY}, \hat{s}, z, C)$ should, in principle, have the same functional character for both the primary beam and the Auger current. A detailed calculation of A would inevitably be complicated. Let us consider a simple attenuation model (Chang, 1974; Bauer, 1972). Consider N electrons passing through a layer of thickness dz and being attenuated by an amount dN. The attenuation should be proportional to N and also to the layer thickness dz. Hence,

$$-\, dN = \alpha N\, dz, \qquad (8.29)$$

α being the constant of proportionality. Evidently, α should have the dimensions of the inverse of a length, i.e.

$$\alpha = \frac{1}{l}.$$

Thus

$$\frac{dN}{N} = -\frac{dz}{l}$$

and

$$N = N_0\, e^{-z/l}, \qquad (8.30)$$

N_0 being the number of electrons incident at $z = 0$. The parameter l may be interpreted as a mean free path for the beam. Then l should depend on the energy E of the beam at the depth z and we can introduce the direction of the beam by writing

$$N = N_0 \, e^{-z/l(E)\cos\theta}.$$

Thus we get

$$A(E_{WXY}, \hat{s}, z, C) = e^{-z/l(E_{WXY})\cos\theta}. \tag{8.31}$$

Such a model effectively assumes that all inelastically scattered electrons are lost from the Auger peak. That is why the beam gets attenuated. As for elastic events, it is implied that as many electrons are lost to the direction θ as might be gained from such processes. No terms need therefore be added to the original beam. Furthermore, diffraction effects are neglected in this simple model.

For the primary beam at depth z, we can write

$$i_p(E_p, \hat{s}_0, E, z, C) = i_0(E_p) \, e^{-z/l(E_p)\cos\theta_0} + i_s(E_p, \hat{s}_0, E, z), \tag{8.32}$$

where the scattering part i_s cannot be neglected because of the much greater intensity of the incident beam $i_0(E_p)$ from which it is ultimately derived. The usual practice is to separate i_s into a forward-scattered i_F and a backscattered component i_B. For near normal incidence, these are the two important components anyway. Because $E_p > E_{WXY}$, $l(E_p) > l(E_{WXY})$, we can write

$$\begin{aligned} i_p(E_p, \hat{s}_0, E, z, C) &\simeq i_0(E_p)\left(1 - \frac{z}{l(E_p)}\right) + i_F + i_B, \\ &= i_0(E_p) - i_0(E_p)\frac{z}{l(E_p)} + i_F + i_B. \end{aligned}$$

Taking $i_F \approx i_0(E_p)(z/l(E_p))$, this gives

$$i_p(E_p, \hat{s}_0, E, z, C) \approx i_p(E_p, z = 0) + i_B(E_p, \hat{s}_0, E, z, C). \tag{8.33}$$

Thus the departure of the total intensity of the primary beam at a given depth within the solid from its incident value at the surface depends on the backscattered current i_B. This is an important factor which is peculiar to a solid and must be taken account of in surface chemical analysis. Sometimes the entire effect of scattering on the Auger current is incorporated by multiplying i_p by a backscattering factor r. For example, Bishop and Rivière (1969a) derived an expression for the Auger current i_A emitted from a monolayer film adsorbed on the surface of a uniform substrate:

$$i_A = i_p N \frac{\Omega}{4\pi} r \operatorname{cosec}\theta_0 \, a\sigma_W. \tag{8.34}$$

Here N is the number of adsorbed atoms per square cm of the substrate and Ω the solid angle over which the Auger current is collected. All other symbols have been already explained. The factor cosec θ_0 explicitly describes the dependence on $\hat{s}_0$. Although good agreement with experiment of this and analogous expressions for several layers (Jacobi and Hölzl, 1971) is yet to be obtained, eqn (8.34) does provide a kind of guideline for experiments. To have a feeling for the experimental situation, consider, for example, the case of oxygen adsorbed on copper (Bishop and Rivière, 1969a). Let the primary energy be about 1600 eV. The *K*-shell ionization energy of oxygen is 532 eV, so that $U \sim 3$. The corresponding value of $(1/U)\ln U$ is 0.55. The backscattering factor calculated by Bishop (1967) for $Z = 29$ and $1/U \sim 0.35$ is 1.5. The value of N here is about 10^{15}, $\sigma_W = 8.8 \times 10^{-20}$ cm^2. Thus $i_A/i_p = 10^{15} \times 1.5 \times 8.8 \times 10^{-20}\, \Theta \operatorname{cosec}\theta_0(\Omega/4\pi) = 1.3 \times 10^{-4}\, \Theta \operatorname{cosec}\theta_0(\Omega/4\pi)$. Θ is the coverage of oxygen on copper, which is defined such that $\Theta = 1$ corresponds to a monolayer deposit. The instrumental factor $\Omega/4\pi$ is typically about 0.17 in a LEED display, where it represents the fraction of solid angle subtended by the screen. The total Auger current striking the screen will, therefore, be about 2×10^{-11} amps.

A word about the mean free path $l(E)$ may be in order here, since it is an important parameter for the surface analyst. Measurements have been made by Palmberg and Rhodin (1968) who looked at the attenuation of Auger electrons in a thin surface layer of silver deposited on gold. For 72 eV electrons $l(E)$ turns out to be two monolayers or 4 Å and for 362 eV electrons, it is four monolayers or 8 Å. Absolute measurements of the Auger current were used by Jacobi and Hölzl (1971) to obtain $l(E)$. Ridgway and Haneman (1971) made thin layer measurements with Fe on Si. Both theory and experiment give values ranging from 3 to 5 Å for energies lying between 20 eV and 100 eV and from 10 to 25 Å for energies around 1000 eV.

6. Experimental technique

The experimental task, then, is to energy-analyse electrons with energies in the range of a few electron volts to about 1500 eV. In this energy region, electrostatic analysers are to be preferred over the more bulky, and usually more costly, analysers of the magnetic type. The latter become more convenient at high energies.

Electrostatic analysers used in AES fall into two classes: (a) the retarding field analyser (RFA), of which the LEED-Auger device based on the LEED electron eptics is the popular example and (b) the dispersive focusing analysers, namely the spherical, the cylindrical 127° sector and the cylindrical mirror analysers (CMA). In a dispersive analyser, only electrons of energy within a certain transmission range, E to $E + \Delta E$, are collected, the energy window ΔE being gradually shifted across the desired energy range to get a spectrum. In the retarding field analyser, all electrons with energies above the potential of a cut-off grid are accepted. Thus the dispersive instrument differentiates—it gives the energy-distribution $N(E)$ of the secondary electrons directly, while the retarding grid analyser gives the integral $\int_{E}^{E_{max}} N(E')\,dE'$, where $E = eV$, V being the voltage applied to the cut-off grid.

We shall confine our discussion to the basic principles underlying the two most important analysers used in Auger electron spectroscopy, namely, the LEED-Auger device and the cylindrical mirror analyser (CMA). The importance of these two instruments can be ascribed to the following:

1. They have high sensitivity.
2. They are relatively simple in construction and easy to operate.
3. They can be combined with other complementary experimental techniques such as LEED, RHEED, the electron microprobe, the scanning electron microscope, etc.

A. The LEED-Auger retarding field analyser (RFA)

A schematic diagram of the LEED-Auger analyser is shown in Fig. 8.18. It consists of a series of four concentric spherical grids followed by a spherical collector, the latter serving as the fluorescent screen in LEED experiments. Three-grid optics, may also be used, but the four-grid arrangement provides better resolution. In the arrangement shown here, the primary beam of electrons passes along the axis of the optics, striking the sample at normal incidence. The sample is placed at the centre of curvature of the grid-system. Ideally, the innermost grid should be midway between the sample and the next set of grids. It is grounded, as are the sample and all neighbouring components, to provide a field-free region between the grid and the sample. This ensures that the secondary electrons produced by the sample will travel radially towards the grid. The second and third grids are electrically connected and serve as the retarding grids for energy analysis. They prevent electrons having an energy less than that corresponding to the retarding voltage from passing the grid-system. The use of two grids instead of one serves to sharply define the radial retarding field. This improves the resolution typically from about 3% to almost 0.3% of V, the applied voltage. To these two retarding grids is also applied the AC modulating voltage which gives electronic differentiation of the energy spectrum. The fourth grid, usually grounded, serves as an AC shield which reduces the capacitative coupling between the collector and the retarding grids caused by the AC perturbation applied to the latter. The collector is biased a few hundred volts positive with respect to the ground. This ensures that most of the secondary electrons produced at the collector are returned to it. It also increases the energy of the electrons striking the collector so that stable collection is possible through adequate bombardment-induced conductivity of the phosphor. The phosphor helps in properly locating the sample and in positioning the electron beam on the sample. This is important,

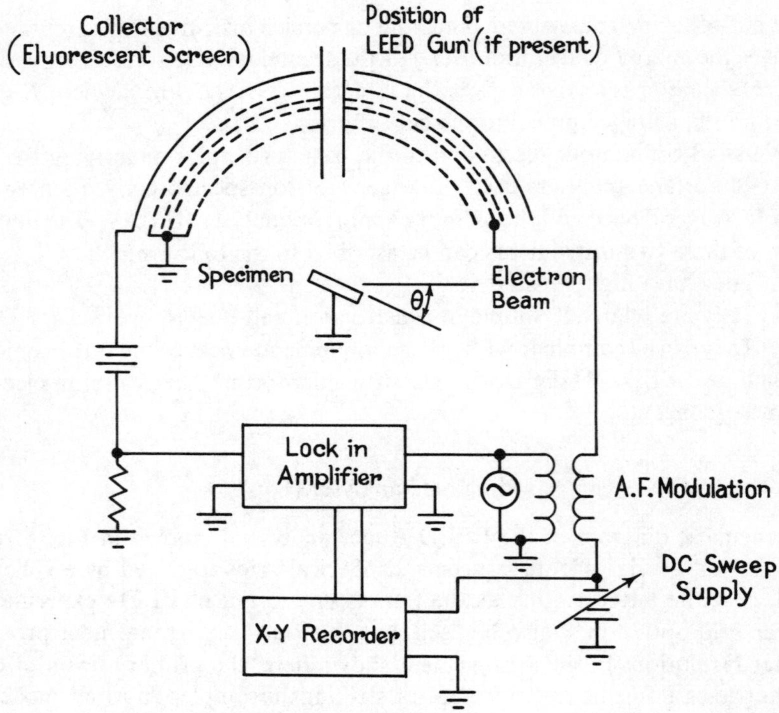

Fig. 8.18. Schematic diagram of the retarding field analyzer (RFA) (from Taylor, 1971, courtesy of Wiley Interscience).

because proper alignment is one of the factors affecting resolution. Other such factors are lack of residual magnetic fields and control over grid warping, i.e. deviation from sphericity.

The sample is excited by bombardment with electrons of energy eV_0 from an electron gun, $-V_0$ being the cathode potential of this gun. The current to the collector is recorded as a function of the retarding field voltage V which is varied from ground potential to $-V_0$. Thus a retarding field plot $I(V)$ versus V is obtained. At any field voltage V, the current

$$I(V) \propto \int_{E}^{E\max} N(E')\,\mathrm{d}E' \quad (E = eV), \tag{8.35}$$

so to obtain the energy distribution we have to differentiate this curve with respect to V, i.e. determine $\mathrm{d}I(V)/\mathrm{d}V$ as a function of V. This is effectively done by applying a small AC modulation voltage to the retarding grids and tuning the detector to the frequency of the modulation. To see how this is done, consider a perturbing voltage $\Delta V = k \sin \omega t$ superposed on the field voltage V

so that the collected electron current is energy modulated. The current $I(V + \Delta V)$ can be written as a Taylor series,

$$I(V + \Delta V) = I(V) + I'(V)\Delta V + \frac{I''(V)}{2}(\Delta V)^2 + \frac{I'''(V)}{3!}(\Delta V)^3 + \frac{I^{\mathrm{IV}}(V)}{4!}(\Delta V)^4 + \cdots, \tag{8.36}$$

where the prime denotes differentiation with respect to V. Collecting the terms corresponding to different harmonics of the modulating frequency, we can write

$$I(V + \Delta V) = I_0(V) + \left[I'(V)k + \frac{I'''(V)}{8}k^3 + \cdots \right] \sin \omega t - \left[\frac{I''(V)k^2}{4} + \frac{I^{\mathrm{IV}}(V)k^4}{48} + \cdots \right] \cos 2\omega t + \cdots$$

$$\sim I_0(V) + I'(V)k \sin \omega t - \tfrac{1}{4}I''(V)k^2 \cos 2\omega t + \cdots \tag{8.37}$$

provided that k^3 and higher order terms can be neglected. The term $I_0(V)$ corresponds to the part which does not vary with time. Equation (8.37) shows that (a) the fundamental frequency term is proportional to $I'(V)$, i.e. to the energy distribution $N(E)$, the proportionality being more precise as the amplitude k becomes smaller and (b) the coefficient of the second harmonic term (2ω) is $-I''(V)$, which gives the derivative of the energy distribution $\mathrm{d}N(E)/\mathrm{d}E$.

Since the second harmonic contains higher powers of k, its proportionality to $I''(V)$ remains valid even when k is not small enough for the k^3 term in the fundamental frequency to be neglected. Thus for a reasonably small modulating voltage, the second harmonic term in the collector current enables us to plot the derivative $\mathrm{d}N(E)/\mathrm{d}E$. This is nice, because the Auger spectrum detectability is enhanced when the derivative of the energy distribution is displayed.

To understand the basic characteristics of the analyser, the ideal Auger peak may be assumed to have a Gaussian shape on the $\mathrm{d}I/\mathrm{d}V$ plot (Taylor, 1969),

$$I'(V) = i\left(\frac{1}{\sigma\sqrt{(2\pi)}}\right) \mathrm{e}^{-V_x^2/2\sigma^2}, \tag{8.38}$$

where i is the total Auger current entering the analyser, V_x is the potential difference as measured from the peak maximum and σ is the standard deviation (Fig. 8.19a). This is equivalent to assuming that the Auger electrons become

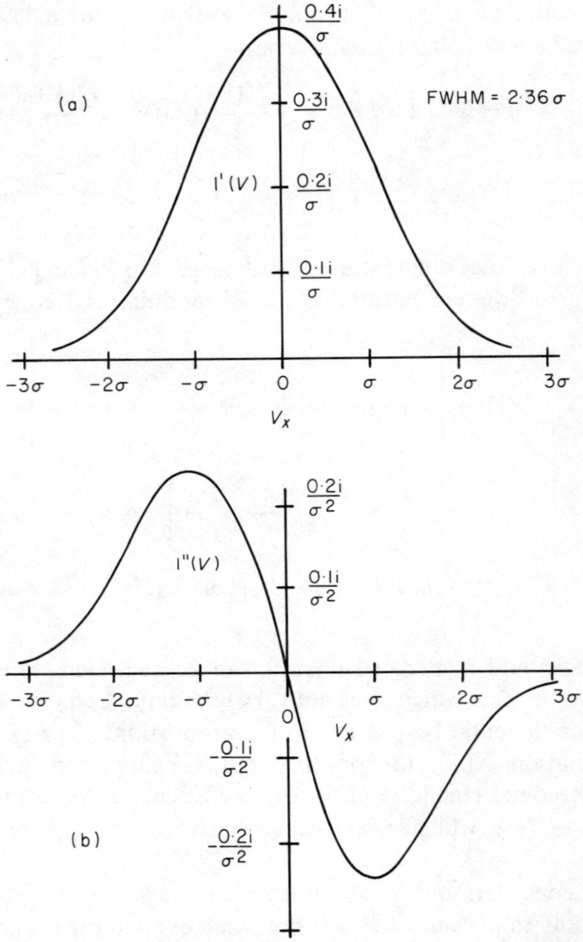

Fig. 8.19 (a) and (b). The Gaussian peak and its derivative.

randomly distributed about a peak energy during their passage to the collector, the randomness being caused by various instrumental limitations. On the d^2I/dV^2 curve, this appears as the combination of a maximum on the low-energy side and a minimum on the high-energy side, as shown in Fig. 8.19(b). Either half of this combination exhibits a greater slope on the side closer to the point $V_x = 0$.

Now let us see what happens to this peak when we actually plot the curves for $I(V)$, $-dI(V)/dV$ and $-d^2I(V)/dV^2$ (Figs. 8.20a,b and c). On the retarding field plot, which simply presents the DC current I versus the retarding potential V, the Auger spectra are barely discernible as kinks on a large background. By tuning the detector to the fundamental, i.e. the modulating frequency, one gets the curve for $-dI/dV$, which yields the energy distribution $N(E)$. On this curve, Auger peaks just appear. Finally, by tuning the detector to the second harmonic, the curve for $-d^2I/dV^2$, i.e. the derivative $dN(E)/dE$ of the energy distribution, is obtained. Here the background slope practically vanishes so that the Auger peak can be more easily detected and a wide energy range can be scanned with high gain. Also, variations of the second derivative

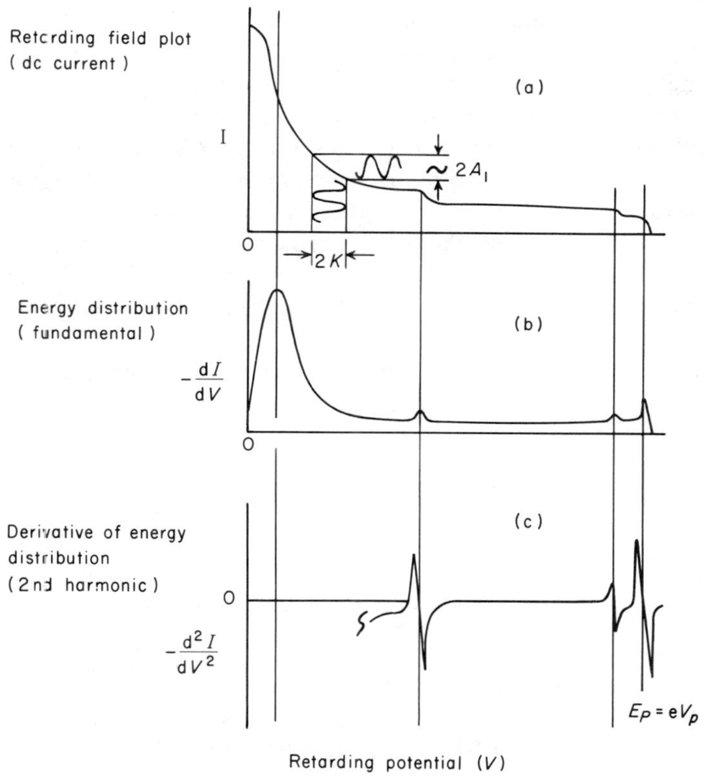

Fig. 8.20 (a), (b) and (c). Schematic diagram showing collector currents in the retarding field analyser in different modes of operation. E_P is the primary beam energy (from Taylor, 1971, courtesy of Wiley Interscience).

are more rapid than the variations of the first derivative, so that Auger peaks are more sharply projected on this plot. As can be seen from Figs 8.20(b) and (c), the actual appearance of the Auger spectra on both the $I'(V)$ and $I''(V)$ curves differs markedly from the idealized shapes given by eqn (8.38). The numerous energy loss processes undergone by the escaping electrons impart to peaks on the $I'(V)$ curve a characteristic tailing structure on the low energy side. This makes it difficult to locate the position of a peak maximum. On the $I''(V)$ curve, i.e. the derivative of the energy distribution, we get a low, broad peak on the low-energy tail side and a large and sharp minimum is easier to locate, its position is customarily used as the energy E_{WXY} of the corresponding Auger peak.

Although a real Auger peak does not have a strictly Gaussian shape, we can use the latter to form an estimate of the desirable range of values for the modulating amplitude k. As we have already indicated, the proportionality of the amplitudes A_1 and A_2 of the fundamental and second harmonic signals to the derivatives $I'(V)$ and $I''(V)$ depends upon k being sufficiently small. As k increases, the amplitude of the fundamental signal to the collector becomes increasingly less than proportional to $kI'(V)$. Similarly, the amplitude of the second harmonic is also no more proportional to $k^2I''(V)$. With the Auger peak described by eqn (8.38), the error terms $(k^3/8)I'''(V)$ for the fundamental and $(k^4/48)I^{\mathrm{IV}}(V)$ for the second harmonic both depend on the ratio k/σ. To minimize these errors, k/σ must be kept small. This is also understandable from the viewpoint of resolution: the Auger peak which we wish to detect by means of the modulating amplitude k has a spread of the order of σ, so that for good resolution we should have $k < \sigma$. Usually we demand that $k = \sigma/2$. Let us see how this affects the ratio A_2/A_1.

For a Gaussian peak with the desired behaviour of the first and second harmonics,

$$A_1 = kI'(V) = i\frac{k}{\sigma}\frac{1}{\sqrt{(2\pi)}}\,\mathrm{e}^{-V_x^2/2\sigma^2} \tag{8.39}$$

and

$$A_2 = \frac{k^2}{4}I''(V) = \frac{i}{4\sqrt{(2\pi)}}\left(\frac{k}{\sigma}\right)^2\frac{V_x}{\sigma}\,\mathrm{e}^{-V_x^2/2\sigma^2}. \tag{8.40}$$

The maximum value of A_1 occurs at $V_x = 0$,

$$(A_1)_{\max} = \frac{i}{\sqrt{(2\pi)}}\frac{k}{\sigma} = 0.4\frac{k}{\sigma}i, \tag{8.41}$$

while for A_2 it occurs at $V_x = \sigma$,

$$(A_2)_{\max} = \frac{i}{4\sqrt{(2\pi)}}\left(\frac{k}{\sigma}\right)^2\mathrm{e}^{-1/2} = 0.06\frac{k^2}{\sigma^2}i. \tag{8.42}$$

Thus with $k = \sigma/2$, the maximum amplitude of current in the fundamental peak will be 20% of i, and that in the second harmonic peak will be 1.5% of i (i being the total current in the Auger peak). Although the maximum amplitude of the second harmonic signal is smaller, it can be more easily detected because of the vanishing background slope.

Another important advantage of tuning the detector to the second harmonic relates to the capacitative coupling between the collector and the retarding grids. Such coupling gives rise to a current with frequency ω which cannot be altogether eliminated by the shielding grid. However, it has no 2ω component, so the detector tuned to the second harmonic ignores it. All we have to do is to see that it is not large enough to saturate the amplifiers.

The electronics for driving the LEED-Auger analyser, shown in Fig. 8.21, consists of a sweep generator which supplies the retarding potential, an audio-

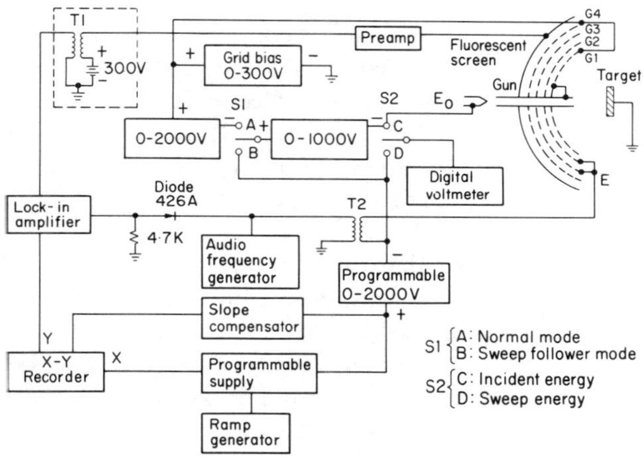

Fig. 8.21. Electronics for driving the LEED-Auger analyser. *G*2, *G*3 are the analysing grids. Their potential is controlled by the ramp generator and programmable supplies. The audio-frequency generator provides the a.c. modulation and the diode arrangement allows for tuning the lock-in amplifier to the second harmonic for electronic differentiation. *G*1, *G*4 are shielding grids. The secondary electrons created at these grids can be "biased out" using the grid bias supply. Electrons are collected by the fluorescent screen and detected with the lock-in amplifier.

frequency generator which provides the modulating voltage and a lock-in amplifier for synchronous detection. Typical excitation voltages range from about 100 V to 3 kV. To measure the energy distribution, the amplifier is tuned to the frequency ω of the modulating voltage, and the retarding voltage and the amplifier output are fed to the X and Y channels of a recorder. To measure the derivative of the energy distribution, the signal channel of the lock-in amplifier is tuned to a frequency 2ω and this double frequency is also applied to the reference channel of the lock-in amplifier.

An important task in this work is to recover the intrinsic Auger distribution from the measured distribution in a LEED-Auger analyser. The latter contains "loss peaks" which we need to distinguish from the actual Auger spectra. A simple scheme is to suppress all loss peaks by applying the modulation to the crystal sample (Gerlach *et al.*, 1970). Alternatively, a deconvolution technique suggested by Mularie and Peria (1971) may be used. In this method, the Auger distribution is recorded along with the energy distribution for an elastic peak of the same energy as the Auger line. From the broadening of the elastic peak one estimates the instrumental broadening function, while measurement of the low-energy tail of the elastic peak gives the broadening produced by inelastic scattering in the solid. From a knowledge of these two functions we can infer the true Auger electron distribution from the measured distribution.

1. Sensitivity of the analyser

We have already noted that the total current from an Auger peak is typically of the order of 10^{-11} amp. From eqn (8.42) we can see that for a total Auger current of 9.3×10^{-11} amp reaching the collector, with $k = \frac{1}{2}\sigma$, the current in the second harmonic is about 14×10^{-11} amp. Thus the potential drop across a 1 MΩ sensing resistor will be about 1.4 μV rms. The ability of a detector to measure this signal will depend on the gain of the detector and the level of noise accompanying the signal.

In spite of all precautions that can be taken to prevent spurious output voltages in the detector, there is always some output without a signal input. This spurious output is generated by the random motion of electrons at the input and is called *noise* (Spangenberg, 1948; Redhead *et al.*, 1968). Electron noise is found in both electronic circuits and devices. Any signal smaller than this electron noise gets masked by it. Thus noise sets a final limit on the sensitivity of a detecting system. Usually, the noise characteristics of electronic devices are specified by the so-called *noise figure*, which may be defined as follows:

$$\text{noise figure} = \frac{\text{signal to noise power ratio of an ideal equipment}}{\text{actual signal to noise power ratio}}.$$

This factor is always greater than unity and accurately describes the noise characteristics only for small signals because of the nonlinear nature of electronic devices. In a LEED-Auger analyser, information is actually sought in the form of the output voltage across a sensing resistor R at the position of an Auger peak. Thus the sensitivity limit of the detecting system can be described by the largest attainable signal-to-noise voltage ratio (S/N) at this peak. An experimental value of the signal S would consist typically of the peak rms value, expressed in microvolts, of the potential drop across R caused by the Auger current. The corresponding sensitivity limit can be estimated if the minimum noise voltage N is known.

Of the possible sources of noise, the following have a bearing on the performance of the retarding field analyser.

(*a*) *Gain modulation or flicker effect noise.* This is caused by the fluctuating output current in the transistor amplifying circuits arising from the small pulsations in electron flow. Under typical operating conditions with a lock-in amplifier this kind of noise is negligible.

(*b*) *Thermal noise* in the input resistor to the lock-in amplifier. The bulk behaviour of electrons in metals is characterized by an average energy depending on the temperature, but they are also subject to random variations in their individual velocities due to thermal agitation, a kin to Brownian motion. These random motions represent small fluctuating currents resulting in minute voltages in the amplifier. This is often called Johnson noise (Johnson, 1928). Thermal noise is not the most important limiting factor in the types of analyser we are concerned with. However, an understanding of the distribution of this noise as a function of frequency can be helpful in discussing the more important type of noise, namely shot noise (also called Schottky noise). We shall, therefore, give here a brief heuristic account based on Spangenberg (1948).

Consider an ideal non-dissipative transmission line of length L with a characteristic resistance R at each end. In a state of thermal equilibrium, i.e. at a steady absolute temperature T, each resistor feeds noise power into the other resistor through the line. Now imagine the line to be shorted at each end. Such a line behaves as a resonator with an infinite number of modes of oscillation with eigenfrequencies $c/2L$, $3c/2L$, $5c/2L$, etc. (c = velocity of light). The spacing between the eigenfrequencies is

$$\Delta\vartheta = \frac{c}{2L}. \tag{8.43}$$

If we now regard this system as an assembly of oscillators obeying the classical Boltzmann statistics, each resonant mode will have energy $2 \times (\frac{1}{2}kT) = kT$ due to the electric and magnetic fields. Thus the energy per unit bandwidth, i.e. per unit frequency range, is

$$E = \frac{kT}{\Delta\nu} = \frac{kT}{c/2L}. \tag{8.44}$$

This is the energy per unit bandwidth which flows into the line from both resistances in a time L/c, the time taken by a wave to pass from one end to the other. Let P be the power per unit bandwidth from one resistance. Then

$$E = 2P\frac{L}{c} = \frac{kT}{c/2L}, \tag{8.45}$$

$$P = kT. \tag{8.46}$$

Hence the power flowing from one resistance into the other due to thermal excitation for a bandwidth B is

$$P_n = kTB. \tag{8.47}$$

Thus the rms noise voltage across the resistor R is given by

$$e_{\mathrm{rms}}^2 = 4P_n R = 4kTRB. \tag{8.48}$$

This is the voltage that would be measured across the resistor by a fictitious infinite-impedance noiseless voltmeter responding to the effective value. In any electronic device we can imagine such a voltmeter to be placed across the resistor in a thought experiment. Expressing kT in electron volts, we get, at a room temperature of 290°K,

$$e_{\mathrm{rms}} = 126(RB)^{1/2} \times 10^{-12} \text{ volts.} \tag{8.49}$$

With a typical one megohm resistor and a bandwidth $B = 0.125$ Hz (corresponding to a time constant $\tau = 1$ s for 12 decibels/octave rolloff) the rms noise voltage is 0.045 microvolts (μV). Corresponding to eqn (8.48), the rms noise current is given by

$$i^2_{\mathrm{rms}} = \frac{4kTB}{R}. \tag{8.50}$$

(*c*) *Shot noise*. Shot noise is a consequence of the fact that electrons arrive at the collector in discrete groups in a random manner. This discreteness in time implies a corresponding spread in frequencies which is represented by the bandwidth B ($B = 1/\tau$, where τ is the time constant, i.e. the decay-time of the random "signal"). If electrons were collected at a uniform rate, B would be infinitely large and the lowest-frequency component would be greater than the highest measurable radio frequency. Under such idealized conditions, no fluctuations would be recorded by the detector. However, we have an inherently fluctuating output even without any contribution to the noise level from the analysing equipment. Shot noise is an essentially statistical phenomenon having nothing to do with instrumental limitations. Because the retarding field analyser is a high-pass filter, i.e. it allows all electrons above a certain threshold energy to enter the detector, it has a significantly higher level of shot noise than a window device like the CMA.

To form an idea of this phenomenon, let us consider a special vacuum diode with two externally shorted cathodes aligned face to face (Spangenberg, 1948). To make things simple, let the cathodes be identical in shape, material composition and temperature. Furthermore, let all electrons emitted from cathode 1 reach cathode 2 and vice versa. Since only high-energy electrons from the tail end of the Fermi–Dirac distribution get emitted, we can assume the velocities of the electrons to follow Maxwell's distribution law, so that the current I_r capable of overcoming a retarding potential V_r is

$$I_r = I_0 \, \mathrm{e}^{-eV_r/kT}, \tag{8.51}$$

where I_0 = emitted current and e = electronic charge. The conductance of the diode is

$$-\frac{\partial I_r}{\partial V_r} = I_0 \frac{e}{kT} \, \mathrm{e}^{-eV_r/kT}.$$

For $V_r = 0$, this becomes

$$G = \frac{eI_0}{kT}. \tag{8.52}$$

For accelerating voltages, the conductance is zero. Thus eqn (8.52) represents the composite conductance in either direction. By hypothesis, I_0 is also the total collected current.

We shall now imagine the diode to be replaced by an equivalent resistor and take the shot noise current to be given by an expression exactly analogous to eqn (8.50) with the conductance G replacing $1/R$:

$$i_{\text{rms}}^2 = 4\,kTGB = 4\,kT\frac{eI_0}{kT}B = 4\,eI_0B. \tag{8.53}$$

This is the noise due to two identical but independent noise sources. Hence for a single cathode we have

$$i_{\text{rms}}^2 = 2\,eI_0\,B. \tag{8.54}$$

The above "derivation" is of course far from being rigorous. However, it has at least the virtue of making the much-used expression (8.54) seem reasonable without going into the intracacies of circuit theory. It may be noted that this expression is independent of the cathode temperature and of the resistance through which the noise current flows. It also does not depend on the type of emitter, but is known to hold more accurately for metal emitters. Since a retarding field analyser (or, for that matter, any electron spectrometer) is a device in which electrons emitted by a sample (cathode 1) are collected by a collecting electrode (cathode 2), it can in principle be replaced by an equivalent diode in so far as noise characteristics are concerned. The fact that not all electrons emitted by the sample are collected by the detector need not be disturbing, since only the total current actually collected occurs in eqn (8.54). In other words, the equivalent diode may be considered to have a low emission current depending on the transmission of the analyser. We shall therefore adopt eqn (8.54) as the expression for the shot noise current I_N in an electron spectrometer for which I_0 is the total collected current and write

$$I_N = (2eI_0\,B)^{1/2}. \tag{8.55}$$

The corresponding rms voltage developed across a 1 MΩ sensing resistor at room temperature is

$$e_{\text{rms}} = 5.6 \times 10^{-4}\,(I_0\,B)^{1/2} \text{ volts.} \tag{8.56}$$

Taking $I_0 = 10\ \mu$amp, $B = 0.125$ Hz, we get $e_{\text{rms}} = 0.63\ \mu$V, which is an order of magnitude greater than the thermal noise.

To see what shot noise does to the performance of a retarding field analyser, we must understand the relation between the bandwidth B and the scanning rate of such a device. In detecting an Auger peak, an important characteristic of the detecting circuit which is brought into play is the ability of the lock-in amplifier to reproduce abrupt changes in the input voltage. This characteristic, usually known as *transient response*, is required of any video amplifier and is commonly expressed in terms of the time it takes for the output voltage to rise from 10% to 90% of its full amplitude. It is an empirical fact that this rise time, usually called the *time constant*,‡ is inversely proportional to the bandwidth of the amplifier. The constant of proportionality is determined by direct observation, depending on the exact definition of the bandwidth. The time constant determines the speed at which the detecting circuit can operate, i.e. continues to register fresh pulses as the voltage is varied over the secondary-electron energy spectrum. The situation is analogous to that in a television receiver, where a rise time greater than about 0.1 μs implies that the fine detail of the picture cannot be received. As for the definition of the bandwidth B, it should by now be clear that this must be directly relevant to the noise characteristics of the amplifier. In defining such a noise bandwidth, we must remember that the response of the amplifier to a given pulse spread over a frequency band drops off near the edges of the band, i.e. the amplification A is a function of the frequency ν. Hence, we can define an equivalent noise bandwidth B_{eq} that has the same noise power as it actually developed by the amplifier output by writing (Terman, 1955),

$$B_{eq} = \frac{\int_0^\infty A^2 \, d\nu}{A_0^2}, \tag{8.57}$$

where A_0 is the maximum amplification. Alternatively, we can also define the noise bandwidth of an amplifier as the frequency width of an idealized band-pass filter having the same area under the power-versus-frequency curve as the actual device and having the same peak value of power (Wallman, 1948). The frequency width is now quantified by giving the roll-off, i.e. the drop-off ratio of the power output from its peak value to the value at either edge of the bandwidth. Usually, this is given in logarithmic units, i.e.

$$\text{No. of decibels} = 10 \log_{10} \frac{P_{max}}{P_{edge}},$$

where P_{max} and P_{edge} are the power values at the peak and the edge of the bandwidth. Since power varies as the square of the voltage or current, we can write:

$$\text{No. of decibels} = 20 \log_{10} \frac{V_{max}}{V_{edge}}$$

$$= 20 \log_{10} \frac{I_{max}}{I_{edge}},$$

‡ Since it depends on the parameters characterizing the detecting circuit.

the notation being self-explanatory. It will suffice to say here that this second definition is the one usually adopted by workers in surface science, the customary roll-off being 12 decibels. With the bandwidth defined in this manner, it turns out that the time constant

$$\tau = \frac{0.125}{B}, \tag{8.58}$$

B being given in Hz and τ in seconds. Thus for $B = 0.125$ Hz, $\tau = 1$ s. Obviously, by reducing B, the amplitude of the noise is reduced so as to increase sensitivity, but the time constant τ is increased so that the rate at which data can be taken is also reduced. The desirable goal in an Auger spectrometer is to obtain a favourable signal-to-noise ratio at a bandwidth such that the time constant is no more than a few seconds. With Auger peaks of typical width (3–20 eV), this implies maximum scan rates of several tens of volts/min. Each pulse from such a peak represents a discrete amount of power, so that the time taken to scan a given voltage range can be obtained by dividing the corresponding area under the I–V curve by this amount and multiplying the result by the time constant τ. For example, with a time constant of 3 s, it takes almost 10 min to get a single trace over an energy range of 200 V.

Using $B = 0.04$ Hz ($\tau \sim 3$ s, 12 decibel rolloff) and $I = 10$ μamp, the rms shot noise voltage turns out to be about 0.1 μV. Experimental values of S for a clean surface can be larger than 500 μV. Thus the maximum value of S/N is about 5×10^3. Let us see what this implies in regard to the limit of detection for surface contaminants. The Auger signal is produced by about 5×10^{15} atoms/sq. cm of the surface. By making a linear extrapolation from the sensitivity limit, i.e. $S/N = 5 \times 10^3$, where a density of 5×10^{15} atoms/cm^2 is detected, we can see that the lowest detectable concentration of surface contaminants, which would correspond to $S/N = 1$, is $(5 \times 10^{15})/(5 \times 10^3) = 1 \times 10^{12}$ atoms/cm^2. In making this extrapolation, we have of course assumed that the signal varies with the surface concentration while the noise level remains fixed. Interestingly, these theoretical limits of detection have already been reached in practice. For example, pure silicon or carbon from graphite produces peaks with a signal-to-noise ratio greater than 5×10^3, using $\tau = 3$ s and a modulation voltage k somewhat greater than 10 V. Once the theoretical limit of detection has been reached with an optimal scan rate, the only way to increase sensitivity is to improve the collection efficiency so that the current I is increased. This is possible up to a point, because the signal is proportional to I whereas the noise goes only as $\sqrt{I}$.

The signal-to-noise ratio, proportional to $i_A/(IB)^{1/2}$ where i_A is the Auger current and I the total collected current, is much smaller for the same i_A in the retarding field analyser than in a dispersive analyser, because of the large current collected. However, this is partly compensated by the very much larger collection angle provided by the hemispherical grid-system. This means that for a given resolution the retarding grid analyser collects a larger Auger current i_A as well. In fact Taylor (1969) has shown that the performance of the retarding grid spectrometer compares quite favourably with that of the 127° electrostatic sector spectrometer.

Huchital and Rigden (1970) showed that it is possible to improve the poor signal-to-noise performance of the retarding grid analyser by allowing only those electrons which just pass through the analysing grid to impinge on the collector. They replaced the conventional large hemispherical collector by a small electrode shielded from direct secondary impingement and used a further grid system after the analyser to focus electrons with energies close to the retarding potential onto the collector. The more energetic electrons pass on and do not contribute to the shot noise. By removing the large collector from the close proximity of the analysing grid, this arrangement also reduces field penetration effects. This analyser combines the advantages of a LEED system with the improved sensitivity and resolution of a dispersive analyser, but it has not yet been used in Auger spectroscopy.

2. Resolution of the analyser

We have already pointed out that even if a monoenergetic beam of electrons of energy E is injected into the analyser, a peak of full-width at half-maximum (FWHM) ΔE is recorded due to various instrumental limitations. ΔE is sometimes called the half-width of the line. For the ideal Gaussian peak, ΔE turns out to be about 2.2σ. The ratio $E/\Delta E$ defines the resolution of the instrument and its reciprocal is expressed as a percentage, i.e. $\Delta E/E \times 100\%$ is called the instrumental line-width (ILW). Of the various limitations on the half-width and resolution, the following are of importance.

(*a*) *Effect of local variations in potential in the vicinity of the retarding grid.* Actually the hemispherical grids are not exactly equipotential surfaces, there being a potential variation across the space between the wires which constitute the grid. Typically the ratio $\Delta V/V$, where ΔV is the potential difference between the centre of a grid aperture and the grid wires, is about 2%. Improvements can be achieved by increasing the grid spacing, using a finer mesh or increasing the depth of the mesh with a honeycomb structure. A more practical solution is to use two retarding grids separated by a short distance. This avoids problems created by secondary electron production at the grids.

(*b*) *Aberrations created by the potential distribution about the first grid.* As the electrons pass through the first grid into the retarding field region, the potential distribution at each of the cells of the mesh causes them to diverge from the normal direction. Let θ be the angle of divergence of electrons having energy $E = eV_0 = \frac{1}{2}mv^2$ (Fig. 8.22). In the approximation of a plane retarding surface which acts only on the component of velocity normal to it, the effective energy of the electrons is $\frac{1}{2}mv^2\cos^2\theta = \frac{1}{2}mv^2 - \frac{1}{2}mv^2\sin^2\theta$. Thus the potential V

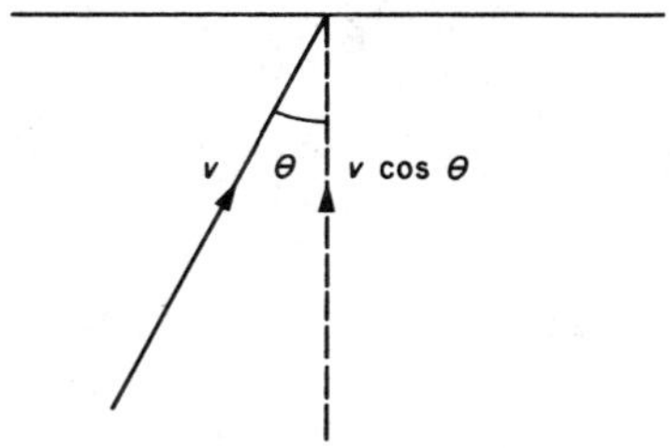

Fig. 8.22. For electrons with energy $\frac{1}{2}mv^2 = eV_0$ incident on a plane retarding surface at an angle θ to the normal direction, the effective energy "seen" by the grid is $eV = \frac{1}{2}mv^2 \cos^2 \theta$. The resulting change in cut-off potential $\Delta V = (1/e)(\frac{1}{2}mv^2 - \frac{1}{2}mv^2 \cos^2 \theta) = (mv^2/2e) \sin^2 \theta$, and $\Delta V/V_0 = \sin^2 \theta$.

required to cut off these electrons is given by

$$V = V_0 - \Delta V$$

where

$$e\Delta V = \tfrac{1}{2}mv^2 \sin^2 \theta.$$

Thus

$$\frac{\Delta V}{V_0} = \sin^2 \theta. \tag{8.59}$$

Now, to estimate θ, Taylor used the expression given by Davisson and Calbick (1931) for the angular divergence of an aperture of radius r with different fields E' and E on either side (Klemperer, 1953):

$$\tan \theta = \frac{r}{4} \frac{(E' - E)}{V_0}. \tag{8.60}$$

In our case, the field E between the sample and the first grid is zero. Thus

$$\tan \theta = \frac{rE'}{4V_0}. \tag{8.61}$$

To minimize θ, r and E' should be small. One way of minimizing E' is to make the radius of the first grid equal to half that of the retarding grid.

(*c*) *Deviations of grids from the ideal spherical geometry, i.e. manufacturing tolerances, wrinkles, grid warping, etc.* These deviations lead to local gross non-radial fields between the grids giving nonradial trajectories, so that cut-off occurs at a reduced retarding voltage. A check on grid uniformity can be made by observing the uniformity of illumination of the fluorescent screen as the retarding grid voltage is taken through cut-off.

(*d*) *Effects of finite beam size and improper positioning of the sample.* These effects cause the beams to enter the retarding field region in an off-normal direction and resolution is reduced by the $\sin^2 \theta$ factor of (b) above. However, it has been seen that a resolution superior to 0.25% can be obtained from a sample of 4 mm diameter positioned to within ±1 mm of the centre of curvature so that its entire surface is flooded by the exciting beam.

Apart from these factors there are some effects which lead to a broadening of the observed peaks in the secondary electron spectrum. For example, the energy spread in an Auger peak depends on the breadth of the electron energy levels of the atoms and also on the scattering losses suffered by the Auger electrons in escaping from the sample. Variations in the work function of the retarding grid also makes a contribution to the half-width ΔE. For grids made of nickel-plated tungsten, this variation is generally less than 0.2 eV. Stray transverse magnetic fields cause angular deviations resulting in an energy spread which can be reduced by modest shielding and field cancellation to less than 0.1 eV.

B. The cylindrical mirror analyser (CMA)

From the sensitivity considerations outlined above, we can see that a device with sensitivity higher than the LEED-Auger system would have to produce the same signal S at a smaller collector current I. This is a feasible proposition, because actually the retarding potential analyser set at a given energy E collects a lot of unwanted current:

$$I(E) \propto \int_E^{\infty} N(E')\, \mathrm{d}E'.$$

The information we really want is $N(E)$ which is contained only at the lower limit of integration. Electrons with energy $E' > E$ create only noise and provide no information. A "high-pass filter" like the LEED-Auger device eliminates noise from electrons below the retardation threshold, but not from those above. For better sensitivity we need a window device or "band-pass filter" which only collects electrons in a narrow energy range ΔE at E.

The cylindrical mirror analyser (CMA), also called the coaxial cylindrical analyser, is a window device of this type. First used by Blauth (1957), it was introduced by Palmberg *et al.* (1969). Of the various types of deflection and mirror electrostatic analysers used so far, the CMA is the best suited for AES. The arrangement, as shown schematically in Fig. 8.23, consists basically of two coaxial cylinders. The inner cylinder is grounded and has two cylindrical gridded apertures suitably positioned along its length. The outer cylinder has a negative voltage V_a applied to it so that secondary electrons leaving a suitably located grounded sample and entering the first of the apertures tend to be reflected towards the second aperture. The sample is usually placed with its centre on the axis of the cylinders. For a particular voltage V_a applied to the outer cylinder, only electrons with energy $E = eV_e$ will pass through the second gridded aperture and come to a focus (strictly speaking, produce a minimum trace) in a small annular region close to the axis. The ratio V_e/V_a is determined by the geometry of the analyser and lies usually between 1 and 2. The electron trajectories follow a toroidal geometry before being brought to a

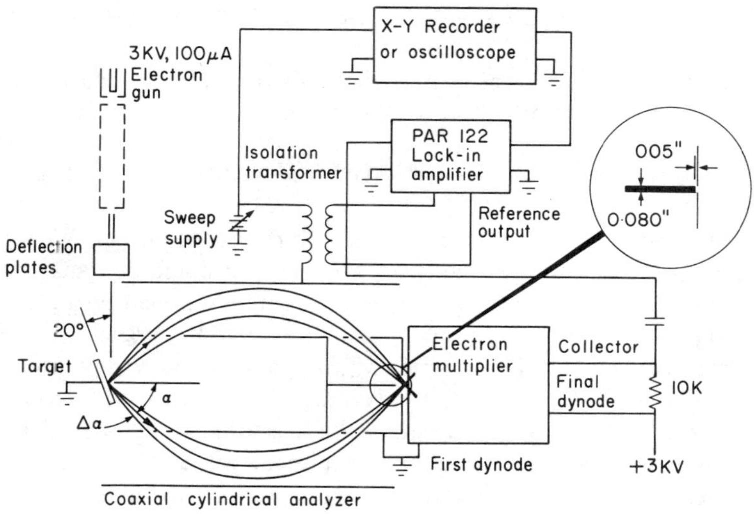

Fig. 8.23. The cylindrical-mirror analyzer (CMA) (from Palmberg *et al.*, 1969, courtesy of the American Institute of Physics).

focus. An annular slit located at the "focal position" allows only electrons with energy values lying in a narrow "energy window" ΔE around the energy E to reach the collector. The current reaching the collector is

$$I(E) = K\Delta E\, N(E) = KREN(E), \tag{8.62}$$

where K is a constant for the geometry used and $R = \Delta E/E$, the resolution. In the absence of stray magnetic fields R is a constant. This analyser can be used in conjunction with an electron multiplier for easier detection. In this case the output of the multiplier becomes $g(V_a)I(E)$, $g(V_a)$ being the gain of the multiplier at the applied voltage V_a. One advantage of using an electron multiplier is that at low energy a combination of small ΔE and low $g(V)$ can suppress the large secondary electron peak. The simplest procedure is to allow the electrons to strike the first dynode of the multiplier with their own energy E. In this case the multiplier output decreases with E, particularly below 50 eV where the output goes to zero. The behaviour of the output is nonlinear in this region, which is a disadvantage in quantitative work. However, this is not really a catastrophe, since the signal otherwise increases rapidly near zero eV. In the LEED-Auger system, the sensitivity has to be reduced by about five orders of magnitude to cope with this problem.

In this analyser the current I can be electronically differentiated by applying a modulation voltage $k \sin \omega t$ to the outer cylinder and detecting the first harmonic of the output current. We write

$$I(E + ek \sin \omega t) = I(E) + eI'(E)\, k \sin \omega t + e^2 I''(E) \frac{k^2 \sin^2 \omega t}{2!} + \cdots \quad (8.63)$$

Neglecting higher order terms, we see that only the $I'(E)$ term corresponds to the first harmonic. The amplitude of this first harmonic current is

$$I(\omega) = ekI'(E) = ekK \frac{\mathrm{d}}{\mathrm{d}E}(REN(E)) = ekKR[N(E) + EN'(E)], \quad (8.64)$$

where the prime indicates differentiation with respect to energy. We see that, unlike in the LEED-Auger analyser, the first harmonic contains $N(E)$ in addition to $N'(E)$. However, because $N'(E)$ gets multiplied by E, the $N(E)$ term happens to be relatively small except at very low energies, i.e. $\lesssim$ 50 eV. There are other modulation schemes which eliminate the $N(E)$ term.

The CMA has two important advantages with regard to focusing:

1. It is a double-focusing analyser, i.e. it focuses in two different directions.
2. It has second order focusing for an entrance angle α of 42.3°. To see what this means, let us expand the focal distance L as a Taylor series about the ideal entrance angle α_0 which gives the best focusing:

$$L(\alpha_0 + \Delta\alpha) = L(\alpha_0) + \left.\frac{\partial L}{\partial \alpha}\right|_{\alpha_0} \Delta\alpha + \left.\frac{\partial^2 L}{\partial \alpha^2}\right|_{\alpha_0} \frac{(\Delta\alpha)^2}{2!} + \left.\frac{\partial^3 L}{\partial \alpha^3}\right|_{\alpha_0} \frac{(\Delta\alpha)^3}{3!} + \cdots \quad (8.65)$$

Here $\Delta\alpha$ denotes the deviation of the secondary electron beam from the ideal angle α_0. An analyser is said to have first order focusing at α_0 if $(\partial L/\partial \alpha)_{\alpha_0} = 0$ and the first non-vanishing partial derivative is $(\partial^2 L/\partial \alpha^2)|_{\alpha_0}$. When we say that the CMA has second order focusing at an entrance angle of 42.3°, we mean that both $(\partial L/\partial \alpha)|_{\alpha_0}$ and $(\partial^2 L/\partial \alpha^2)|_{\alpha_0}$ vanish at $\alpha_0 = 42.3°$ and the first non-vanishing derivative is $(\partial^3 L/\partial \alpha^3)|_{42.3°}$. Thus for electrons leaving at an angle $\Delta\alpha$ from the ideal angle of 42.3°, the CMA creates an image of the source at a distance proportional to $(\Delta\alpha)^3$ from the focal point instead of $(\Delta\alpha)^2$ as would happen for first order focusing. Hence a larger value of $\Delta\alpha$ can be tolerated for the same resolution, resulting in high collection efficiency.

The major improvement resulting from the use of this analyser is reduction in shot noise. Because electrons in only a narrow energy window are collected, the current I is typically about 10^{-4} times that in a retarding field device. Since $N \propto (IB)^{1/2}$, shot noise is reduced 100 times. Operating at the same

S/N ratio, the bandwidth can increase 10^4 times and the scanning rate can become as high as 10 kV/s. This allows a continuous oscilloscope display. Scans of this rapidity are particularly useful in kinetic studies of adsorption, catalytic reactions and surface segregation of impurities. Alternatively, such sensitivity can be traded off for higher resolution or for lower beam currents.

The highest attainable resolution in this type of analyser depends on $\Delta\alpha$ and can be varied by changing the width of the entrance or exit slit. For good resolution the secondary electrons should travel only in axial planes from a specific point on the axis. The incident electron beam and the sample should be positioned to about $\pm\frac{1}{2}$ mm of this point so as to maintain a resolution of 1 % in analysers with an inner cylinder having a diameter of about 2.5 cm.

The electronics for this system are broadly similar to that for the LEED-Auger device. It should be noted that the collected current is here proportional to $N(E)$, so that detection at ω gives the most generally useful $N'(E)$ curve. The resulting signal is four times larger and is proportional to k instead of k^2, in contrast to the $N'(E)$ curve obtained in the LEED-Auger system by tuning to the second harmonic (2ω). Figure 8.24 shows oscilloscope traces of an

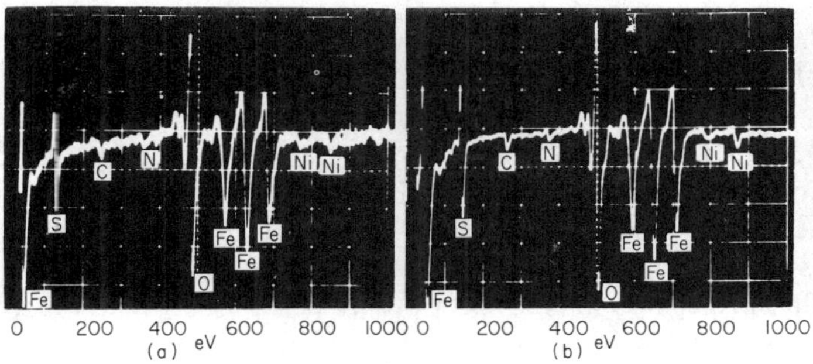

Fig. 8.24. Oscilloscope traces of Auger spectrum taken by Palmberg, Bohn and Tracy (1969) at scanning speeds of (a) 20 000 V/s and (b) 2000 V/s. The bandwidths for (a) and (b) were 2.5 kHz and 0.25 kHz respectively (from Palmberg *et al.*, 1969, courtesy of the American Institute of Physics).

Auger spectrum taken by Palmberg *et al.* (1969) on the CMA with scanning speeds of (a) 20 000 V/s and (b) 2000 V/s.

In addition to the LEED-Auger analyser and the CMA, a high resolution Auger electron spectrometer using a hemispherical electrostatic deflector as the energy dispensing element has been described by Bassett *et al.* (1972). This instrument is based on the design considerations enunciated by Simpson (1964) and Kuyatt and Simpson (1967). An important feature of this design is that there are no physical apertures in the entrance and exit planes of the electrostatic deflector. Instead, these planes are imaged by electrostatic lenses onto physical apertures at much higher potentials than exist in the hemispherical deflector. In the instrument used by Basset *et al.*, both the input and output lenses operate in a decelerating mode so that secondary electrons produced at the physical input slit are prevented from reaching the deflector and secondary electrons produced at the output end do not reach the detector. This instrument achieves an energy resolution of 1200 and a transmission efficiency above 40% for Auger electrons in the energy range 300–1000 keV. Because it accepts a cone of emitted electrons with a small half-angle, it can be adapted for studies of angular distribution. At the same time, it achieves a high collection efficiency by concentrating the primary beam on a small area of the sample surface. The signal-to-noise ratio is very favourable as compared to the LEED-Auger system—about seven for the carbon Auger peak at 263 eV with a primary energy of 2 keV. For an Auger peak with 3 eV full width at half maximum, these authors report the very satisfactory scanning speed of 1000 eV/s using a bandwidth of 300 Hz. Another important advantage of this spectrometer is that because of its imaging optics it allows a working distance of 60 mm between the sample and the input surface of the analysing system as compared to the 6 mm space available in the current commerical CMA. This is convenient where experiments have to be performed on the sample.

7. Surface chemical analysis with AES

The use of Auger electron spectroscopy for the identification and quantitative estimation of chemical species present in solids is, in practice, a fairly complicated task. In this section we shall only outline some of the basic principles involved.

The first problem is to isolate the Auger peaks from energy loss spectra. Plasma loss peaks associated with the elastic peak may be detected by applying a potential to the sample. This will shift the Auger peaks and the energy loss peaks associated with them, but not the elastic peak and its plasma associates. Alternatively, only the latter group may be shifted by altering the primary beam energy. This is the easier course and is applicable to insulators as well as metals.

One must also decide whether small peaks on the low energy side of a major Auger peak are plasma loss peaks or additional Auger peaks or satellites. This is particularly important because the intensity of any given peak depends on so many different factors. The method used by Mularie and Rusch (1970) was to reduce the primary beam energy to that of the large Auger peak in question. The relative positions and magnitudes of the Auger peak and its "satellites" are now compared with those of the elastic peak and its associated plasma loss peaks. If these are similar, the "satellite" peaks may be attributed to Auger electrons suffering one or more plasmon losses.

Ionization loss peaks resulting from incident electrons occur at a fixed energy relative to the elastic peak and can be distinguished from Auger peaks.

Actual identification of elements present in a sample hinges on a knowledge of the factors discussed in Section 5, namely, energies of the Auger peaks, their widths, shapes and, to an extent, their relative intensities. For reasons already enumerated, low-energy Auger spectra below 1500 eV are usually employed. The characteristic spectra of most elements in this region are well known (Palmberg *et al.*, 1972). However, we have seen how relative intensities and widths are modified in a solid, thereby making identification more difficult. The situation is further complicated by the absence of selection rules for Auger transitions. Uncertainties exist with regard to the labelling of peaks even when the element involved is known. With the presence of a number of unknown impurities, the experimentalist has to be cautious about interpreting his data.

A standard method of measuring the Auger current is to use the peak-to-peak height in the differentiated spectrum $\mathrm{d}N(E)/\mathrm{d}E$. As we have seen for a LEED-Auger analyser

$$I(V) = \int_E^\infty N(E')\,\mathrm{d}E', \tag{8.66a}$$

so that

$$\frac{\mathrm{d}I(V)}{\mathrm{d}V} = N(E) \quad \text{and} \quad \frac{\mathrm{d}^2 I(V)}{\mathrm{d}V^2} = \frac{\mathrm{d}N(E)}{\mathrm{d}E}.$$

Thus

$$\frac{\mathrm{d}^2 I(V)}{\mathrm{d}E^2} \bigg/ \frac{\mathrm{d}I(V)}{\mathrm{d}E} = \frac{\mathrm{d}N(E)}{\mathrm{d}E} \bigg/ N(E). \tag{8.66}$$

Now consider a plausible shape, e.g. the Gaussian peak. Then, from our discussion in the last section,

$$\frac{\mathrm{d}N(E)}{\mathrm{d}E} \bigg/ N(E) = -\frac{V_x}{\sigma^2}, \tag{8.67}$$

where V_x is the voltage difference with the peak maximum and 2σ the FWHM. If we assume that the width of the peak remains unchanged as we change the surface concentration of the sample, V_x/σ^2 remains constant for a given V_x. Hence the peak-to-peak deflection in the differentiated spectrum should be proportional to $N(E)$, the number of atoms present.

The above reasoning applies only if the amplitude k of the modulating signal is less than half the rms width of an Auger peak. For a larger amplitude, the series expansion introduced in Section 6 cannot be used. Instead, the undifferentiated Auger current can be approximated by a step of height i_A. The AC component of the current collected by the screen appears as a square wave of changing width as the scanning voltage passes through the peak. The amplitude of the second harmonic is now given by the corresponding Fourier coefficient in the expansion for this square wave. This amplitude turns out to be constant and equal to i_A/π (Bishop and Rivière, 1969a). The proportionality with the Auger current holds as the concentration is changed provided that the total AC signal remains a square wave.

If the shape of the peak changes with concentration, it is more convenient to use the total current in the Auger peak which is proportional to the number of atoms present. This current may be obtained by integrating twice through the differentiated peak. Background nulling techniques may also be used if the undifferentiated spectrum is recorded (Musket and Ferrante, 1970).

As for the position of the Auger peak, it is customary to identify this with the high-energy turning point in the curve for $dN(E)/dE$. This point is easier to detect than the crossing point with the energy axis, which corresponds to the maximum of $N(E)$, particularly if the peak is asymmetric or there is a large background. However, it needs to be remembered that the turning point is a point of inflection of the $N(E)$ curve and its position depends on instrumental broadening effects in low-energy spectrometers as well as on the amplitude "k" of the modulating voltage if this is large (Gallon and Matthew, 1972).

In spite of the difficulties involved, Auger electron spectroscopy has become a valuable analytical tool in those areas of research and industry where the chemical composition of surfaces is of importance. Its applications range from fundamental surface studies to semiconductor technology and from the temper embrittlement of steels to the study of gain fatigue in channel electron multipliers. It provides a method for detailed analysis of the transition region between the bulk of a solid and its surface and for the study of thin films. What follows should be regarded as no more than a sampling of these applications.

A. Qualitative analysis

1. Surface cleanliness and evaluation of surface cleaning methods

Auger electron spectroscopy has proved very useful in the monitoring of surface cleanliness and the determination of the efficiency of different methods for producing a clean surface. It was shown by Weber and Peria (1967) that a surface impurity in a quantity as small as one-tenth of a monolayer can be detected and identified with a LEED-Auger system. They observed Auger electron spectra from clean and alkali-covered germanium and silicon surfaces, the alkali atoms deposited being potassium and cesium. The amount deposited is described by the coverage Θ, which is defined such that $\Theta = 1.0$ when there is one adsorbate atom on the surface for each substrate atom in a perfect atomic plane for the surface under consideration. Obviously, $\Theta = 1.0$ corresponds to the deposition of a monolayer. The fabrication of integrated circuit devices involves cleaning, polishing, controlled diffusion and thin-film deposition of metals on silicon and germanium substrates. All these steps can be evaluated by AES. For example, conventional methods for cleaning silicon surfaces involved degreasing, chemical etching in HF and washing. It was shown by Gallon *et al.* (1972) that the Auger spectra in this case contained a large peak corresponding to carbon and smaller peaks for oxygen and phosphorus. However, several cycles of a treatment consisting of flashing to 1100°C for about 15 s and annealing at about 600°C for 10 min led to the disappearance of all except traces of phosphorus and carbon.

2. *Impurities and LEED patterns*

AES has also elucidated the question of possible correlations between observed LEED patterns and the presence or absence of impurities on surfaces. Numerous impurities have been revealed on surfaces previously assumed to be clean because of their good LEED patterns. On the other hand, some surfaces with poor LEED patterns have been shown to contain no impurities within the detection limit (Ignatiev *et al.*, 1972). More importantly, the influence of impurities on the formation of superstructures on clean surfaces is being better understood. For example, the Si(111) surface is known to form a large number of superstructures depending upon the nature and amount of impurities present. It was suggested that the (7×7) LEED pattern was caused by the presence of submonolayer quantities of iron or nickel. It has now been established that Fe is not essential for the formation of this structure (Bauer, 1969; Thomas and Francombe, 1971; Taylor, 1969). In some experiments, Fe may play a scavenging role by removing impurities which suppress the (7×7) pattern. On the other hand, Ni does induce the $(\sqrt{19} \times \sqrt{19})$ pattern. These observations underline the necessity of using AES and LEED as complementary techniques.

3. *Surface segregation*

Another problem in the preparation and maintenance of clean surfaces is the surface segregation of impurities. During their experiments with *n* type silicon, Gallon *et al.* (1972) observed that the concentration of phosphorus at the Si surface was much greater than that in the bulk. This led to the conclusion that phosphorus had diffused from the bulk to the surface. Earlier, during his examination of the Auger spectrum of Ni before and after heating, Harris (1968b) found that a large peak had appeared at 150 eV corresponding to sulphur which had segregated to the surface from the bulk.

Because silicon wafers are widely used in integrated circuits, surface segregation of impurities like phosphorus or boron is of considerable importance. In the presence of an oxidizing atmosphere, phosphorus segregates at the oxide–vacuum and oxide–silicon interfaces. Oxidation also lowers the energy of the silicon Auger peak and broadens it. Oxygen atoms act as donors of electrons and the resulting charge transfer shifts the peak by about 14 eV.

Segregation of impurities to the surface of single-crystal specimens can give rise to new ordered structures. Auger electron spectroscopy can be useful in deciding whether a new structure evidenced by the LEED pattern is due to the reconstitution of a clean surface or to the ordering of an impurity which has segregated to the surface. Bishop and Rivière (1969b) doped the surface region of a silicon single crystal with gold by ion implantation and observed the changes caused by heating with both LEED and AES. A series of surface structures resulted, with corresponding changes in the Auger signal from gold.

4. Grain boundary segregation

Some metals and alloys, when subjected to a critical load, exhibit two possible types of fracture: (a) ductile fracture or (b) brittle fracture. Brittle fracture occurs at lower temperatures, whereas ductile fracture corresponds to higher temperatures. Usually, each such metal has a characteristic temperature at which transition takes place from the brittle to ductile fracture. The yield stress for ductile fracture is usually much greater than that for brittle fracture, so that for industrial applications a high transition temperature is a desirable feature. One way to achieve this is to subject the metal to heat treatment, when it is said to be temper embrittled. For a temper embrittled metal below the transition temperature, fracture occurs along grain boundaries. Above this temperature, fracture takes place through grain boundaries. Before AES came to be used in such studies, the assumption was that embrittlement, i.e. fracture along grain boundaries below the heightened transition temperature, was caused by the segregation of impurities at grain boundaries during the tempering treatment. But there was no direct evidence for this surmise. AES has confirmed the segregation at grain boundaries. For example, Marcus and Palmberg (1969) showed that there is a considerable segregation of antimony to fracture surfaces in embrittled low alloy steel containing as little as 0.03 atoms % of antimony, whereas no antimony could be detected at non-brittle fracture surfaces in the same steel.

B. Quantitative methods

For a discussion of the quantitative methods employed in AES, it is necessary to characterize the various parameters on which the observed Auger current $i_{Aj}(E_{WXY})$ from a sample j depends. Several of these parameters were introduced in Section 5. Here we shall complete the list by writing down, after Chang (1974), the symbolic equation

$$i_{Aj}(E_{WXY}) = A i_p X_j \rho_j D_j r \sigma_W a_j(E_{WXY}, C) RT. \tag{8.68}$$

This is a symbolic equation in the sense that not all factors are mutually independent. We may also note that the directional dependence has been integrated out. The symbols not defined so far are as follows. A is the area of the material irradiated with the primary beam, X_j is the concentration of element j in the surface region, so that $0 \leqslant X_j \leqslant 1$ and R is a surface roughness factor. On a rough surface, the angle of incidence of the primary beam will vary from point to point so that only a part of the beam is effective in causing ionization. For normal incidence this may not be too much of a problem. But for grazing incidence the valleys in a rough surface may be partly "shaded" by the hills. T is the instrumental transmission factor. ρ_j represents the average atomic density, i.e. number of atoms/cc, of the element j. The parameter D_j relates to the escape probability $P(E_{WXY})$ of an Auger electron through the surface. Actually it should be a complex function of the surface energy barrier. However, in a simplified treatment it may be described by an escape depth (or ejection depth) D_j. This depth specifies the detected volume from which Auger electrons are actually collected by the analyser. In this viewpoint, we can write $D_j = l_j(E_{WXY})$, where l_j is the mean free path of the Auger electron. It should be noted, however, that l_j is less than the range of the electron in the material. The latter represents the average distance the electron can travel in the material before coming to rest. For the purposes of Auger spectroscopy, an electron which suffers energy loss greater than the half-width of the Auger peak does not contribute to the peak. Thus the detected volume is always smaller than the volume in which Auger electrons are excited. Because energy loss occurs mainly through plasmon excitations in the valence band, the escape depth depends critically on the structure of the band. Also, even with the simplest choice of an escape probability function, e.g. the exponential attenuation model, the escape probability decreases rapidly with the depth z. As a result, if a sample does not have a uniform composition within the detected volume, Auger measurements are weighted heavily in favour of the species in the topmost layers.

Fortunately for the experimentalist, most of the quantitative methods employed so far are based on empirical calibration procedures. Thus the diverse factors enumerated above need not all be known for each measurement.

1. Calculation of surface composition

Consider a solid surface with a homogeneous distribution of elements j with concentrations X_j (Chang, 1974). We now write eqn (8.68) in the form

$$b_j i_{Aj} = X_j \bar{D}(E_j), \tag{8.69}$$

where for brevity we use E_j to denote the Auger energy corresponding to the transition WXY, $\bar{D}(E_j)$ represents an average escape depth for all the elements and b_j incorporates the factors not explicitly shown in this equation. For a pure element, $X_j = 1$ and the corresponding equation is

$$b_j^0 i_{Aj}^0 = D_j(E_j). \tag{8.70}$$

From these two equations we get

$$X_j = b_j i_{Aj} D_j(E_j)/a_j^0 i_{Aj}^0 \bar{D}(E_j). \tag{8.71}$$

We must now relate the average escape depth $\bar{D}(E_j)$ to the individual escape depths $D_j(E_j)$ for the different constituents. To do this we define the electron stopping power for element j

$$S_j(E_j) = \frac{1}{D_j(E_j)}. \tag{8.72}$$

Stopping power is by definition an intensive quantity, i.e. its value for a mixture is obtained by averaging rather than summing over the constituents. Thus the total stopping power of the mixture is

$$\bar{S}(E_j) = \sum_j X_j S_j(E_j) = \sum_j \frac{X_j}{D_j(E_j)} = \frac{1}{\bar{D}(E_j)}. \tag{8.73}$$

This gives us a way of finding $\bar{D}(E_j)$ provided that the $D_j(E_j)$ are known.

As a first approximation, let us now set $b_j \approx b_j^0$. Also, if we know the Auger current I_{As}^0 for a pure standard element s, then we can define an inverse Auger sensitivity factor α_j for the element j by writing

$$\alpha_j i_{Aj}^0 = i_{As}^0. \tag{8.74}$$

The name inverse sensitivity factor is justified by noting that the larger the Auger current i_{Aj}^0 is compared to the current i_{As}^0 for the standard element, the smaller will α_j be. Then, substituting this in eqn (8.71),

$$X_j = \frac{i_{Aj} D_j(E_j)}{i_{Aj}^0 \bar{D}(E_j)} = \frac{\alpha_j i_{Aj} D_j(E_j)}{i_{As}^0 \bar{D}(E_j)}. \tag{8.75}$$

Now adding all X_j's and setting $\sum_j X_j = 1$, we get

$$\sum_j \frac{\alpha_j i_{Aj} D_j(E_j)}{i_{As}^0 \bar{D}(E_j)} = 1. \tag{8.76}$$

Hence

$$X_j = \frac{\alpha_j i_{Aj} D_j(E_j)}{i_{As}^0 \bar{D}(E_j)} \Bigg/ \sum_j \frac{\alpha_j i_{Aj} D_j(E_j)}{i_{As}^0 \bar{D}(E_j)}$$

$$= \frac{\alpha_j i_{Aj} D_j(E_j)}{\bar{D}(E_j)} \Bigg/ \sum_j \frac{\alpha_j i_{Aj} D_j(E_j)}{\bar{D}(E_j)}, \tag{8.77}$$

because i^0_{As} is a constant quantity and may be factored out of the summation. The concentration can be found in principle from this expression, provided that the average escape depth $\bar{D}$ is known. However, the latter is not easy to determine. To bypass this difficulty we define a relative emittivity parameter of the element j for Auger electrons:

$$\eta_j = \frac{\text{Escape depth for element } j \text{ at energy } E_j}{\text{Escape depth for standard element } s \text{ at the same energy}}$$

$$= \frac{D_j(E_j)}{D_s(E_j)}. \tag{8.78}$$

Then obviously $\eta_s = 1$. Note that η_j depends not only on the element j but also on the energy E_j of the Auger electron. Now

$$\bar{D}(E_j) = \frac{1}{\bar{S}(E_j)} = \frac{1}{\sum_j \frac{X_j}{D_j(E_j)}} = \frac{1}{\sum_j \frac{X_j}{\eta_j D_s(E_j)}} = \frac{1}{\frac{1}{D_s(E_j)} \sum_j \frac{X_j}{\eta_j}}$$

$$= \frac{D_s(E_j)}{\sum_k \frac{X_k}{\eta_k}}. \tag{8.79}$$

Here we have assumed that η_j is independent of the Auger energy, so that each element is described by one η_j. Substituting this in eqn (8.77)

$$X_j = \frac{\alpha_j i_{Aj} D_j(E_j)\left(\sum_k \frac{X_k}{\eta_k}\right)\Big/ D_s(E_j)}{\sum_j \alpha_j i_{Aj} D_j(E_j)\left(\sum_k \frac{X_k}{\eta_k}\right)\Big/ D_s(E_j)} = \frac{\alpha_j i_{Aj} \eta_j}{\sum_j \alpha_j i_{Aj} \eta_j}. \tag{8.80}$$

To see what the emittivity η implies, let us consider a binary mixture of elements 1 and 2 with equal concentrations, i.e. $X_1 = X_2$, and equal inverse sensitivity, i.e. $\alpha_1 = \alpha_2$. Let $X_1 = X_2 = X$ and $\alpha_1 = \alpha_2 = \alpha$. Let element 1 be the standard element, i.e. $s = 1$. Then

$$X = \frac{\alpha i_{A1} \eta_1}{\alpha i_{A1} \eta_1 + \alpha i_{A2} \eta_2} = \frac{\alpha i_{A2} \eta_2}{\alpha i_{A1} \eta_1 + \alpha i_{A2} \eta_2}.$$

$$\therefore \alpha i_{A1} \eta_1 = \alpha i_{A2} \eta_2.$$

But by definition $\eta_1 = 1$, so that $i_{A1} = i_{A2} \eta_2$ or, $i_{A1} = (D_2/D_1) i_{A2}$. In other words, the material with the shorter escape depth gives the larger Auger peak.

Equation (8.80) is the basic equation relating Auger currents to concentrations. For a mixture of elements with $\eta = 1$, it assumes the simple form

$$X_j = \frac{\alpha_j i_{Aj}}{\sum_j \alpha_j i_{Aj}}. \tag{8.81}$$

In the more general situation, eqn (8.81) is replaced by

$$X_j = \frac{M_j \alpha_j i_{Aj}}{\sum_j M_j \alpha_j i_{Aj}}, \tag{8.82}$$

where M_j are called the "matrix effect" parameters. Then, writing $M_j \alpha_j = \beta_j$, we get

$$X_j = \frac{\beta_j i_{Aj}}{\sum_j \beta_j i_{Aj}}, \tag{8.83}$$

where the new coefficients β_j are obtained by calibration against a known standard. In effect this means that the coefficients α_j in the simpler eqn (8.81) are replaced by the more general coefficients β_j. The mapping between these two sets of coefficients is accomplished by the diagonal matrix M_j. If $M_i = M_j$ for all $i \neq j$, then the matrix M is a multiple of the identity matrix and we can replace eqn (8.83) by eqn (8.81) even if $\alpha_j \neq \beta_j$. This is the meaning of the approximation $\eta = 1$.

It must be clear that even with the approximations made above, we need to calibrate measurements by means of some known standard. In fact any accurate determination of the concentration or the absolute amount of material detected depends on such calibration. The method of calibration will depend on the form in which the material occurs on the surface. Once this has been done, the Auger current i_A from a given peak for the element j is the basic information needed. For a Gaussian peak, we can see from eqns (8.38) and (8.66a) that the Auger current i may be obtained from the area under the $N(E)$ curve. However, assuming that the shape of the peak does not change with concentration, we can write

$$\frac{dN(E)}{dE} = -\frac{V_x}{\sigma^2} N(E) = -\frac{V_x}{\sigma^2}\frac{dI(V)}{dV} = -\frac{V_x}{\sigma^3}\frac{1}{\sqrt{2\pi}} e^{-V_x/2\sigma^2} i,$$

so that the Auger current may equally well be obtained from the peak-to-peak height of the $dN(E)/dE$ curve.

Unfortunately, changes do occur in the shape of the peak with composition in many cases. For example, atoms nearer the surface produce sharper peaks and removal of a single monolayer of surface impurities results in all the peaks from the substrate becoming sharper. Care must be taken in handling such situations.

2. Submonolayer surface deposits

The simplest situation in quantitative surface analysis occurs when a material A forms a uniformly flat deposit on a substrate B and less than a monolayer is present. In this case the Auger signal i_A from A should vary linearly with coverage Θ. For $\Theta = 1$ we have a complete monolayer. Let i_{A_0} be the corresponding Auger signal. Then

$$i_A = \Theta i_{A_0}, \quad 0 < \Theta < 1. \tag{8.84}$$

That the actual behaviour of a submonolayer deposit is well described by this assumption can be seen from Fig. 8.25 which gives the results obtained by

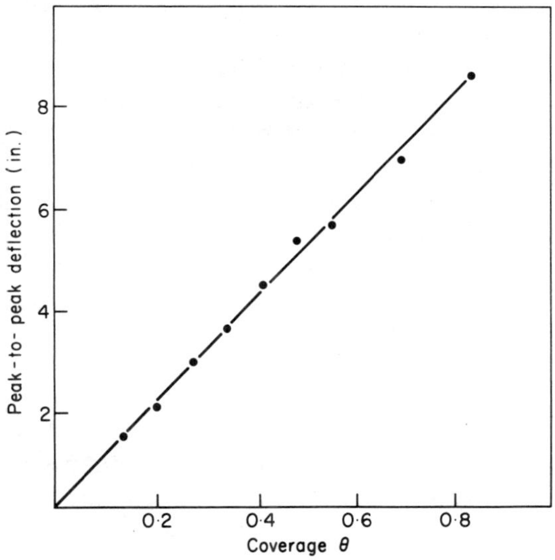

Fig. 8.25. Height of the principal Auger peak in potassium as a function of coverage Θ (from Weber and Johnson, 1969, courtesy of the American Institute of Physics).

Weber and Johnson (1969) for potassium on Ge(111). They deposited known amounts of potassium on germanium and measured the peak to peak height of the K Auger signal as a function of coverage. Linearity holds so long as all spectrometer parameters are kept constant. The coverage was checked independently by means of the LEED pattern. In LEED the spots arising from the deposit form ordered structures. The intensity of these spots is measured as a function of coverage. The agreement was good but, however, the procedure is not necessarily unique, because the same LEED pattern was found to occur over a wide range of coverage. Other methods used for calibration are (a) ion counting, (b) radiotracer techniques, (c) weighing in a microbalance, (d) measurement of adsorption energy by mass spectrometer and (e) ellipsometry. Calibrations using one or the other of these methods have been made for (a) silver on gold (Palmberg and Rhodin, 1968), (b) alkali atoms on germanium silicon and molybdenum (Weber and Johnson, 1969), (c) sulphur on nickel (Perdereau, 1971), (d) phosphorus on silicon (Levenson *et al.*, 1972), (e) oxygen on silicon (Vrakking and Meyer, 1971) and (f) oxygen on tungsten (Musket and Ferrante, 1970).

Once a calibration has been made, unknown values of Θ can be found from i_A. The important restriction is that the sample to be investigated should have the same lateral and depth distribution as the standard used for calibration. Other experimental conditions should also be similar.

Because of the linearity, the accuracy of this method depends ultimately on the current sensitivity of the analyser. It was shown by Weber and Johnson that about 2% of a monoatomic layer of caesium or potassium on Si could be detected with a LEED-Auger analyser. Musket and Ferrante (1970) observed that with the same type of analyser they could detect less than 2% of a monolayer of oxygen on tungsten. They used the area under the $N(E)$ curve as a measure of the Auger current, using a square wave signal to null the background input to the tuned signal amplifier. To calibrate the instrument, they had to know i_{A_0}. For this, they exposed the tungsten surface to large quantities of oxygen and assumed that saturation coverage implied a monolayer deposit. Interestingly, they found that the shape of the oxygen peak did not change with coverage.

3. *Layer growth*

Another simple system from the viewpoint of quantitative analysis is one in which a material is deposited on a solid surface and grows layer by layer. A simple Darwin-type model was proposed by Gallon (1969) for the study of such systems. The model pictures the crystalline deposit as being made up of a number of monoatomic plane layers parallel to the surface of the substrate. The Auger current from the deposit is regarded as the sum of contributions from these parallel planes. It is also assumed that an Auger electron emitted in any given layer has a fixed probability of penetrating the layer immediately above which does not depend on where these layers are located. A quantitative estimate of the thickness of the surface deposit may be made from the Auger spectrum, provided that the ratio of the Auger intensity from a monolayer to that of the bulk material, i.e. material with a very large number of layers, is known.

Let P be the probability that a primary electron causing ionization is scattered at one of these layers, so that $(1 - P)$ is the probability that it penetrates to the next layer. Let A be the probability that (a) a scattered primary electron ionizes an atom and (b) the resulting ionized atom emits a characteristic Auger electron in the backward (i.e. upward) direction. Finally, let E be the probability that this Auger electron penetrates without energy loss the layer immediately above.

Let N be the total number of primary electrons incident normally on the surface/s. Then the number of primary electrons scattered by the nth layer is $N(1-P)^{n-1}P$. This produces $N(1-P)^{n-1}PA$ Auger electrons in the upward direction. Out of these, only $N(1-P)^{n-1}PAE^{n-1}$ reach the surface and escape so as to be recognized as Auger electrons belonging to the characteristic peak in question. The number of such Auger electrons from the different layers may be obtained by assigning the values 1, 2, 3, ... to n. Thus the total Auger current i_n from n such layers is given by

$$i_n = NPA\{1 + (1-P)E + \cdots + (1-P)^{n-1}E^{n-1}\}$$

$$= NPA\left\{\frac{1-(1-P)^n E^n}{1-(1-P)E}\right\}. \tag{8.85}$$

For a single monoatomic layer, $n = 1$, and

$$i_1 = NPA.$$

For the bulk material, $n \to \infty$,

$$i_\infty \equiv \lim_{n\to\infty} i_n = NPA\left\{\frac{1}{1-(1-P)E}\right\},$$

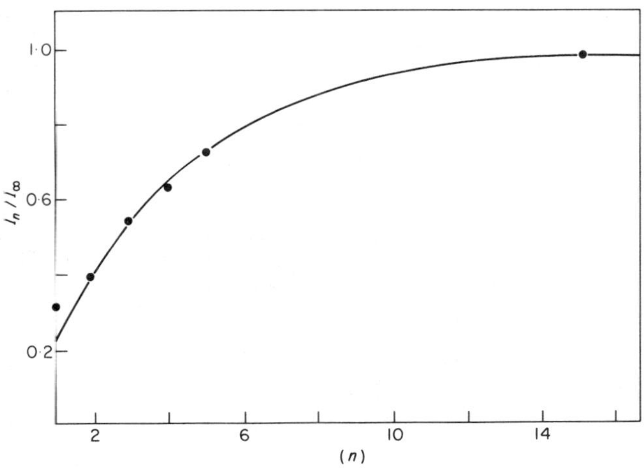

Fig. 8.26. Variation of Auger signal with thickness corresponding to layer-by-layer growth. Experimental points are from the data obtained by Palmberg and Rhodin (1968) and the curve represents the prediction made by Gallon (1969) (from Gallon and Matthew, 1972, courtesy of the Institute of Physics).

because the product $(1-P)^n E^n$ becomes vanishingly small as $n \to \infty$. Using these values of i_1 and i_∞ in eqn (8.85),

$$i_n = i_\infty \left\{1 - \left(1 - \frac{i_1}{i_\infty}\right)^n\right\}. \tag{8.86}$$

This relation gives the ratio i_n/i_∞ as a function of the number of layers, provided that the ratio i_1/i_∞ is a known parameter. In most experiments, the Auger current is not measured directly. However, Gallon showed that i_n/i_∞ plotted as a function of n fitted the experimental curves of Palmberg and Rhodin (1968), obtained with Ag on Au, for $n = 2$ (Fig. 8.26). For i_∞, Gallon used the Auger current for $n = 15$. It can be seen that the current reaches its saturation value for this value of n.

For quasicontinuous growth (Gallon and Matthew, 1972), eqn (8.86) may be written in the form

$$i_Z = i_\infty\{1 - e^{-Z/Z_0}\}, \tag{8.87}$$

where

$$Z_0 = d \ln^{-1}\left(\frac{i_\infty}{i_\infty - i_1}\right),$$

d being the layer spacing and Z the depth of the layer.

References

Åberg, T. (1969). *Ann. Acad. Sci. Fenn. Ser.* **A VI**. *Physica*, 308.

Åberg, T. (1972). *In* Proceedings of International Symposium on X-ray Spectra and Electronic Structure of Matter, Munich.

Albridge, R. G. and Hollander, J. M. (1961). *Nucl. Phys.*, **27**, 554.

Amelio, G. F. (1970). *Surface Science*, **22**, 301.

Archard, G. D. (1963). Private Communication.

Archard, G. D. (1964). Unpublished observations.

Asaad, W. N. and Burhop, E. H. S. (1958). *Proc. Phys. Soc.* (*London*), **71**, 369.

Asaad, W. N. (1959). *Proc. Roy. Soc.* (*London*), **A249**, 555.

Asaad, W. N. (1963a). *Nucl. Phys.*, **44**, 399.

Asaad, W. N. (1963b). *Nucl. Phys.*, **44**, 415.

Asaad, W. N. (1965a). *Nucl. Phys.*, **63**, 337.

Asaad, W. N. (1965b). *Nucl. Phys.*, **66**, 494.

Asaad, W. N. (1967). *Z. Phys.*, **203**, 362.

Asaad, W. N. and Mehlhorn, W. (1968). *Z. Phys.*, **217**, 304.

Bambynek, W., Crasemann, B., Fink, R. W., Freund, H. U., Mark, H., Swift, C. D., Price, R. E. and Rao, P. V. (1972). *Rev. Mod. Phys.*, **44**, 716.

Bank, W., Barz, A., Kocian, P., Nöller, H. G., Polaschegg, H. D., Schillalies, H., Spohr, R. and Wischnewski, K. (1972). *Vacuum*, **22**, 497.

Bargmann, V. (1936). *Z. Phys.*, **99**, 576.

Barkla, C. C. (1918). *Phil. Trans. Roy. Soc.* (*London*) *Ser.*, **A217**, 315.

Bassett, P. J., Gallon, T. E. and Prutton, M. (1972). *J. Phys. E: Scientific Instruments*, **5**, 1008.

Bauer, E. (1969). *In* "The Structure and Chemistry of Solid Surfaces" (Ed. G. A. Somorjai). Wiley Interscience, New York.

Bauer, E. (1972). *Vacuum*, **22**, 539.

Bearden, J. A. and Burr, A. F. (1967). *Rev. Mod. Phys.*, **39**, 125.

Bergström, I. and Hill, R. D. (1954). *Ark. Fys.* **8**, 21.

Bergström, I. and Nordling, C. (1965). *In* "Alpha, Beta and Gamma Spectroscopy Vol. 2" (Ed. K. Siegbahn). North Holland, Amsterdam.

Bethe, H. A. and Jackiw, R. W. (1968). "Intermediate Quantum Mechanics". W. A. Benjamin, Inc., New York.

Bhalla, C. P. (1970a). Relativistic effects in the KLM Auger Spectrum. *J. Phys.*, **B3**, L29.

Bhalla, C. P. (1970b). *Phys. Rev.*, **A2**, 722.

Bhalla, C. P. (1970c). *J. Phys.*, **B3**, 916.

Bhalla, C. P. and Ramsdale, D. J. (1970a). *J. Phys.*, **B3**, L14.

Bhalla, C. P. and Ramsdale, D. J. (1970b). *Z. Phys.*, **239**, 95.

Bhalla, C. P., Rosner, H. R. and Ramsdale, D. J. (1970). *J. Phys.*, **B3**, 1232.

Bhalla, C. P. (1973). *Phys. Lett.*, **44A**, 103.

Biedenharn, L. C. (1962). *Phys. Rev.*, **126**, 845.

Biedenharn, L. C. (1962). Lecture notes on Group Theory, "Summer Institute

for Theoretical Physics" (Ed. W. E. Brittin and B. W. Downs). Boulder, Colorado.
Biedenharn, L. C. and Swamy, N. V. V. J. (1964). *Phys. Rev.*, **133**, B1353.
Biedenharn, L. C. and Brussaard, P. J. (1965). "Coulomb Excitation". Oxford University Press, London.
Bishop, H. E. (1967). *Brit. J. Appl. Phys.*, **18**, 703.
Bishop, H. E. and Rivière, J. C. (1969a). *J. Appl. Phys.*, **40**, 1740.
Bishop, H. E. and Rivière, J. C. (1969b). *J. Phys. D: Appl. Phys.*, **2**, 1635.
Bishop, H. E. and Rivière, J. C. (1970). *Appl. Phys. Lett.*, **16**, 21.
Blauth, E. (1957). *Z. Phys.*, **147**, 228.
Bohm, D. and Pines, D. (1951). *Phys. Rev.*, **82**, 625.
Bohm, D. and Pines, D. (1953). *Phys. Rev.*, **92**, 609.
Breuckmann, B. and Schmidt, V. (1974). *Z. Phys.*, **268**, 235.
Briancon, C. (1970). *Ann. Phys.*, **5**, 151.
Briand, J. P., Chevallier, P., and Tavernier, M. (1970). *J. Phys.* (*Paris*) (suppl. 10), **32**, C4–165.
Briand, J. P., Frilley, M., Chevallier, P., Johnson, A., Rozet, J. P., Tavernier, M., and Touati, A. (1973). *In* Proceedings of the Int. Conf. on Inner Shell Ionization Phenomena, p. 1930. U.S. A.E.C. Report No. CONF-720404.
Broyles, C., Thomas, D. and Haynes, S. (1953). *Phys. Rev.*, **89**, 715.
Brueckner, K. A. (1955). *Phys. Rev.*, **97**, 1353.
Burhop, E. H. S. (1935). *Proc. Roy. Soc.*, **A 148**, 272.
Burhop, E. H. S. (1937). Ph.D. Thesis, Trinity College, Cambridge (unpublished).
Burhop, E. H. S. (1952). "The Auger Effect and Other Radiationless Transitions". Cambridge University Press, Cambridge.
Burhop, E. H. S. (1955). *J. Phys. Radium*, **16**, 625.
Burhop, E. H. S. and Asaad, W. N. (1972). *In* "Advances in Atomic and Molecular Physics", Vol. 8. Academic Press, New York and London.
Byrne, J. and Howarth, N. (1970). *J. Phys.*, **B3**, 280.
Callan, E. J. (1961). *Phys. Rev.*, **124**, 793.
Callan, E. J. (1962). *Bull. Am. Phys. Soc.*, **7**, N416.
Callan, E. J. (1963). *In* "Role of Atomic Electrons in Nuclear Transformations", Proceedings of International Conference, Vol. 3, p. 419. Nuclear Energy Information Centre, Warsaw.
Callan, E. J., Nikolai, P. and McDavid, W. L. (1964). Proceedings of the 3rd International Conference on the Physics of Electronic and Atomic Collisions, London, 1963. North Holland, Amsterdam.
Callan, E. J. (1968). Private communication.
Carlson, T. A. (1973). *In* Proceedings of the International Conference on Inner Shell Ionization Phenomena, p. 2274. U.S. Atomic Energy Commission Report No. CONF-720404.
Carriere, B., Deville, J. P. and Goldsztaub, S. (1972). *Vacuum*, **22**, 485.
Castaing, R. (1951). Thesis, Univ. of Paris, publ. O.N.E.R.A. No. 55.
Castaing, R. (1960). *Advan. Electron. Electron Phys.*, **13**, 317.
Chacon, E., Moshinsky, M. and Novaro, O. (1971). *Phys. Rev.*, **A3**, 166.

Chang, C. C. (1971). *In* Modern Methods of Surface Analysis, p. 53, *Surface Science*, **25** (1), March.
Chang, C. C. (1974). *In* "Characterization of Solid Surfaces" (Ed. P. F. Kane and G. B. Larrabee). Plenum Press, New York.
Chase, R. L., Kelly, H. P. and Köhler, H. S. (1971). *Phys. Rev.*, **A3**, 1550.
Chattarji, D. (1967). *Il Nuovo Cimento*, **A48** (Series X), 524.
Chattarji, D. and Talukdar, B. K. (1968). *Phys. Rev.*, **174**, 44.
Chattarji, D. and Talukdar, B. (1973). *Phys. Rev.*, **A7**, 42.
Chen, M. H., Crasemann, B. and Kostroun, V. O. (1971). *Phys. Rev.*, **A4**, 1.
Chung, M. F. and Jenkins, L. H. (1970). *Surface Science*, **22**, 479.
Cleff, B. and Mehlhorn, W. (1969). *Z. Phys.*, **219**, 311.
Coad, J. P. and Rivière, J. C. (1971). *Z. Phys.*, **244**, 19.
Condon, E. U. and Shortley, G. H. (1959). "Theory of Atomic Spectra". Cambridge University Press, Cambridge.
Crasemann, B., Chen, M. H. and Kostroun, V. O. (1971). *Phys. Rev.*, **A4**, 2161.
Darwin, C. G. (1928). *Proc. Roy. Soc.* (*London*), **A118**, 654.
Davisson, C. and Germer, L. H. (1927). *Phys. Rev.*, **30**, 705.
Davisson, C. J. and Calbick, C. J. (1931). *Phys. Rev.*, **38**, 585.
Delbrück, M. (1930). *Proc. Roy. Soc.*, **129**, 686.
De Shalit, A. and Talmi, J. (1963). "Nuclear Shell Theory". Academic Press, New York and London.
Dick, C. E. and Lucas, A. C. (1970). *Phys. Rev.*, **A2**, 580.
Dirac, P. A. M. (1958). "The Principles of Quantum Mechanics" (4th edn). Oxford University Press, London.
Duke, C. B. and Park, R. L. (1972). *Physics Today*, **25** (8), p. 23.
Dyson, F. J. (1949a). *Phys. Rev.*, **75**, 486.
Dyson, F. J. (1949b). *Phys. Rev.*, **75**, 1736.
Dyson, F. J. (1951). Advanced Quantum Mechanics. Lectures given at Cornell University, Fall.
Edmonds, A. R. (1957). "Angular Momentum in Quantum Mechanics". Princeton University Press.
El Ibayari, S. N., Asaad, W. N. and McGuire, E. J. (1972). *Phys. Rev.*, **A5**, 1048.
Englefield, M. J. (1972). "Group Theory and the Coulomb Problem". Wiley Interscience, New York.
Erdelyi, A., Magnus, W., Oberhettinger, F. and Tricomi, F. G. (1953). "Higher Transcendental Functions", Vol. I. McGraw-Hill, New York.
Erman, P. and Sujkowski, Z. (1961). *Ark. Fys.*, **20**, 209.
Estrup, P. J. and McRae, E. G. (1971). *In* Modern Methods of Surface Analysis, p. 1, *Surface Science*, **25** (1), March.
Fahlman, A. and Nordling, C. (1964). *Ark. Fys.*, **26**, 248.
Fahlman, A., Nordberg, R., Nordling, C. and Siegbahn, K. (1966). *Z. Phys.*, **192**, 476.
Farnsworth, H. E., Schlier, R. E., George, T. H. and Burger, R. M. (1958). *J. Appl. Phys.*, **29**, 1150.
Fink, R. W., Jopson, R. C., Mark, H. and Swift, C. D. (1966). *Rev. Mod. Phys.*, **38**, 513.

Fock, V. (1936). *Z. Phys.*, **98**, 145.

Froese, C. University of British Columbia, 1966 (unpublished).

Gallon, T. E. (1969). *Surface Science*, **17**, 486.

Gallon, T. E. (1972). *J. Phys. D: Appl. Phys.*, **5**, 822.

Gallon, T. E., Prutton, M. and Wray, L. (1972). Private communication.

Gallon, T. E. and Matthew, J. A. D. (1972). *Rev. Phys. Techn.*, **3** (1), p. 31.

Gaspar, R. (1954). *Acta Phys. Acad. Sci. Hung.*, **3**, 263.

Gaunt, J. A. (1930). *Phil. Trans. Roy. Soc. London*, **A229**, 163.

Geffrion, C. and Nadeau, G. (1959). Air Force Office of Scientific Research, Report TR 59-145 (unpublished).

Gerlach, R. L., Houston, J. E. and Park, R. L. (1970). *Appl. Phys. Lett.*, **16**, 179.

Goldstone, J. (1957). *Proc. Roy. Soc.* (*London*), **A239**, 267.

Grivet, P. (1972). "Electron Optics" (2nd English edn). Pergamon Press, Oxford.

Gryzinski, M. (1959). *Phys. Rev.*, **115**, 374.

Gryzinski, M. (1965). *Phys. Rev.*, **138A**, 305.

Haas, T. W. and Grant, J. T. (1970). *Appl. Phys. Lett.*, **16**, 172.

Haas, T. W., Grant, J. T. and Dooley, G. J. (1972). *J. Appl. Phys.*, **43**, 1853.

Hagstrum, H. D. (1966). *Phys. Rev.*, **150**, 495.

Hammermesh, M. (1962). "Group Theory and its Applications to Physical Problems". Addison-Wesley Publishing Company, Inc., Reading, Massachusetts.

Harris, L. A. (1968a). *J. Appl. Phys.*, **39**, 1419.

Harris, L. A. (1968b). *J. Appl. Phys.*, **39**, 1428.

Harrower, G. A. (1956). *Phys. Rev.*, **102**, 340.

Hartree, D. R. (1928). *Proc. Cambridge Phil. Soc.*, **24**, 89, 111.

Hartree, D. R. (1957). "The Calculation of Atomic Structures". Wiley Interscience, New York.

Helmer, J. C. and Weichert, N. H. (1968). *Appl. Phys. Lett.*, **13**, 266.

Herman, F. and Skillman, S. (1963). "Atomic Structure Calculations". Prentice Hall, Englewood Cliffs, N.J.

Herman, F. and Schwarz, K. (1972). *In* "Computational Solid State Physics" (Ed. F. Herman, N. W. Dalton and T. R. Koehler). Plenum Press, New York.

Hillig, H., Cleff, B., Mehlhorn, W. and Schmitz, W. (1974). *Z. Phys.*, **268**, 225.

Hink, W. and Ziegler, A. (1969). *Z. Phys.*, **226**, 222.

Hink, W. and Paschke, H. (1971). *Z. Phys.*, **244**, 140.

Hörnfeldt, O. (1963). *Ark. Fys.*, **23**, 235.

Hörnfeldt, O., Fahlman, A. and Nordling, C. (1963). *Ark. Fys.*, **23**, 155.

Huchital, D. A. and Rigden, J. D. (1970). *Appl. Phys. Lett.*, **16**, 348.

Hulme, H. R. (1936). *Proc. Roy. Soc.* (*London*), **A154**, 487.

Huzinaga, S. and Arnau, C. (1970). *Phys. Rev.*, **A1**, 2285.

Ignatiev, A., Jones, A. V. and Rhodin, T. N. (1972). *Surface Science*, **30**, 573.

Jacobi, K. and Hölzl, J. (1971). *Surface Science*, **26**, 54.

Jackson, J. D. (1962). "Classical Electrodynamics". Wiley Interscience, New York.

Johnson, J. B. (1928). *Phys. Rev.*, **32**, 97.
Johnson, F. A. and Foster, J. S. (1953). *Can. J. Phys.*, **31**, 469.
Kammlott, G. W. (1971). *In* Modern Methods of Surface Analysis, p. 120, *Surface Science*, **25** (1), March.
Kaufman, B. (1966). *J. Math. Phys.*, **7**, 447.
Kelly, H. P. Private communication.
Kleima, D., Holman, W. J. III, and Biedenharn, L. C. (1968). *In* "Group Theory and Its Applications" (Ed. E. M. Loebl). Academic Press, New York.
Klemperer, O. (1953). "Electron Optics". Cambridge University Press.
Kohn, W. and Sham, L. J. (1965). *Phys. Rev.*, **140**, A1133.
Körber, H. and Mehlhorn, W. (1966). *Z. Phys.*, **191**, 217.
Kostroun, V. O., Chen, M. H. and Crasemann, B. (1971). *Phys. Rev.*, **A3**, 533.
Krause, M. O., Carlson, T. A., Vestal, M. L. and Johnston, W. H. (1964). *Phys. Rev.*, **133**, A385.
Krause, M. O., Stevie, F. A., Lewis, L. J., Carlson, T. A. and Moddeman, W. E. (1970). *Phys. Lett.*, **31A**, 81.
Krause, M. O., Carlson, T. A. and Moddeman, W. E. (1971). *J. Phys.* (*Paris*), *Colloque*, **C4**, 139.
Krisciokaitis, R. J. and Haynes, S. K. (1967). *Nucl. Phys.*, **A104**, 466.
Kuyatt, C. E. and Simpson, J. Arol (1967). *Rev. Sci. Instr.*, **38**, 103.
Lander, J. J. (1953). *Phys. Rev.*, **91**, 1382.
Larkins, F. P. (1971a). *J. Phys. B: Atom. Molec. Phys.*, **4**, 1.
Larkins, F. P. (1971b). *J. Phys. B: Atom. Molec. Phys.*, **4**, 14.
Leder, L. B. and Simpson, J. A. (1958). *Rev. Sci. Instr.*, **29**, 571.
LeJeune, E. J., Dixon, R. D. (1972). *J. Appl. Phys.*, **43**, 1998.
Lenz, W. (1932). *Z. Phys.*, **77**, 713.
Levenson, L. L., Davis, L. E., Bryson, C. E., Melles, J. J. and Kon, W. H. (1972). *J. Vac. Sci. Technol.*, **9**, 608.
Listengarten, M. A. (1960). *Izv. Akad. Nauk SSSR, Sev. Fiz.*, **24**, 1041 (*Bull. Acad. Sci. USSR, Phys. Ser.*, **24**, 1050 (1960)).
Listengarten, M. A. (1961). *Izv. Akad. Nauk SSSR, Ser. Fiz.*, **25**, 803. (*Bull. Acad. Sci, USSR, Phys. Ser.*, **25**, 803 (1961)).
Listengarten, M. A. (1962). *Izv. Akad. Nauk SSSR, Ser. Fiz*, **26**, 182 (*Bull. Acad. Sci. USSR, Phys. Ser.*, **26**, 182 (1962)).
Löfgren, H. and Wallden, L. (1973). *Solid State Comm.*, **12**, 19.
Loucks, T. L. and Cutler, P. H. (1964). *Phys. Rev.*, **133**, A819.
Lyubarskii, G. Ya. (1960). "The Application of Group Theory in Physics". Pergamon Press, Oxford.
Manson, S. T. and Cooper, J. W. (1968). *Phys. Rev.*, **165**, 126.
Marcus, H. L. and Palmberg, P. W. (1969). *Trans. Metal. Soc. AIME*, **245**, 1664.
Mark, Peter (1971). *In* Modern Methods of Surface Analysis, p. 192, *Surface Science*, **25** (1), March.
Massey, H. S. W., and Burhop, E. H. S. (1936). *Proc. Roy. Soc.* (*London*), **A153**, 661.
McDonnell, L. and Woodruff, D. P. (1972). *Vacuum*, **22**, 477.

McGuire, E. J. (1967). *Phys. Rev.*, **161**, 51.
McGuire, E. J. (1968). *Phys. Rev.*, **175**, 20.
McGuire, E. J. (1969a). *Phys. Rev.*, **185**, 1.
McGuire, E. J. (1969b). Sandia Lab. Research Report SC-RR-69-137 (unpublished).
McGuire, E. J. (1970). *Phys. Rev.*, **A2**, 273.
McGuire, E. J. (1971a). *Phys. Rev.*, **A3**, 587.
McGuire, E. J. (1971b). *Phys. Rev.*, **A3**, 1801.
McGuire, E. J. (1972a). *Phys. Rev.*, **A5**, 1043.
McGuire, E. J. (1972b). *Phys. Rev.*, **A5**, 1052.
McGuire, E. J. (1974a). Auger and Coster-Kronig transitions. *In* "Inner Shell Atomic Processes" (Ed. B. Crasemann). Academic Press, New York and London.
McGuire, E. J. (1974b). *Phys. Rev.* (In press.)
McGuire, E. J. (1974c). *Phys. Rev.* (In press.)
McGuire, E. J. (1974d). *Phys. Rev.*, **A9**, 1840.
McVoy, K. W. (1965). Symmetry groups in physics. *Rev. Mod. Phys.*, **37**, 84.
Mehlhorn, W. and Albridge, R. G. (1963). *Z. Phys.*, **175**, 506.
Mehlhorn, W. and Asaad, W. N. (1966). *Z. Phys.*, **191**, 231.
Mehlhorn, W., Stalherm, D. and Verbeck, H. (1968). *Z. Naturf.*, **23A**, 287.
Mehlhorn, W. (1969). The Auger Effect. Notes based on lectures given at the University of Nebraska (1969) (unpublished).
Mehlhorn, W., Stalherm, D. and Schmitz, W. (1972). *Z. Phys.*, **252**, 399.
Metropolis, M. and Reitz, J. (1951). *J. Chem. Phys.*, **19**, 555.
Mladjenovic, M. and Slätis, H. (1955). *Arkiv. Fys.*, **9**, 41.
Møller, C. (1931). *Z. Phys.*, **70**, 786.
Mott, N. F. and Massey, H. S. W. (1965). "The Theory of Atomic Collisions". Oxford University Press, Oxford.
Mularie, W. M. and Rusch, T. W. (1970). *Surface Science*, **19**, 469.
Mularie, W. M. and Peria, W. T. (1971). *Surface Science*, **26**, 125.
Müller, E. W., Panitz, F. A. and McLane, S. B. (1968). *Rev. Sci. Instr.*, **39**, 83.
Musket, R. G. and Ferrante, J. (1970). *J. Vac. Sci. Techn.*, **7**, 14.
Musket, R. G. and Fortner, R. J. (1971). *Phys. Rev. Lett.*, **26**, 80.
Nall, J. C., Baird, Q. L. and Haynes, S. K. (1960). *Phys. Rev.*, **118**, 1278.
Nestor, C. W., Tucker, T. C., Carlson, T. A., Roberts, L. D., Malik, F. B. and Froese, C. (1966). Oak Ridge National Lab. Report–4027 (UC-32), December.
O'Connell, R. F. and Carroll, C. O. (1965). *Phys. Rev.*, **138**, B1042.
Oppenheimer, J. R. (1929). *Z. Phys.*, **55**, 725.
Packer, M. E. and Wilson, J. M. (1971). "Auger Transitions between 0 and 1500 eV for elements up to $Z=88$". The Institute of Physics, London.
Palmberg, P. W. and Rhodin, T. N. (1968). *J. Appl. Phys.*, **39**, 2425.
Palmberg, P. W., Bohn, G. K. and Tracy, J. C. (1969). *Appl. Phys. Lett.*, **15**, 254.
Palmberg, P. W. (1972). *In* "Electron Spectroscopy" (Ed. D. A. Shirley). North Holland, Amsterdam.

Palmberg, P. W., Rich, G. E., Weber, R. E. and MacDonald, N. C. (1972). "Handbook of Auger Electron Spectroscopy". Physical Electronics Ind., Inc., Edina, Minnesota.

Pauli, W. (1926). *Z. Phys.*, **36**, 336.

Perdereau, M. (1971). *Surface Science*, **24**, 239.

Pincherle, L. (1935). *Nuovo Cimento*, **12**, 81.

Pines, D. and Bohm, D. (1952). *Phys. Rev.*, **85**, 338.

Pines, D. (1955). *Sol. St. Phys.*, **1**, 367.

Pines, D. (1964). "Elementary Excitations in Solids". W. A. Benjamin, New York.

Prutton, M. (1971). *Metall. Rev.*, **5**, 57.

Racah, G. (1951). Group Theory and Spectroscopy (unpublished). Lectures given at the Institute for Advanced Study, Princeton.

Ramberg, E. G. and Richtmeyer, F. K. (1937). *Phys. Rev.*, **51**, 913.

Ramsdale, D. J. (1969). Ph.D. Thesis, Kansas State University, Manhattan, Kansas (unpublished).

Redhead, P. A., Hobson, J. P., Kornelsen, E. V. (1968). "The Physical Basis of Ultra High Vacuum". Chapman and Hall, London.

Reuter, W. (1971). *In* Modern Methods of Surface Analysis, p. 80, *Surface Science*, **25** (1), March.

Richard, P., Hodge, W. and Moore, C. F. (1972). *Phys. Rev. Lett.*, **29**, 393.

Ridgway, J. W. T. and Haneman, D. (1971). *Surface Science*, **24**, 451.

Rose, M. E. (1955). "Multipole Fields". Wiley Interscience, New York.

Rose, M. E. (1957). "Elementary Theory of Angular Momentum". Wiley Interscience, New York.

Rose, M. E. (1961). "Relativistic Electron Theory". Wiley Interscience, New York.

Rubenstein, R. A. (1955). Ph.D. Thesis: University of Illinois (unpublished).

Rubenstein, R. A. and Snyder, J. N. (1955). *Phys. Rev.*, **97**, 1653.

Rudge, M. R. H. and Schwartz, S. B. (1966). *Proc. Phys. Soc. (London)*, **88**, 563.

Sagawa, T. (1968). *In* Proc. Conf. on Soft X-ray Spectrometry and the Band Structure of Metals and Alloys (Ed. D. J. Fabian). Academic Press, New York and London.

Schmidt, V. (1973). *In* Proceedings of the Intern. Conf. on Inner Shell Ionization Phenomena, vol. 2, p. 548. U.S.-AEC Conference Report No. CONF-720404.

Scofield, J. H. (1969). *Phys. Rev.*, **179**, 9.

Seah, M. P. (1972). *Vacuum*, **22**, 475.

Sewell, P. B. and Cohen, M. (1967). *Appl. Phys. Lett.*, **11**, 298.

Siegbahn, K. *et al.* (1967a). "Atomic, Molecular and Solid State Structure Studied by Means of Electron Spectroscopy, ESCA". Almqvist and Wiksell, Uppsala.

Siegbahn, K., Nordling, C., Fahlman, A., Nordberg, R., Hamrin, K., Hedman, J., Johansson, G., Bergmark, T., Karlsson, S. E., Lindgren, I. and Lindberg, B. (1967b). *Nova Acta Reg. Soc. Sci. Upsal. Ser. IV*, **20**, Uppsala.

Silverstone, H. J. and Yin, M. L. (1968). *J. Chem. Phys.*, **49**, 2076.

Simpson, J. A. (1964). *Rev. Sci. Instr.*, **35**, 1698.
Slater, J. C. (1930). *Phys. Rev.*, **36**, 57.
Slater, J. C. (1951). *Phys. Rev.*, **81**, 385.
Slater, J. C. (1960). "Quantum Theory of Atomic Structure", Vols I and II. McGraw-Hill, New York.
Smith, David P. (1971). *In* Modern Methods of Surface Analysis, p. 171, *Surface Science*, **25** (1), March.
Socha, A. J. (1971). Modern Methods of Surface Analysis, p. 147, *Surface Science*, **25** (1), March.
Spangenberg, K. K. (1948). "Vacuum Tubes". McGraw-Hill, New York.
Staib, P. (1972). *Vacuum*, **22**, 481.
Stalherm, D. (1968). *Diplomarbeit Inst. Kernphysik, Univ. Münster.*
Sujkowski, Z. (1961). *Arkiv Fysik*, **20**, 243.
Swirles, B. (1935). *Proc. Roy. Soc.* (*London*), **A152**, 625.
Talukdar, B. K. (1970). Ph.D. Thesis, Visva-Bharati University, India (unpublished).
Talukdar, B. and Chattarji, D. (1970). *Phys. Rev.*, **A1**, 33.
Taylor, H. M. and Mott, N. F. (1933). *Proc. Roy. Soc.* (*London*), **A142**, 215.
Taylor, N. J. (1969). *Rev. Sci. Instr.*, **40**, 792.
Taylor, N. J. (1971). *In* "Techniques of Metals Research" (Ed. R. F. Bunshah), Vol. VII, p. 118. Wiley Interscience, New York.
Terman, F. E. (1955). "Electronic and Radio Engineering" (4th edn). McGraw-Hill, New York.
Tharp, L. N. and Scheibner, E. J. (1967). *J. Appl. Phys.*, **38**, 3320.
Thomas, L. (1954). *J. Chem. Phys.*, **22**, 1758.
Thomas, R. N. and Francombe, M. H. (1971). *Surface Science*, **25**, 357.
Thomson, G. M., Laudieri, P. C., Smith, W. W. and Russek, A. (1971). *Phys. Rev.*, **A3**, 2028.
Volz, D. J. and Rudd, M. E. (1970). *Phys. Rev.*, **A2**, 1395.
Vrakking, J. J. and Meyer, F. (1971). *Appl. Phys. Lett.*, **18**, 226.
Vriens, L. (1969). *Case Studies in Atomic Collision Physics*, **1**, 337.
Wallman, H. (1948). *In* "Vacuum Tube Amplifiers" (Ed. G. E. Valley), Radiation Laboratory Series, Vol. 18. McGraw-Hill, New York.
Walters, D. L. and Bhalla, C. P. (1971a). *Phys. Rev.*, **A3**, 519.
Walters, D. L. and Bhalla, C. P. (1971b). *Phys. Rev.*, **A3**, 1919.
Walters, D. L. and Bhalla, C. P. (1971c). *Phys. Rev.*, **A4**, 2164.
Weber, R. E. and Peria, W. T. (1967). *J. Appl. Phys.*, **38**, 4355.
Weber, R. E. and Johnson, A. L. (1969). *J. Appl. Phys.*, **40**, 314.
von Weizsäcker, C. F. (1934). *Z. Phys.*, **88**, 612.
Wentzel, G. (1927). *Z. Phys.*, **43**, 524.
Werme, L. O., Bergmark, T. and Siegbahn, K. (1972). *Physica Scripta*, **6**, 141.
Werme, L. O., Bergmark, T. and Siegbahn, K. (1973). *Physica Scripta*, **8**, 149.
Williams, E. J. (1935). *Mat. Fys. Medd.*, **13**, No. 4.
Worthington, C. R. and Tomlin, S. C. (1956). *Proc. Phys. Soc.*, **A69**, 401.

Subject Index

Accidental degeneracy (*l*-degeneracy), 31, 66
Adiabatic switching, 173
  theorem, 173
Algebra,
  Lie, 3, 5, 34, 36
  Racah, 97
  Wigner, 37, 65, 68, 92, 105, 130
Angular momentum,
  orbital, 7, 32, 65, 69
  addition of, 11, 126
  total, 12, 65
  non-integer, 47, 48, 62, 63
Annihilation operator, 69
  fermion, 23, 198
  for hole, 173
Antisymmetric wave function, 69, 93
  (see also Slater determinant)
Antisymmetry, 128, 129, 173
Anticommutation relation, 172
Atom probe, 181
Auger
  current, 221, 231, 246
  effect, 13
  electron, 14, 243
  yield, 28, 59, 61, 101, 158, 210
    in solids, 211
Auger transition, 13, 61
  probability, 14, 62
  Pauli (NRS) rate, 59
    calculation of, 39
  Dirac-Coulomb calculation of, 51, 52
  unscreened nonrelativistic rate, 59
  total transition rate, 100
  transition rate in *jj* coupling, 93
    in *LS* coupling, 104
  transition rate in intermediate coupling, 109
  transition rate with correlation, 177
Auger electron, spectroscopy (AES), 61, 181, 183, 193
  peak, 184
Average energy of multiplet, 83

Backscattering factor, 217
Band-pass filter, 234
Bandwidth, 230
Bergstrom and Hill ansatz for screening, 120, 188
Binding energies, 83, 118, 119, 120, 188, 191, 202
Bohr orbit, radius (first), 36, 40
Brueckner-Goldstone many body perturbation theory, 172

Casimir invariant (invariant operator), 8, 33, 34, 52, 69
Central field (potential), 30, 126
  Auger calculation, 91
  approximation, 127, 130
Characteristic Auger spectra, 200
Charge density
  non-local, 132, 133
  transition, 21, 25, 54, 198
Chemical shift, 203, 207
Chung–Jenkins ansatz for screening, 189, 192
Clebsch–Gordan coefficient, 11, 41, 99, 113
  symmetry relation, 97
Clebsch–Gordan series, 112
Closed shell (complete shell), see configuration
Closure, 3, 133
Collective oscillation (of plasma), 199
Commutation relations, 5, 33
Concentration of impurities, 243
Conduction band, 191
Configuration, 68, 69, 79, 81
  closed-shell (complete shell), 68, 126
  open shell (incomplete shell), 127
  two-electron, 68, 69, 74

Configuration interaction (mixing), 79, 80, 81, 84, 117, 118, 121, 139, 170
in Auger satellites, 160
Confluent hypergeometric function, 31, 39, 42, 137
Continuum
electron, 17
states, 31, 33
eigenvalues of, 33
wave function for NRS system, 38
hole, 86, 91, 163, 164
Correlation effects, 118, 171
diagram, 174
energy, 171, 174
Coster-Kronig transition, 14, 27, 28, 29, 101, 102, 203
probability, 27
super, 28
Coulomb
field, 30
potential, 30
wave function, 30, 31
wave function, screened (hydrogenic), 30, 51, 56, 60, 61, 116
symmetry of the, 31, 64
relativistic wave function, 31
non-relativistic wave function (NR), 31, 34
Creation operator, 69
fermion, 23, 198
for hole, 173
Current-current interaction, 22, 45, 51
Current density transition, 21, 25, 54
Cylindrical mirror analyser (CMA), 194, 218, 234

Decay rate for Auger transition in many-body theory, 175
Deconvolution technique, 206, 225
Degeneracy, 68, 76
(see also "accidental degeneracy")
Density of states, 203
Determinantal wave function, 127
(see also Slater determinant)
Diagram Auger line, 159, 168
Dipole
selection rule, 19
transition, 16
Dirac
delta function, 21, 24, 25, 26, 134
equation for planewave, 46
basis function, 23
operator, 37, 38
quantum number, 26, 31, 37
wave function, 21
velocity vector, 21
matrices, 46
spinor, 52, 143
Dirac–Coulomb
field, 37
problem, 45
wave equation, 47, 50, 146
bound-state radial eigenfunctions, 50
coupled radial equations, 51, 146
Direct matrix element (term), 24, 91, 92, 95, 97
Direct product
group, 9, 66, 67
space, 37
Direct radial integral, 97, 105
relativistic, 144
Discrete
eigenvalues, 33
states, 36
wave function for the NR system, 38
for the Pauli NRS system, 39
Double-focusing analyser, 212, 236
Dyson chronological operator, 23

Effective incremental charge, 188
Effective nuclear charge, 30, 46, 56, 116
Effective potential (energy), 133
averaged, 134
Ejection depth, see escape depth
Electrical differentiation (of secondary electrons), 194, 196, 236
Electron probe microanalysis, 183, 184
Electron stopping power, 244
Electrostatic interaction, 1, 64, 66, 69, 76, 92
Energy of the Auger electron, 119, 121
Energy-loss, electron, 186
peak, 194, 197
processes, 210
Energy window, 218, 235
Equivalent electron, 75, 98
vacancy, 94
ESCA, 182, 184, 185
Escape depth, 185, 243
average, 243
Exchange
matrix-element, 24, 91, 92, 95, 98
potential, 132, 133

universal potential, 133
relativistic potential, 144
Exchange radial integral, 98, 105
relativistic, 144

Fermi energy, 148
level, 193
momentum, 148
Feynman graph, 24, 172
Field operator, electron and positron, 23
Field electron microscopy, 184
Field ion microscopy, 184
Finite nuclear size effects, 146
Fluorescence yield, 14, 27, 28, 29, 59, 61, 158
for K-shell, 107, 140, 148
Fold of a function, 206
Fractional parentage coefficient, 164
Frozen atom approximation, 160
Full width at half maximum (FWHM), 191

Gaussian hypergeometric function ($_2F_1$), 31, 43, 44
elimination method, 44
recurrence relations for (see recurrence relations)
Gaussian peak, 221
Gegenbauer polynomials, 34
Green's function, scalar, 21
Ground-state energy of many-particle system, 131
Group, concept of, 2
property, 3
Lie, 3, 4
finite, 3
direct product, 9
generators, 5, 8, 34, 67
Abelian, 3
semi-simple, 7
elements, 3
multiplication, 3
continuous, 4
structure constants of, 5
local properties of, 5
O_3, 5, 7, 35, 36, 37
O_4, 34, 36, 37
$O_{3,1}$, 34, 35
SU_2, 6, 37
rotation, 5, 66, 67

Hamiltonian, 7, 8
non-relativistic, 32, 66
symmetric Dirac–Coulomb, 47, 48
many-particle, 80
Hartree–Fock, equation (non-relativistic), 65, 132, 135
(relativistic), 142
calculation, 59
energy, 171
potential, 172, 175
Hartree–Fock–Slater, equation (non-relativistic), 86, 134, 136
calculation of Auger transition, 136, 139, 140
Hartree–Fock–Slater calculation (relativistic) of Auger transition, 146, 155
Herman–Skillman, 136
potential, 137, 147
McGuire's approach to, 137
unit of length, 151
High-pass filter, 228, 234
Hilbert space, 7, 82
Hole (many-body theory), 173
(see also vacancy)
Hydrogenic model, 30, 59
screened, 30, 56, 60, 65
wave functions, 30, 56, 60, 100, 116
Hypersatellite Auger lines, 160
satellites of, 160

Impurities, 241
Independent-particle, approximation, 128
potential, 172
Instrumental line width (ILW), 191, 232
Intensities of Auger lines and groups, see relative intensities
Intensities of *KL–LLL* satellites, 161
Intermediate coupling, 76, 78, 81
wave function, 77, 80, 110
transition probabilities with, 109
energy of *K*-Auger spectrum with, 121
Interconfiguration term, 84
Internal conversion
nuclear, 15
atomic, 15
coefficient, 16
Invariance, 1, 7
of Hamiltonian under permutation, 128
Invariant subspace, 7, 80, 84
irreducible, 7

Inverse Auger sensitivity factor, 244
Ion, microprobe, 184
neutralization spectroscopy, 184
scattering, 184
Ionization
cross-section, 213
inner shell, 13, 120
potential (energy), 13, 182
Ionization loss peak, 200

jj coupling, 41, 68, 72, 73, 76, 77, 79, 87, 88, 89, 100, 163
wave function, 73
Auger transition probabilities in, 93, 98, 116, 117
Jacobi identity, 5

KLL Auger, spectrum, 66, 76, 81, 159
transition, 66, 70, 86
transition probability (rate), 100, 106, 115, 177
energies in intermediate coupling, 123
energies with correlation, 179
KL–LLL Auger satellite lines, 158, 159
Schmidt's theoretical treatment of, 160
Kepler problem
relativistic, 45
non-relativistic, 32
Kummer's transformation, 42, 49

Ladder operator, 38
Lagrange parameter, 131
Laguerre polynomials (associated), 40
Layer growth, 248
LEED–Auger analyser, 188, 218, 219, 225, 237, 239
Legendre operator, 46
Lie, algebra, 3, 5, 34, 36
group, 3, 4, 5, 6, 7, 8, 9, 66
Line width, 203
Linear group, 129
Local potential, 132, 133
Lorentz
group, 35
invariance, 25
Low energy electron diffraction (LEED), 185, 186, 187, 188, 194, 195, 218, 219, 225, 226, 232, 236, 237, 240, 241, 242
L-shell Auger transitions, 101, 102, 103, 104
Coster–Kronig transition, 101, 102, 103
LM–MMM satellites, 159

Matrix effect parameter, 246
Matrix element
non-relativistic Coulomb Auger, 31, 41, 59
non-relativistic dipole Coulomb, 60
relativistic Dirac-Coulomb, 54
of single-particle operators, 71, 172
of two-particle operators, 71, 172, 173, 174
of electrostatic interaction, 112, 122
of spin–orbit interaction, 73, 114
spin–angular part, 95
McGuire
approximate SCF method, 101, 137
Auger transition probabilities with, 108, 139, 140
theory of partially filled shells, 163
Mean free path (in solids), 217, 218
Mixed coupling, 86, 87, 88, 89, 90, 168
Mixing coefficients
for configuration interaction, see configuration interaction
for intermediate coupling, 110, 112, 114, 116
Modulating voltage, 220, 221
Møller formula, 20
Monatomic layer (monolayer), 181, 249
M-shell satellites, 159
Multiplet, 7, 8, 64, 65, 67, 74, 75, 76
structure, 34
Multiple ionization, 158
Multiple vacancy, 163
Multipole, 93
expansion of electrostatic interaction, 93, 94

Nine-*j* (9-*j*) symbol, 74, 164
Noise, 226
figure, 226
flicker-effect, 226
shot, 228, 236
thermal, 227
Noise bandwidth, 230
Non-local potential, 132, 133
Non-degenerate ground state, 173
Normalization
of continuum electron, 25
on energy scale, 25, 26, 49
to one electron per unit time, 26

Orbital, single-particle, 71, 127, 129, 130, 131, 132
Orbital angular momentum, see angular momentum
Orthonormalization, 111, 131

Parity, 15, 38, 80, 92
Partially filled shells, 163
Passive shells, 168
Pauli
matrices, 5
particle, 36
spinor, 37, 52, 93
central-field spinor, 37, 52
NRS system, Coulomb field, 36
transition probability for, 39
exclusion principle, 126, 129, 133, 148, 173
Peak-to-peak height, 239
Permutation operator, 128, 129, 143
Perturbation, energy (of many-body system), 174
series, 173, 174
Plasma, excitation, 184, 194
loss peak, 200
oscillation, 196, 197
Plasmon, 198
Poisson bracket, 32
Poisson's equation, 149, 151
Population distribution following Auger decay, 170
Potential, spherically symmetric, 7, 146
Primary excitation, 195
Principal quantum number, 14, 30, 36
apparent, 50
Principal value integration, 175
Projection operator, 175
Propagator, photon, 23

Qualitative chemical analysis, 240
Quantitative chemical analysis, 242

Racah coefficients, 41, 96, 99
Radial wave function
non-relativistic Coulomb bound state, 36
Dirac–Coulomb continuum, 49
bound-state, 50
self-consistent-field (Hartree-Fock)
non-relativistic, 66, 135
relativistic, 144
Radial integral, see Slater integral
Radiationless transition, see Auger transitions
Radiative transitions, see transitions
Raising and lowering operators, 35
Reciprocal lattice, 185
Recombination process, 182
Recurrence relation (recursion relation)
for Gaussian hypergeometric functions, 32, 43
for Coulomb wave functions, 31, 34, 36, 38
for special functions, 34
for Auger radial integrals, 32
Reduced matrix element, 114
Reflection high energy electron diffraction (RHEED), 185, 195
Relative emittivity, 245
Relative intensities (Auger)
for groups, 117, 138
for lines, 117, 138
in solids, 212
Residual interaction (perturbation), 80
Retarded interaction (retardation term), 18, 21, 51, 55, 147, 151
Retarding field analyser (RFA), 191, 218, 219
(see also LEED-Auger analyser)
Retarding grid, 219, 224, 225
Rotation
of a rigid body, 3
group, 4, 66, 67
infinitesimal, 4
Rotation matrix, 129
Rotational symmetry, see symmetry
Runge–Lenz vector, 33, 38
Russell–Saunders coupling, 59, 68, 76, 84, 87, 90, 163, 164, 168
wave function, 69, 70, 77, 80, 81, 110, 112
Auger transition rates in, 104, 105, 115, 116, 117, 159

Satellite Auger lines, 158
spectra, 158
Scanning speed (rate), 194, 231, 237
Scattering matrix, element of, 22, 24
operator, 22
Schmidt orthogonalization procedure, 111
Screening, 56, 57, 58, 59, 61, 120, 137
(see also Herman-Skillman)

(see also Bergstrom and Hill, Chung–Jenkins)
Screening constant (parameter), 30, 56, 57, 58, 59
  Hartree's definition of, 57
  mean radius for, 57, 59
  median radius for, 57
  ultimate, 58, 60
Second quantization, 172, 198
Secondary electrons, 188, 195, 196
Secular equation, 83, 110, 121
Selection rule
  for orbital angular momentum, 19, 76, 93
  for parity, 41, 93
Self-consistency, 130
Self-convolution, 206
Self-consistent-field (SCF)
  wave function, 56, 57
  (see also Hartree)
Shake-off, 158
  statistically distributed, 170
Shell
  atomic, 14
  inner, 14, 16
  major, 14
  sub-, 14, 53
  outer, 14
Signal-to-noise ratio, 226, 231
Silverstone-Yin-Huzinaga-Arnau potential, 175
Similarity transformation, 121
Six-*j* (6-*j*) symbol, 167
Slater determinant, 71, 126, 129, 130, 132, 143
  general integral, 72
  direct integral, 72, 113, 117, 123, 124
    Z-dependence of, 124
  exchange integral, 72, 113, 117, 123, 124
    Z-dependence of, 124
  relativistic integral, 144
SLJM representation, 70, 74, 77, 81, 105, 164
Sommerfeld, rosette motion, 46
  parameter, 35, 36, 59, 168
    relativistic analogue of, 47
Spherical Bessel functions, 55
  Hankel functions, 55
Spherical harmonic, 7, 35, 52, 65, 95, 127
  addition theorem for, 41, 94
Spherical symmetry, see symmetry
Spherically averaged potential, 127, 136
Spin, 37, 67
  dynamically independent, 37
Spin-angle functions, 37
Spinors, 143
  (see also Pauli and Dirac spinors)
Spin–orbit coupling (interaction), 1, 64, 67, 68, 69, 72, 73, 74, 76, 84, 103, 118, 121, 127
  energy splitting, 73, 83, 103
  matrix element of, 73, 114
Subgroup, 3, 67
Submonolayer surface deposit, 247
Sudden approximation, 158
Statistical weight, 165
Sticking probability (coefficient), 181
Supermultiplet, 65, 67
Surface chemical analysis, 181, 238
Surface cleaning methods, 240
Surface layer, 181
Surface plasmon, 199
Surface segregation, 195, 241
Surface sensitivity, 185, 189, 195
Symmetric Hamiltonian (Biedenharn–Swamy), 47
Symmetry, 1
  breaking, 31, 45, 65, 66, 68, 77
  rotational, 126
  spherical, 66
  of many-electron atom, 128
  O_4, 36, 45, 64, 65, 66, 140
  O_3, 35, 45, 65, 66, 67
  particle-hole, 69
  relation for Clebsch-Gordan coefficients, 96

Thomas precession, 73
Thomas–Fermi equation, 149
  model, 149
Thomas–Fermi–Dirac, equation, 149, 151
  model, 63, 104, 133, 148
  calculation of Auger transitions, 151
  function, 151
Time-dependent perturbation theory, 173
Transition amplitude, Auger, 59, 116, 118
Transition probabilities
  Auger, 14, 27, 31, 60, 84, 85, 86, 91, 98, 105, 106, 109
  radiative, 16, 27, 60, 102, 140

Triangle rule, 41, 99, 113
Transformations
  group of, 3
  set of, 3
  unitary, see unitary transformation
  similarity, 121
Transient response, 230

Unitarity, 6
Unitary transformation, 6, 74, 76, 77, 145

Vacancy
  inner shell, 13, 14, 15, 163, 169
  lifetime of, 203
  multiple, 13, 163
  primary, 13, 14, 195
  states, 69, 92
Vacuum (ground state of many-body system), 173
Vacuum level, 192, 206
Valence band, 191, 204, 207
  effects, 203
  transitions, 204
Variation principle, 130, 131
Vector spherical harmonic, 147
Virtual
  photon, 14, 18, 19, 20, 95
  state, 19
  quanta, 18, 19, 45

Wigner–Eckart theorem, 41, 55, 95, 105, 113
Work function, 192, 193, 196

X-ray transitions, see radiative transitions

Z-coefficient, 99